Polymer photophysics and photochemistry

Polymer photophysics and photochemistry

AN INTRODUCTION TO THE STUDY OF PHOTOPROCESSES IN MACROMOLECULES

JAMES GUILLET

Professor of Chemistry, University of Toronto

CAMBRIDGE UNIVERSITY PRESS

Cambridge

New York New Rochelle

Melbourne Sydney

Published by the Press Syndicate of the University of Cambridge
The Pitt Building, Trumpington Street, Cambridge CB2 1RP
32 East 57th Street, New York, NY 10022, USA
10 Stamford Road, Oakleigh, Melbourne 3166, Australia

First published 1985
First paperback edition with corrections 1987

Printed in Great Britain at the University Press, Cambridge

Library of Congress catalogue card number: 83-18838

British Library cataloguing in publication data
Guillet, J. E.
Polymer photophysics and photochemistry
1. Polymers and polymerization
2. Photochemistry
I. Title
547.7'0455 QD381

ISBN 0 521 23506 5 hard covers
ISBN 0 521 34783 1 paperback

MCP

To E.C.G.
who showed the way
R.G.W.N.
who gave the light
H.B.G.
who made it possible

Contents

Preface

This book was originally intended to be a comprehensive survey of the field of macromolecular photophysics and photochemistry. However, the rapid expansion of the literature in this area of science in recent years and the publisher's limitations on the size of the book have made it necessary to attempt a more modest objective, namely an outline of the more fundamental aspects of the subject. Since this is a monograph it represents my personal view of the field and is of necessity biased towards those areas with which I am most familiar. For this reason, the experimental studies discussed herein relate primarily to the photochemistry of synthetic polymers, and ignore the extensive literature on biological photosystems. The only justification for this is that the structures of synthetic polymer molecules are usually better understood and hence the results are simpler to interpret unambiguously.

Furthermore, a number of the more applied topics, such as conventional photopolymerization, photo-oxidation and photodegradation are not included in this volume. The last two topics have been the subject of books by Rånby and Rabek (*Photo-oxidation, photostabilization and photodegradation of polymers*, Wiley, 1977) and by McKellar and Allen (*Photochemistry of man-made polymers*, Applied Science, 1979) which are commended to the attention of those interested in the more applied aspects of polymer photochemistry.

For economy of space, details of experimental procedures have generally not been included, since these are readily available by reference to the original papers, or in standard textbooks of photochemistry or polymer chemistry. Furthermore, the advent of lasers and microcomputers has made many of the older procedures obsolete.

The book is designed primarily as a reference text suitable for two types of scientists: (*a*) those with training in polymer chemistry who wish to apply their knowledge to polymer photochemistry, and (*b*) photochemists who wish to study photo-processes in polymeric systems. The applications of such studies may include the synthesis of photoresists for microcircuitry, polymers for xerography and photomechanical reproduction, photocuring of paints and resins and photodegradation and stabilization of commercial plastics and synthetic fibers, to name only the most prominent. A new area of great potential significance is the use of polymers in various solar energy conversion devices. Stability to solar radiation often becomes a limiting factor on the design of such devices, and it is even possible that polymeric molecules may become the key to the successful imitation of plant photosynthesis (see Chapter 9).

In preparing this book, I reviewed more than 1000 original papers and review articles. It is an occupational hazard of the writer's profession that the significance of some work will be misunderstood, or attributed to another author, or even ignored completely. To those who perceive themselves to have been slighted in this way, I can only offer my apologies and a promise to make amends in subsequent publications.

I am indebted to a number of people who in one way or another made this book possible. My father, Edwin C. Guillet, himself the author of 22 scholarly books, brought me early to appreciate the fascination of the printer's art, at a time when books were printed with lead type and set by hand.

Professor R. G. W. Norrish, Nobel Laureate, distinguished teacher and photochemist, who was wise enough to recognize the future possibilities in combining photons with polymers, introduced me to the arcane rules of photochemistry. Professor Howard Purnell, an editor of the Cambridge Monographs in Physical Chemistry, persuaded me to undertake this book, and was undismayed by the long delays, as the length of the book expanded to the point where it could no longer fit into the format of the series. His encouragement and careful editing of the preliminary drafts are much appreciated.

A number of colleagues and associates have read chapters of the book and have given valuable suggestions and corrections. These include Professor Alastair M. North, Professor Mitchell Winnik, Dr David A. Holden and Dr James R. McCallum. They should not, however, be blamed for the errors which may remain, which are strictly my responsibility.

In a less direct way, I have benefited throughout my career from scientific discussions in the area of polymers and photochemistry with Professors Peter Yates, Herbert Morawetz, Michael Szwarc and Paul Flory.

Portions of this book were written in libraries in a number of locations. I am particularly grateful to the Chairman of the Chemistry Department at the University of Toronto, Professor Keith Yates, for approving the study leaves which made it possible to work undisturbed on the manuscript. Professor Murray Goodman was kind enough to provide me with an office and access to the library at the University of California at San Diego during a sabbatical leave. This has led to a continuing, if informal, relationship with that famous institution. I am also grateful to Joan Siereki and Karen Feeney of the USCD library for introducing me to the wonders of computer methods for searching Chemical Abstracts.

A study leave made possible by the Guggenheim Foundation allowed me to spend three-month periods at the University of Mainz, West Germany, and at the University of California, Berkeley. I am much indebted to my hosts, Professor Helmut Ringsdorf and Professor Melvin Calvin, for stimulating my imagination during those brief visits. Professor George Pimentel at Berkeley was also helpful in many ways.

Karen Ott and Carolyn Guillet helped with the typing of many drafts. My secretary, Susan Arbuckle, has helped in many ways, almost impossible to acknowledge, from drawing to final drafts, references, indexes and proofreading. My work is made so much easier by her competence and unfailing good humour.

This book is in large measure a creation of my many students and research associates. Their interest and enthusiasm has kept the whole project going, and I am deeply indebted to them for teaching me much of what I know about the subject. I hope that they are suitably acknowledged by name in the chapter references.

Finally, I must recognize the able assistance of my wife, Helen Bircher Guillet, who has not only helped with the onerous task of tracking down innumerable literature references, but also taught herself to use our Apple computer to help in indexing, literature searching and other tasks. More particularly she has furnished the life-support system which provides the only means of survival for an academic in the increasingly hostile environment of the 1980s. For these and other blessings I am truly grateful.

Toronto, May 1984 James Guillet

1

Introduction to photochemistry and photophysics

1.1 The nature and properties of light

Photochemistry deals with those processes by which light interacts with matter in such a way as to induce chemical reactions. However, as we study the subject, we find that the chemical reactions are usually competitive with photophysical processes which involve luminescence and conversion of electronic and other forms of energy to heat. Consequently, the subject of photochemistry will include topics which might otherwise have been considered to be photophysical. A number of excellent texts have been published on the general subject of the photochemistry of organic molecules (Calvert and Pitts 1966, Turro 1978).

For the purposes of this book, we will consider light as being that part of the electromagnetic spectrum of radiation characterized by radiation of wavelengths in the range 750 to about 100 nm. Light, in common with other electromagnetic radiation, has a velocity which is independent of the velocity of its source, given by $c = 2.9979 \times 10^{10}$ cm s^{-1} in vacuum. However, in a medium containing any form of matter, light will move at a different velocity, c', given by the equation

$$c' = c/(\varepsilon\mu)^{\frac{1}{2}} \qquad (1.1)$$

where ε is the dielectric constant of the medium and μ is its magnetic permeability.

For our purposes, electromagnetic radiation can be described in terms of a transverse plane wave involving associated electric and magnetic fields. The electric vector E and the magnetic vector H, which describe the relative field strengths are aligned in planes at right angles to each other, and both are perpendicular to the direction of propagation of the wave. In Cartesian coordinates, these vectors are oscillating functions with time, and are given by the following equations:

$$E_y = A \sin 2\pi[(x/\lambda) - \nu t] \tag{1.2}$$

$$H_z = (\varepsilon/\mu)^{\frac{1}{2}} A \sin 2\pi[(x/\lambda) - \upsilon t] \tag{1.3}$$

In these equations, E_y is the electric field strength vector lying in the xy plane, increasing along the y axis, and H is the magnetic field strength vector lying in the xz plane and increasing along the z axis. A is the amplitude of the electric vector, and the intensity of the wave is proportional to A^2. In vacuum, ε and μ are both equal to unity and they are approximately the same value in air. The wavelength of the wave is given by λ which is the distance between adjacent maxima in the wave or, in fact, any comparable point on subsequent waves and is measured at any instant along the direction of wave propagation, that is, the x axis, while ν is the frequency or number of cycles of vector position change per second. The frequency ν and the wavelength λ are related by the equation

$$\lambda = c/\nu \tag{1.4}$$

where c is the velocity of the radiation in the medium.

Normal light is unpolarized, that is to say, it is made up of a great number of electromagnetic waves, each of which has its electric and magnetic vectors undulating in fixed planes, but the planes of the various waves are randomly oriented with respect to one another. Such light is normally given out by incandescent bodies and unoriented emitting systems.

Plane polarized light, on the other hand, is presumed to consist of many waves, all of which have their electric vectors oriented in the same direction. Plane polarized light may be made by deflection from plane surfaces or by transmission through doubly refracting crystals. The degree of polarization of the light p is usually expressed by the equation

$$p = \frac{I_{\parallel} - I_{\perp}}{I_{\parallel} + I_{\perp}} \tag{1.5}$$

where $I_{\parallel}$ and $I_{\perp}$ are the intensities of light transmitted by an analyser parallel and perpendicular to the plane of maximum polarization of the incident beam.

Elliptically polarized light is usually described in terms of a light wave compounded from two plane waves of equal wavelength and frequency, and having identical directions of propagation but electric vectors which are perpendicular to one another and out of phase. The resultant electric vector of the total wave is that of an ellipse if viewed along the line of propagation. The equation of an elliptically polarized light wave can be obtained by vector addition of the components, thus the electric vector is given by (1.6) and (1.7) where τ represents the fraction of the cycle by which one wave leads the other. The components of the resulting electric vector along the y and z axes are

$$E_y = A \sin 2\pi \frac{x}{\lambda} \tag{1.6}$$

$$E_z = B \sin 2\pi \left(\frac{x}{\lambda} + \tau \right) \tag{1.7}$$

Circularly polarized light is a special case of electric polarization where $A = B$ and $\tau = 1/4$ cycle. Analysis of these equations shows that in circularly polarized light the resultant electric vector is seen to undergo a helical path, that is, it will appear to rotate clockwise, or counter-clockwise when viewed along the axis of propagation of the wave. The importance of this is that it will interact in a different way with molecules which are either right-handed (dextro) or left-handed (laevo) in optical activity.

1.2 Quantum-mechanical description of light

The simple wave theory described in the previous section is quite useful in describing a number of the properties of light. However, in order to understand the emission and absorption of light by molecules, it is necessary to introduce a second concept for the nature of light in which light is considered to be a beam of photons whose energies are quantized. A detailed description of quantum mechanics and spectroscopy is beyond the scope of this book and therefore only the major conclusions which are necessary for treatment of experimental data will be included here.

In the quantum theory of radiation, any electromagnetic radiation can be considered to be a beam of particles called photons, each of which has an energy E given by the Planck relationship $E = h\nu = hc/\lambda$, where h is the proportionality constant, called Planck's constant, equal to 6.6256×10^{-27} erg s quantum^{-1}. The quantum theory has successfully explained a large number of phenomena relating to the interactions of light with matter. One of the earliest of these is the photoelectric effect, which was discovered by Hertz, who found that short wavelength radiation incident on a metal surface caused the instantaneous emission of photoelectrons. Classical electromagnetic wave theory failed completely to account for the observed phenomena but in 1905, Einstein derived an equation which showed that when a quantum of light of energy $h\nu$ is absorbed by the metal, an electron is emitted and the energy relationships were given by the following equation:

$$\frac{m_e v^2}{2} = h\nu - \omega_0 \tag{1.8}$$

where ω_0 is the minimum energy necessary to remove an electron, and $m_e v^2/2$ is the kinetic energy of the electron emitted.

A further effect which was satisfactorily explained by the new quantum theory is the Compton effect, which describes the encounter of a monochromatic beam of X-rays with any scattering medium. The scattered rays are found to consist of a new longer wavelength radiation, as well as that of the incident rays, and an electron is emitted. When an X-ray is absorbed, the change in wavelength of the X-ray corresponds exactly to the energy transmitted to the electron. That is to say, the interaction can be described as a collision in which energy and momentum are conserved. The change in wavelength of the X-ray is given by

$$\lambda - \lambda_0 = 0.0242(1 - \cos\theta) \tag{1.9}$$

where λ is the wavelength of the scattered radiation at angle θ with respect to the direction of the incident beam of wavelength λ_0.

1.3 Interaction of light with matter

In photochemistry, the quantum theory also is obeyed. After absorption of a quantum of light, a number of chemical or physical events may occur by which the energy of the quantum is ultimately dissipated. The quantum yield of any given process is defined

$$\phi_i = \frac{\text{number of primary events of type } i}{\text{total number of quanta absorbed}} \tag{1.10}$$

and the law of conservation of energy requires that

$$\phi_{\text{total}} = \sum_i \phi_i = 1 \tag{1.11}$$

That is, the sum of the primary quantum yields of all processes is equal to unity. In most photochemical processes we will not be dealing with reactions involving only a single photon, but large numbers of photons, and it is convenient to define a mole of photons as 1 einstein, that is 6.023×10^{23} photons. The energy of each photon is equal to $E = hc/\lambda$ in which h is 6.6256×10^{-27} erg s quantum^{-1} and $c = 2.9979 \times 10^{10}$ cm s^{-1}. In molar terms, however, E (in kJ einstein^{-1}) $= 11.962 \times 10^4/\lambda$ (in nm).

Photochemists commonly use one or the other of a variety of units to express energy. A selection of the most commonly used units and their conversion factors is given in table 1.1. Using these relationships, it is easy to derive the energy in kJ einstein^{-1} corresponding to the wavelength of the radiation. Photons with wavelengths of 700 nm correspond to an energy of 171 kJ einstein^{-1} while those of 200 nm correspond to 598 kJ einstein^{-1}. These energies cover the range of covalent bond energies in most chemical compounds and for this reason radiation of

Table 1.1. *Energy–frequency–wavelength relations*

		kcal mol^{-1}	kJ mol^{-1}
Wave number $\bar{\nu}$	cm^{-1}	$2.86 \times 10^{-3}\,\bar{\nu}$	$11.95 \times 10^{-3}\,\bar{\nu}$
Wavelength λ	Å	$\dfrac{2.86 \times 10^{5}}{\lambda(\text{Å})}$	$\dfrac{11.95 \times 10^{4}}{\lambda(\text{Å})}$
Wavelength λ	nm	$\dfrac{2.86 \times 10^{4}}{\lambda(\text{Å})}$	$\dfrac{11.95 \times 10^{4}}{\lambda(\text{nm})}$

this type, which spans the visible and ultraviolet regions, is particularly effective in inducing chemical reactions.

1.4 Laws of photochemistry

There are two fundamental laws of photochemistry, the first of these, that formulated by Grotthus and Draper, states that 'only the light which is absorbed by a molecule can be effective in producing photochemical change in that molecule'. The second law states that 'the absorption of light by a molecule is a one-quantum process so that the sum of the primary process quantum yield must be unity'. Thus, $\sum_i \phi_i = 1$ where ϕ_i is the quantum yield of the ith primary process. These primary processes may include dissociation, isomerization, fluorescence, phosphorescence, radiationless transitions, and indeed all other reaction paths which lead to destruction or deactivation of the excited state of the molecule. Multiphotonic processes, which are now becoming more commonly observed with the advent of high intensity light sources such as lasers, can be included in this law by the assumption that absorption occurs in discrete one-photon steps to the final excited state.

The absorption of a monochromatic beam of light by homogeneous absorbing systems is usually described by the Beer–Lambert law in the form

$$I/I_0 = 10^{-\varepsilon cl} \tag{1.12}$$

where I_0 represents the incident light intensity per unit time (or number of quanta per unit time) at the front of a column of length l centimeters of a single absorbing species of concentration c mol $liter^{-1}$ and I is the light intensity per unit time emerging from the column. ε is the molar decadic extinction coefficient in liter mol^{-1} cm^{-1}, and is a measure of the probability that a particular molecule will absorb a quantum of light during its interaction with a photon. When more than one species of an absorbing compound is present in a homogeneous mixture, the Beer–Lambert law can be written in the form:

$$I/I_0 = 10^{-(\varepsilon_1 c_1 + \varepsilon_2 c_2 + \cdots + \varepsilon_i c_i \cdots)l} \tag{1.13}$$

The experimentally observed absorption is usually expressed in terms of the absorbance A given by:

$$A = \log_{10}(I_0/I) = \varepsilon cl \tag{1.14}$$

the concentration c being expressed in mol liter^{-1} and the length l in centimeters.

1.5 Absorption and molecular excitation

When light is absorbed by any molecule, the photon must interact with an atom or group of atoms and promote transitions between quantized states. For absorption to occur, two conditions must be satisfied. First, for an initial molecular state m with energy E_m there must be a state n of higher energy E_n such that $h\nu = E_n - E_m$ and second, there must be a specific interaction between the electric component of the incident radiation and the chromophore which results in a change in the dipole moment of the molecule during the transition. That is to say, the transition moment integral, which is defined as

$$R_{mn} = \langle \psi_m | \hat{R} | \psi_n \rangle \tag{1.15}$$

must be non-zero. In this equation, ψ_m and ψ_n are the wavefunctions of states m and n, respectively, and $\hat{R}$ is the electric dipole operator which has the form

$$\hat{R} = e \sum_i \mathbf{r}_i \tag{1.16}$$

In this equation e is the electronic charge and $\mathbf{r}_i$ is the vector which corresponds to the dipole moment operator of electron i. If these two conditions are met, then the probability of absorption would be proportional to $|R_{mn}|^2$. Electronic transitions with high absorption probability are said to be 'allowed'. If the transition moment integral is small or zero, the transition is said to be 'forbidden'.

The energy states that a molecule may have are characterized by the set $\{\psi_n\}$ of solutions to the Schrödinger equation

$$\hat{H}\psi_n = E_n \psi_n \tag{1.17}$$

where $\hat{H}$ is the complete quantum-mechanical Hamiltonian. Exact solutions to the Schrödinger equation are extremely difficult to obtain, even for very simple molecules, so that for larger molecules it is customary to invoke the Born–Oppenheimer approximation, which is

$$\psi \approx \phi_{(\text{orbit})} \chi_{(\text{vibration})} S_{(\text{spin})} \tag{1.18}$$

where ϕ is the electronic part of the complete wavefunction, χ is the nuclear wavefunction, and S is the spin wavefunction. This allows us to separate the transition moment integral into separate orbital, vibrational and spin components, as in

$$R_{mn} \approx \langle \phi_m | R | \phi_n \rangle \langle \chi_m | \chi_n \rangle \langle S_m | S_n \rangle \tag{1.19}$$

Using this equation, certain selection rules can be derived to predict whether the electronic transition is allowed or forbidden. These rules can be summarized as follows.

1.5.1 *Symmetry*

Certain symmetry properties are possessed by ϕ_m and ϕ_n in addition to those corresponding to $\hat{R}$. Transition is symmetry forbidden when the product of the symmetry portions of these functions produces an integrand which is an odd function of its spatial coordinates since this results in a vanishing small transition moment integral.

1.5.2 *Overlap*

Transitions are said to be overlap or spatially forbidden when the two orbitals ϕ_m and ϕ_n involved in the transition do not simultaneously possess large in-phase amplitudes in the same region of space. This lack of overlap leads to zero value for R_e

$$R_e = \langle \phi_m | \hat{R} | \phi_n \rangle \tag{1.20}$$

1.5.3 *Spin*

The transition moment integral will also vanish if the spin multiplicity is not conserved during the electronic transition. A transition for which S_n is not equal to S_m is then said to be 'spin forbidden'.

Although these selection rules are very useful guidelines in predicting the probability of absorption, they may in some cases be relaxed in large molecules because of the approximate nature of ψ in the Born–Oppenheimer treatment, and also because of perturbations of the system caused by spin-orbital coupling. Spin is the most rigid selection rule, and a transition which involves a change in spin is 10^4 to 10^6 times less probable than a totally allowed transition. Similarly, transition probabilities are reduced by a factor of 10 to 1000 for space-forbidden transitions and 10 to 100 for symmetry-forbidden transitions, respectively. Electrons in single bonds are usually held too tightly to be excited by near ultraviolet radiation. Hence, in this region of the electromagnetic spectrum absorption corresponds to promotion of the labile π and n non-bonded electrons of chromophores.

1.6 Molecular orbitals

Ground and excited states of molecules can be described in terms of molecular orbitals (MO). A complete treatment of molecular orbital theory is beyond the scope of this book. For our purposes, two chromophores will be used to illustrate the types of electronic transitions occurring in molecules. These will be the transitions associated with the carbonyl group, and in aromatic rings, respectively. Transitions between different electronic states will be denoted in conventional photochemical usage by listing the ground state first and the excited state second, with a hyphen between them. For example, two types of electronic transitions occur in the carbonyl groups of aldehydes and ketones, corresponding to n-π^* and π-π^* excitation. The molecular orbital states are shown below.

n-π^*

$$S_0\{\cdots(\pi_{CO})^2(n_O)^2(\pi^*_{CO})^0(\sigma^*_{CO})^0\}\xrightarrow{h\nu_1}S_1\{\cdots(\pi_{CO})^2(n_O)^1(\pi^*_{CO})^1(\sigma^*_{CO})^0\}$$

π-π^*

$$S_0\{\cdots(\pi_{CO})^2(n_O)^2(\pi^*_{CO})^0(\sigma^*_{CO})^0\}\xrightarrow{h\nu_2}S_2\{\cdots(\pi_{CO})^1(n_O)^2(\pi^*_{CO})^1(\sigma^*_{CO})^0\}$$

It is now known from both theoretical calculations and experimental evidence that in simple carbonyl systems the highest energy occupied molecular orbital is a non-bonding n orbital and the lowest unoccupied molecular orbital is a π^* anti-bonding orbital delocalized across the C–O bond. These energy levels are illustrated for formaldehyde in figure 1.1. The weak absorption observed around 280 nm in aliphatic ketones is therefore attributed to the n-π^* transition. However, the n and π^* orbitals are orthogonal to each other and hence the n-π^* transition violates both symmetry and the orbital selection rules. It is therefore relatively weak, having an extinction coefficient typically less than 100. In contrast the π-π^* transitions with absorption maxima around 180 nm are completely allowed and are therefore much more intense, with absorption extinction coefficients of the order of 10 000.

In aromatic carbonyl systems the n-π^* absorption is shifted to longer wavelengths, around 320 nm, with a corresponding increase in the extinction coefficient to about 100 to 150. This is a consequence of the inductive effect of the π system of the phenyl ring, which causes a lowering of the π^* energy level of the carbonyl and mixing of the CO n-π^* state with the π-π^* state of the phenyl system.

In planar aromatic compounds such as benzene, the simplest MO picture represents the planar skeleton of the aromatic ring as being constructed by a combination of the sp^2 hybrid orbitals of carbon with the 1s orbitals of hydrogen, as shown diagramatically in figure 1.2*a*. The

p_z atomic orbitals of each carbon atom which are unchanged by the sp^2 hybridization project perpendicularly from the plane of the ring. The six p_z atomic orbitals of each carbon atom interact to form a π-electron cloud above and below the plane of the ring, as shown in figure 1.2*b*. The additional stabilization energy associated with π electrons reduces the carbon–carbon bond length in the aromatic ring from the 1.54 Å associated with the conventional σ-bond to 1.39 Å in benzene. These π electrons are delocalized in benzene and other aromatic molecules and it

Figure 1.1. Molecular orbitals and their approximate energy levels in formaldehyde. (Adapted from Kasha 1960.)

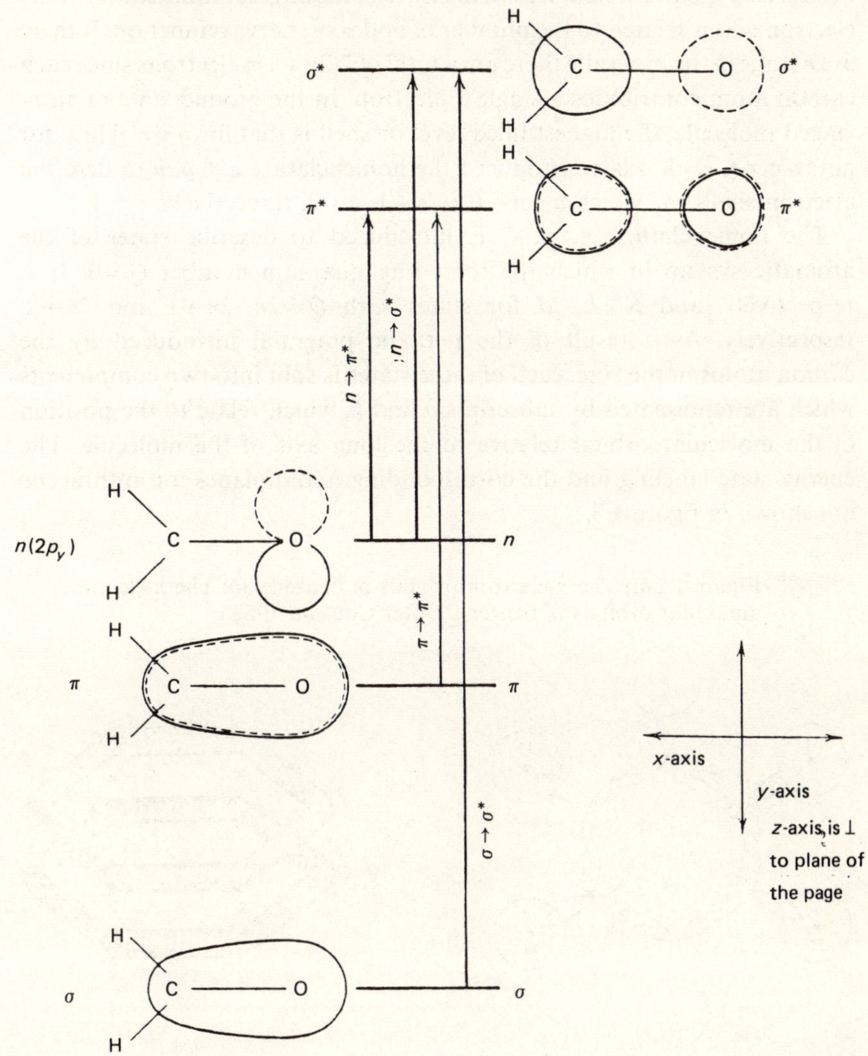

is the excited states of these π electrons systems which are usually those of lowest energy and thus whose transitions and interactions are of most interest in photochemistry.

The excited states associated with π-electrons in aromatic systems are usually described in terms of the perimeter free electron orbital model (PFEO) (Platt 1949, Platt 1964). In this model, the π orbitals are treated as the orbitals of free electrons travelling in a loop around the molecular perimeter of the aromatic system. Each electronic state is characterized by a total ring quantum number Q, which is the algebraic sum of the orbital ring quantum numbers, q, of the individual π electrons. This orbital ring quantum number is a measure of the angular momentum of the electron and is related to the number of nodes of the wavefunction. If there are n rings in the molecule there are a total of $2(2n+1)\pi$ electrons since each carbon atom contributes a single π electron. In the ground state of an n-ringed molecule, the highest filled level or shell is that for $q=n$. Thus, for anthracene, $n=3$. Platt introduced the nomenclature e, f, g, h to describe electron shells for which $q=n-1$, n, $n+1$, $n+2$, respectively.

The nomenclature A, B, C is introduced to describe states of the aromatic system in which the total ring quantum number $Q=0$, 1, 2, respectively, and K, L, M for states with $Q=2n$, $2n+1$ and $2n+2$, respectively. As a result of the periodic potential introduced by the carbon atoms in the ring, each of these states is split into two components which are represented by subscripts a and b, which relate to the position of the molecular orbital relative to the long axis of the molecule. The energy state labeling and the corresponding nodal planes for anthracene are shown in figure 1.3.

Figure 1.2.(*a*) The σ-electron orbitals of benzene. (*b*) The π-electron molecular orbitals of benzene. (After Coulson 1952.)

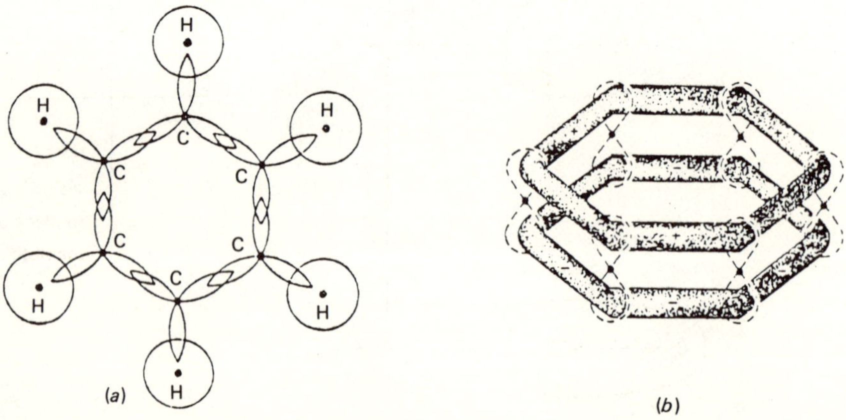

In general, for aromatic compounds in which the benzenoid rings are arranged in a linear fashion the lowest excited state is triplet L_a, followed by singlet L_a or singlet L_b. As indicated in the diagram, the B_a state corresponds to a strong dipole oscillation in the molecular plane, polarized perpendicular to the long molecular axis, while the B_b state corresponds to a strong oscillation polarized parallel to this axis. Both the L_a and L_b states correspond to weak dipole oscillations with polarizations perpendicular and parallel to the long molecular axis, respectively. For further details of the MO treatment of aromatic compounds, the reader is referred to Platt (1964).

1.7 Singlet and triplet states

The ground states of most molecular systems are *singlets*, that is states in which all the electrons are paired. Direct excitation from a singlet ground state to the excited *triplet* state is spin forbidden and is thus negligible for n-π^* transitions. However, the *triplet* state, that is one with an unpaired electron in the upper excited state, can be populated by spin flipping of an electron in the excited singlet state as a result of spin-orbit interactions. This is an example of a process known as *intersystem*

Figure 1.3. The perimeter free-electron orbital model. Classification of the L_a, L_b, B_a and B_b states of anthracene.

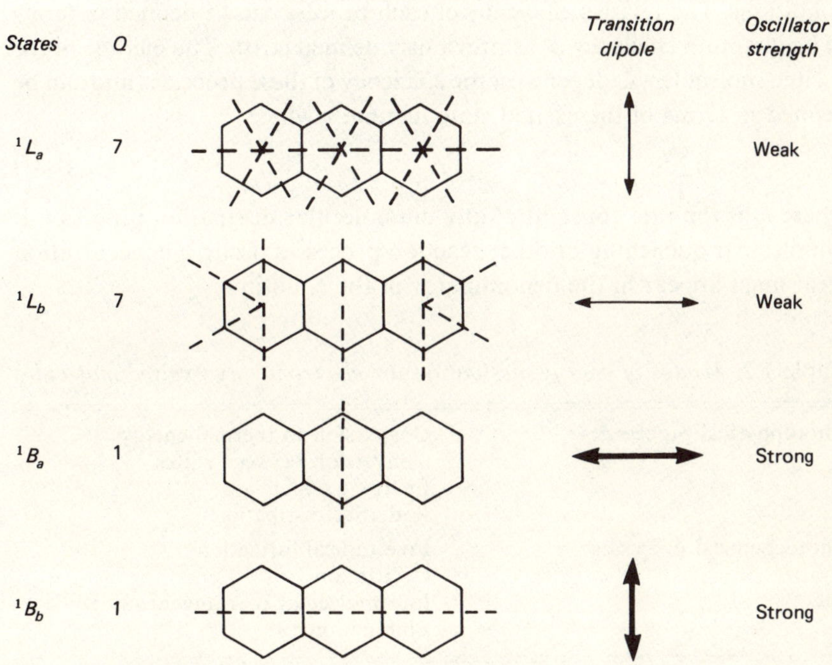

crossing, which will be discussed later. According to Hundt's rule the electronic configuration with the higher spin multiplicity is more stable, other things being equal. For this reason, the triplet state of a given electronic configuration usually has a lower energy than the corresponding singlet state. If the two electrons have parallel spin, they cannot, according to the Pauli exclusion principle, occupy the same molecular orbital and consequently they will exist in different regions of space, thereby reducing the electron–electron repulsion and lowering the energy level of the state correspondingly.

1.8 Dissipation of excitation energy

After absorption of a photon a molecule exists in an excited state, and because it is not in thermal equilibrium with its surroundings, it will, in general, have only a short lifetime since a number of processes can contribute to the deactivation of the excited molecule to the lower state. These include both chemical and physical processes and are summarized in table 1.2. If the process gives rise to chemically distinguishable products, it is said to be photochemical; otherwise, it is labeled as photophysical. The alternate pathways compete with each other for energy dissipation and the dominant routes are usually determined by factors such as the structure and environment of the molecule, the nature of the excited state, and temperature, pressure, and other experimental conditions. The relative efficiency of each process can be defined in terms of its quantum efficiency ϕ_i as previously defined (1.10). The lifetime of the excited molecule will depend on the efficiency of these processes and can be defined in terms of the excited state lifetime τ where

$$\tau = 1/\sum_i k_i \tag{1.21}$$

where k_i is the rate constant of any unimolecular dissipation process i. If bimolecular quenching or other reaction processes occur, a concentration term must appear in the denominator of the equation.

Table 1.2. *Modes of energy dissipation for electronically excited molecules*

Photophysical processes:	Conversion to thermal energy Conversion between states Energy transfer Radiative dissipation
Photochemical processes:	Free radical formation Cyclization Intramolecular rearrangement Elimination

A convenient description of the modes of energy dissipation from possible excited states is given by the Jablonski diagram (figure 1.4). In this diagram only the two lowest energy excited states, S_1 and T_1, are shown. There will, of course, be higher energy singlet and triplet states, but in condensed phases these two are most likely to be involved in photochemical reactions.

According to the Franck–Condon principle, the time required for absorption of a photon and resultant transition of an electron to an excited state is very short, of the order of 10^{-15} s. This is very small compared to the period of vibration of most molecules, which is of the order of 10^{-13} s, so that during the absorption process the nuclei do not alter their relative positions or their kinetic energies appreciably. As a result of this, absorption of a photon usually produces a state which is excited rotationally and vibrationally as well as electronically. In highly dilute gases, where collisional deactivation is slow, these excited states can actually be observed spectroscopically. However, in solids or liquids, the combination of a dense distribution of excited states and high collision frequencies between the molecules causes rapid relaxation of vibrational and rotational modes to occur very efficiently, in about 10^{-11} s, so that only molecules in the lowest vibrational level of the first excited electronic state persist long enough to be important photochemically, although there are some notable exceptions to this rule, e.g.,

Figure 1.4. Energy level diagram showing deactivation pathways for a typical organic chromophore. ABS = absorption, VR_1 = vibrational relaxation, IC = internal conversion, F = fluorescence, ISC = intersystem crossing, P = phosphorescence, S_0 = ground state singlet, S_1 = first excited singlet, T_1 = first triplet.

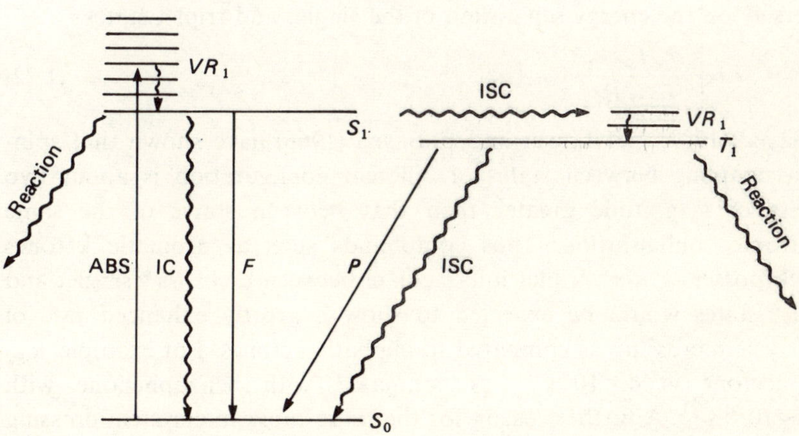

azulene, which shows fluorescence due to an S_2-S_0 transition (Beer and Longuet-Higgins 1955).

If the chromophore is excited to a higher electronic singlet state (say S_2), there is an extremely rapid process called *internal conversion* (IC) which occurs with a rate constant of 10^{11} to $10^{13}\ s^{-1}$, which converts the higher singlet state to the lowest excited singlet and is a result of frequent crossings or near crossings of excited state potential surfaces above the S_1 state. The probability of this occurring is inversely proportional to the energy gap between the states. For this reason, the large energy gap between S_0 and S_1 results in a considerably slower rate of conversion of S_1 to S_0 by internal conversion.

In general, electronic conversions between states of identical multiplicity are known as *internal conversion* (IC), whereas those between states of differing multiplicity are known as *intersystem crossing* (ISC). These processes are similar in that they involve a conversion of electronic energy to vibrational energy (heat) which is followed by rapid relaxation to the lowest vibrational level of the lowest excited state. They occur without emission of a photon of light and hence are known as *radiationless processes.*

1.9 Spin-orbit coupling

Intersystem crossing is, of course, spin forbidden. However, it is facilitated by *spin-orbit coupling* (SO). This coupling arises from magnetic interactions between the orbital motion of an electron and its spin magnetic moment. Spin-orbit coupling results in a mixing of two or more manifolds of differing multiplicity, thus relaxing the spin selection rules and permitting effective intersystem crossing. The magnitude λ of this mixing depends directly on the nature of $\hat{H}_{SO}$, the spin-orbit operator and inversely on the energy separation of the singlet and triplet states.

$$\lambda = \left| \frac{\langle \psi_S \hat{H}_{SO} \psi_T \rangle}{E_S - E_T} \right| \tag{1.22}$$

El-Sayed (1963) and Lower and El-Sayed (1966) have shown that spin-orbit coupling between states of different configuration is about two orders of magnitude greater than that between states of the same electronic configuration. Thus compounds such as aromatic ketones which possess a π-π^* triplet intermediate between their n-π^* singlet and triplet states would be expected to show a greatly enhanced rate of intersystem crossing as compared to aliphatic ketones. For example, k_{ISC} for acetone is $4 \times 10^7\ s^{-1}$ as compared with benzophenone with $k_{ISC} \sim 10^{10}\ s^{-1}$. A further reason for the more rapid intersystem crossing in aromatic ketones is the smaller energy gap between S_1 and T_1 compared to aliphatic ketones. This smaller energy difference increases the Franck–Condon overlap factors between the excited states. The energy gap between the states involved in intersystem crossing also affects the rate at which this transition occurs because it affects the Franck–Condon overlap factors which are inversely related to this gap. Since the energy difference for the S_1 and T_1 states is only roughly 20 kJ mol^{-1} for ketones, as compared to typical values of greater than 100 kJ mol^{-1} for S_0 and T_1 states, k_{ISC} for $S_1 \rightarrow T_1$ will be much larger than the corresponding rate constant for $T_1 \rightarrow S_0$. This dependence upon the energy gap involved has been demonstrated by Siebrand (1967) for aromatic molecules in solid solution and their $T_1 \rightarrow S_0$ solution transitions. Changes in the Franck–Condon overlap factors have also been held responsible for the decrease in intersystem crossing rates upon deuteration since the latter only affects the position of the vibrational levels.

1.10 Radiationless decay

Radiationless decay is essentially the conversion of electronic excitation into rotational and vibrational motions of the molecule itself or of its surroundings. It can be considered as a failure of the Born–Oppenheimer approximation. The overall process may be thought of as consisting of two steps, first the intramolecular conversion of electronic excitation to vibrational energy and second, the transfer of that vibrational energy to the environment. Two approximations have been used in the theoretical analysis of the problem. The most common assumption is that in condensed media, the first step is rate determining, and that the role of the medium is restricted to that of a sink for thermal energy. For example, the Robinson–Frosch theory (Robinson and Frosch 1963; Robinson 1974) postulates a resonance interaction between weakly coupled initial and final electronic states. The theory relies on the determination of the Franck–Condon factors and stresses the formal analogy to the theory of radiative transitions. It has been successful in predicting the temperature independence of radiationless decay, and the effect of deuteration in aromatic hydrocarbon systems.

In an alternative approach, Gouterman (1962) examined the other limiting situation in which the transfer of energy to the surrounding medium is the slow step. In the absence of a dense medium, no transition would occur. A consequence of this theory is a strong temperature dependence for non-radiative transition, a phenomenon which has been shown to occur for transition metal complexes, but not with aromatics. Although the effect of the medium is important, as evidenced by the heavy atom effect, examples exist which show that its presence is not essential to the conversion of electronic energy to vibrational energy. Such results have produced a number of theories which attempt to account for radiationless transitions without requiring the presence of a medium.

Hammond (1969) has proposed that chemical reactions of excited states are special forms of radiationless decay. For unimolecular reactions, the approximate solution of the total rate constant $k_{A^* \to B}$ for conversion of A to a group of nearly isoenergetic vibrating states of B was obtained by applying radiationless decay theory:

$$k_{A^* \to B} = \frac{2\pi \beta_{\mathrm{el}}^2 \sum_n \langle \phi_{nB} | \phi_{0A^*} \rangle^2}{\hbar \alpha} \tag{1.23}$$

where β_{el} is the interaction between the electronic wavefunctions of A^* and B, ϕ_{nB} is the nth vibrational wavefunction for B to which transition initially occurs, ϕ_{0A^*} is the vibrational function for thermally equilibrated A^* and α is the average energy of the interaction between the molecular vibrational modes and the environment.

1.11 Radiative processes

The dissipation of electronic excitation by radiation can be very efficient in certain types of aromatic hydrocarbons. For example, scintillator molecules have been designed, for which the quantum yields of fluorescence emission are close to unity. In other types of chromophores, for example, the ketone carbonyl, where the n-π^* transition is both symmetry and orbital forbidden, the quantum yield for radiative decay is usually very low. In many other chromophores, however, emission of radiation can represent a major method of dissipation of electronic excitation.

1.11.1 *Fluorescence*

Fluorescence is defined as the emissive transition between states of identical multiplicity. Rapid vibrational relaxation and internal conversion in the solid or liquid state enables the emission to take place, usually from the lowest vibrational level of the first excited singlet state. This is known at Kasha's rule (Kasha 1950). The consequence of this relaxation is that the longest absorbed and shortest emitted wavelength maximum usually corresponds to the 0-0 transition. A mirror image relationship then usually exists between absorption and emission spectra. An example of this relationship is shown in figure 1.5.

A theoretical treatment of the rate of emission from excited singlet states gives the inherent radiative lifetime ${}^1\tau_0$ of this state as

$$^{1}\tau_0 = \frac{3.5\times10^8}{\nu_m^2 \int \varepsilon(\nu)\mathrm{d}\nu} = \frac{1}{k_F} \tag{1.24}$$

where ν_m is the mean frequency for the absorption band, and the integral $\int\varepsilon(\nu)\mathrm{d}\nu$ is the experimental extinction coefficient integrated over the width of the absorption band. $^{1}\tau_0$ represents the lifetime in the absence of any other deactivation pathway. If the absorption band is symmetrical, this equation can be approximated by the relation

$$^{1}\tau_0 = \frac{3.5\times10^8}{\nu_m^2 \varepsilon_m \Delta\nu_{1/2}} \tag{1.25}$$

where ε_m is the extinction coefficient at the band maximum and $\Delta\nu_{1/2}$ is the band half-width. The radiative lifetime can thus be computed from spectroscopic information. Fluorescence is a comparatively fast process because it is a spin allowed transition and the radiative fluorescence lifetime $^{1}\tau_0$ is usually in the range of 10^{-6} to 10^{-9} s. However, it should be pointed out that this is longer than the measured fluorescence lifetime $^{1}\tau$ which is the reciprocal of the sum of all rate constants for processes depleting the excited state.

1.11.2 *Phosphorescence*

Phosphorescence emission involves a transition between states of differing multiplicity, usually T_1 and S_0. Since direct excitation to the first

Figure 1.5. Absorption and emission spectra for benzene (3 ml/l) in cyclohexane (λ_{ex} = 254 nm) showing approximate mirror image relationship. (After Berlman 1971)

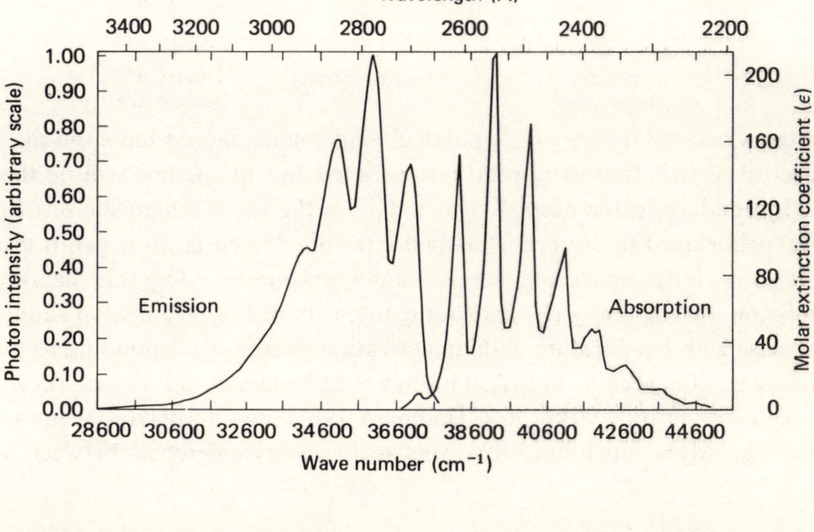

excited triplet T_1 is forbidden by all three selection rules outlined in §1.5, this state is usually populated only by intersystem crossing from the first excited singlet state. The radiative lifetime is given by

$$^3\tau_0 = \frac{3.5 \times 10^8}{\nu_m^2 \int \varepsilon(\nu)\mathrm{d}\nu} \frac{g_3}{g_1} \tag{1.26}$$

where g_3 and g_1 are the degeneracies of the triplet and singlet states, respectively. Such forbidden transitions have very low extinction coefficients ε for absorption, and hence it is obvious from (1.24) that the triplet state will usually possess a long radiative lifetime. This is equivalent to the statement that a state which is difficult to populate by direct excitation is equally difficult to depopulate by radiative emission. For many chromophores the phosphorescence lifetime may range from 10^{-3} s to several minutes, considerably longer than that of the excited singlet state. For this reason the triplet state often plays a central role in organic photochemistry since, other factors being equal, it persists for a much longer time and consequently has a greater probability of undergoing photochemical reaction. Furthermore it might be expected that the triplet and singlet will show different chemical reactivity.

1.11.3 *Delayed fluorescence*

Two other types of emission may deplete triplet population. *Delayed fluorescence* can be caused by intersystem crossing from the first excited triplet T_1 back to the first excited singlet S_1 when the energy gap between them is small enough to allow the thermal repopulation of the excited singlet:

$$\begin{array}{ccccc} S_0 \xrightarrow{h\nu} S_1 & \overset{\text{ISC}}{\rightsquigarrow} & T_1 & \underset{\text{ISC}}{\overset{\Delta}{\rightsquigarrow}} & S_1 \\ \downarrow & & \downarrow & & \downarrow \\ \text{normal fluorescence} & & \text{phosphorescence} & & \text{delayed fluorescence} \end{array} \tag{1.27}$$

Fluorescence of this newly populated singlet state should have the same spectral distribution as normal fluorescence, but its lifetime will be that corresponding to the excited triplet. This is the usual diagnostic test for 'E-type' delayed fluorescence (named after the dye eosin, in which it was first observed). Since the singlet states responsible for this delayed emission are thermally populated, the intensity of E-type emission should increase with temperature with an activation energy corresponding to the energy gap between S_1 and T_1. This has been shown for fluorescein (Lewis *et al.* 1941), eosin (Parker and Hatchard 1961), and acriflavin (Wen and Hsu 1963) dyes which have relatively small energy differences between S_1

and T_1. However, when the energy gap exceeds about 40 kJ mol^{-1}, E-type emission is insignificant at normal temperatures.

The second process arises when the triplet lifetimes of two excited species, *A* and *B*, or the same species, are sufficiently long that an encounter occurs between them. This is termed *triplet–triplet annihilation* and can be represented by

$$T_1 + T_1 \rightsquigarrow S_1^v + S_0^v \rightsquigarrow \underset{\substack{\downarrow \\ \text{delayed} \\ \text{fluorescence}}}{S_1} + S_0 \qquad (1.28)$$

where S_1^v and S_0^v denote vibrationally excited states of S_1 and S_2, respectively. The overall result of this sequence of encounters is also delayed fluorescence emission, but now the lifetime τ will be significantly longer than that for normal fluorescence. In general τ_{DF} will be shorter but of the same order of magnitude as τ for phosphorescence. This is a biphotonic process, and hence the intensity of the delayed emission should be linearly dependent on the square of the excitation intensity. This is the principal diagnostic test for P-type emission, so called because it was first shown experimentally with solutions of pyrene.

An analysis of the kinetic scheme implied in (1.28) results in an emission intensity which decays with a single-exponential law characterized by a lifetime

$$\tau_{DF} = \tfrac{1}{2}\tau_P \qquad (1.29)$$

where τ_P is the phosphorescence lifetime. This also distinguishes P-type emission from E-type where $\tau_{DF} = \tau_P$. However, as pointed out by Parker (1968) this relationship only holds when the excitation intensity is low so that other types of triplet–triplet interactions do not interfere.

1.12 Energy transfer processes

Energy transfer is defined as the donation of excitation energy from one molecule to another or from one chromophore to another. While the transfer of vibrational and rotational energy between molecules can be understood quite well in terms of quantized momentum transfer, additional theory is required to explain the phenomenon of electronic energy transfer. One of the most general forms of electronic energy transfer can be represented by

$$D^* + A \rightarrow D + A^* \qquad (1.30)$$

where D represents a donor molecule, A represents an acceptor molecule and * denotes electronic excitation. This equation assumes that at some initial time, electronic excitation may be associated with D and at some later time it is unambiguously associated with A. The energy transfer process may involve two steps with the intermediacy of a photon (this is known as radiative transfer) or may be a one-step radiationless process requiring direct interaction of the excited donor with the acceptor. Not all energy transfer processes follow (1.30) since there are some well known examples of energy transfer in which the excitation energy originally on the donor does not result in electronic excitation of the acceptor. Such a process could be represented by

$$D^* + A \rightarrow D^v + A^v \tag{1.31}$$

In this case, the electronic excitation is presumably converted into heat, that is, into the rotations and vibrations of donor and acceptor molecules.

Radiative or 'trivial' transfer involves the reabsorption of the donor emission by the acceptor, a process which can be represented by

$$\begin{aligned} &D^* \rightarrow D + h\nu \\ &h\nu + A \rightarrow A^* \end{aligned} \tag{1.32}$$

This type of energy transfer may even occur when D and A are similar molecules in spite of the small overlap of the emission and absorption spectra although the effect is usually small. On the other hand, radiationless transfer of electronic energy involves the simultaneous deexcitation of the donor and excitation of the acceptor, which requires a direct interaction between D and A and a low energy virtual transition between the two. Depending on the nature of the interaction between donor and acceptor, different mechanisms have been suggested to explain the phenomenon of radiationless energy transfer.

1.13 Resonance transfer

Resonance transfer is a long-range transfer of electronic excitation which occurs when the separation of the donor and the acceptor is several times greater than the sum of their van der Waals radii. At this distance only Coulombic force interactions are important and electronic exchange effects are negligible. Expanding the Coulombic term of the Hamiltonian as a power of $1/r$, one can write

$$H_1 = e^2/r^3 + e^2/r^4 + e^2/r^5 \tag{1.33}$$

The first term in this equation describes dipole–dipole interactions, the second term dipole–quadrupole interactions, and so on. For allowed transition, the dipole–dipole interaction term dominates the expression. So-called 'forbidden' energy transfer processes may result from dipole–quadrupole or higher order interactions.

Förster (1948) developed a quantitative relation expressed in terms of experimentally obtainable parameters for the rate of resonance energy transfer originating in dipole–dipole interactions between an isolated donor and acceptor pair.

$$k_{D^*\to A} = \frac{8.8\times 10^{-25}\kappa^2\phi_D}{n^4\tau_D R^6}\int_0^\infty F_D(\nu)\varepsilon_A(\nu)\frac{d\nu}{\nu^4} \tag{1.34}$$

where $k_{D^*\to A}$ is the rate constant for energy transfer, R is the donor–acceptor separation, κ^2 is an orientation factor, usually assumed to be 2/3 for a random distribution of D and A, ϕ_D is the quantum yield for emission from donor, n is the refractive index of the solvent, τ_D is the mean lifetime of the donor excited state, $F_D(\nu)$ is the normalized spectral distribution of donor emission, and $\varepsilon_A(\nu)$ is the molar extinction coefficient of the acceptor as a function of ν. If one defines a critical distance R_0 as that distance between the donor and the acceptor at which energy transfer occurs with 50% probability, it follows that

$$R_0^6 = \frac{8.8\times 10^{-25}\kappa^2\phi_D}{n^4}\int_0^\infty F_D(\nu)\varepsilon_A(\nu)\frac{d\nu}{\nu^4} \tag{1.35}$$

For fully allowed transitions, that is with ε_{max} of the order of 10^4 in the acceptor and with significant spectral overlap and donor emission yields of 0.1 to 1, R_0 values of 5–10 nm arc prcdictcd, and experimentally confirmed by Bowen and Livingston (1954). Efficient transfer only occurs when there is significant overlap between donor emission and acceptor absorption spectra, the same requirement as for trivial transfer, i.e., donor emission followed by acceptor absorption. However, this can usually be neglected for many chromophores if experiments are carried out at high dilution. However, in concentrated solution with chromophores of high extinction coefficient, trivial transfer does become an experimental factor and must be taken into account (figure 1.6). Its occurrence is usually identified by variation of the emission spectrum with path length of the experimental cell.

The requirement for a strong overlap between emission absorption spectra of D and A is very important. However, in certain cases chromophores with only small overlaps can still demonstrate resonance energy transfer if the value of ϕ_D, the quantum yield for donor fluorescence is large so that energy transfer can compete with other deactivating processes for the excited state. Thus forbidden transitions in the donor do not preclude the observation of an electronic energy transfer by these mechanisms, however, if the transition is forbidden in the acceptor the mechanism always predicts a low efficiency for the transfer step.

Figure 1.6. Singlet–singlet energy transfer from *p*-terphenyl (2.17×10^{-2} M) to tetraphenylbutadiene (TPB) in toluene solutions. Energy transfer quantum efficiency f_{YM} against TPB concentration $[^1Y]$. Radiative component, f_R; radiationless component, f_{NR}. Solution thickness, 13 nm. (After Birks and Kuchela 1961.)

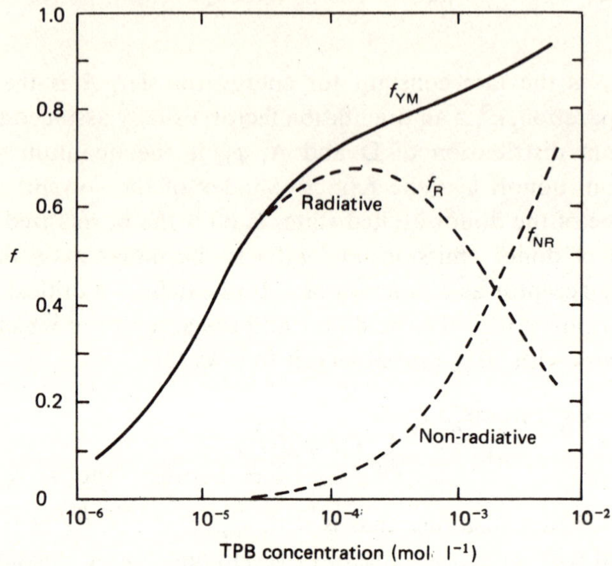

1.14 Exchange energy transfer

A second important mechanism for radiationless energy transfer involves the exchange of electronic excitation when the donor and acceptor are close enough so that their electron clouds overlap. In the region of overlap the electrons are indistinguishable, so that an excited electron on D* may also be on A. Since the densities of the electron clouds fall off rapidly as the separation of D* and A increase, exchange interaction is a short range interaction. Dexter (1953) has derived a rate expression for exchange energy transfer given by

$$k_{D^* \to A} = (2\pi/\hbar)Z^2 \int f_D(\nu) f_A(\nu) d\nu \tag{1.36}$$

where

$$Z^2 = K^2 \exp(-2R/L) \tag{1.37}$$

where K is a constant with the dimensions of energy, L is a constant called the effective average Bohr radius, $f_D(\nu)$ is the donor emission spectrum, $f_A(\nu)$ is the acceptor absorption spectrum, and R is the donor-acceptor distance. The rate constant is clearly more dependent on R than in the case of Förster transfer.

Allowable transitions via this mechanism are governed by the Wigner spin rule that if S_A and S_D are the initial spin quantum numbers of acceptor and donor, respectively, the total spin quantum numbers of the resultant species must have one of the values

$$|S_A+S_D|, |S_A+S_D-1|, |S_A+S_D-2| \ldots |S_A-S_D|$$

For this reason both singlet–singlet and triplet–triplet energy transfer can occur by exchange interaction. However, the latter process cannot occur via resonance transfer because it involves forbidden transitions in both D* and A. Triplet–triplet transfer is possible by the exchange mechanism because the forbiddenness of the acceptor transition is compensated for by the long lifetime of the donor. It is an important process used to determine the multiplicity of excited states in quenching studies, and in sensitizing various types of reaction proceeding from photo-excited intermediates.

1.15 Non-classical energy transfer

Both the resonance and the exchange energy transfer mechanisms require a non-vanishing overlap between the spectrum of emission and absorption of D and A, respectively, and as is obvious from the equations, the rate of energy transfer will be zero if the overlap integral is zero. There are, however, numerous examples of intermolecular deactivation of electronic excitation energy which cannot be explained by the above classical theories. This is usually because of the absence of sufficient low-lying excited states of the quencher or acceptor. An example of this is the fluorescence quenching of aromatic hydrocarbons by dienes. In these systems the spectral overlap of the donor emission and the acceptor absorption spectrum is zero. In these cases it is proposed that a complex, or *exciplex* (Birks 1967) is formed between the excited molecule of the quencher and the acceptor which subsequently undergoes fast deactivation processes. The simplest form of such a mechanism is shown below:

$$\begin{array}{ccccc} D^* + A \rightleftharpoons [D\cdots A]^* \rightarrow & D^*_{(vib)} & + & A^*_{(vib)} & \\ \downarrow \quad & \downarrow & & \downarrow & \\ D+A+h\nu \quad & D & & A & \text{or chemical product} \end{array} \tag{1.38}$$

Since it seems clear that exciplexes are involved as intermediates in a number of fluorescence quenching processes, it now seems likely that they can also be intermediates in many bimolecular photochemical transformations. Further discussion of excimers and exciplexes is given in chapter 7.

1.16 Kinetics of energy transfer

Rate constants for energy transfer processes are of particular interest to photochemists and there are several experimental means for determining them. The first of these, a direct method, involves exciting a chromophore with a rapid light pulse followed by monitoring the dynamics of the excited species, such as by observing the decay of the emission or absorption spectrum. The second approach involves the competition for the excited state between emission or chemical reaction and deactivation by an added quencher or acceptor. The kinetics, following Stern and Volmer (1919), are given in the following equations:

$$D^* \rightarrow \text{product or } h\nu \tag{1.39}$$

$$D^* + A \xrightarrow{k_q} D + A^* \tag{1.40}$$

The experimental quantum yields of the process of interest is related to the molar concentration of quencher A by the *Stern–Volmer equation*,

$$\phi_0/\phi = 1 + k_q\tau[A] \tag{1.41}$$

where ϕ_0 and ϕ are the quantum yields for the process in the absence and presence of A, respectively, and τ is the lifetime of the donor in the absence of the acceptor. Adherence to a Stern–Volmer relationship is usually considered to be evidence for a mechanism in which there is competition for the deactivation of the donor between a bimolecular quenching process, such as energy transfer, and a unimolecular process such as emission. A plot of ϕ_0/ϕ versus [A] gives a straight line of slope $k_q\tau$ enabling the estimation of either quantity to be made, should the other one be known by an independent method.

If several competing unimolecular decay pathways are present for a given excited species, Stern–Volmer plots of each of the competing reactions should have the same slope, provided that all products come from the same excited state and the quencher A does not react with any of the product or intermediates. Deviations from linearity of the Stern–Volmer plot with increasing acceptor concentration are often used to determine the relative importance of different electronically excited states in a reaction. A well known example is the photoreaction of aliphatic carbonyls in n-π^* states where a triplet state is formed irreversibly from the singlet state and where many photoprocesses proceed from both states. In this case, if the triplet state is selectively quenched by species A, the Stern–Volmer plot of the total quantum yield will curve downward, as expressed by

$$\frac{\phi_0}{\phi} = 1 + \frac{k_q\tau_3[A]}{1 + (\phi_1^0/\phi_3^0)(1 + k_q\tau_3[A])} \tag{1.42}$$

where τ_3 is the lifetime of the ketone triplet, ϕ_1 and ϕ_3 are the quantum yields for processes from the singlet and triplet state, respectively, and the other symbols have their usual significance. A detailed discussion of Stern–Volmer kinetics in solution has been presented by Wagner (1972).

In solution, the upper limit to the value of k_q, the bimolecular quenching constant, corresponds to diffusion-controlled quenching where energy transfer occurs at every encounter of the donor and acceptor molecules. An approximate relation for the rate constant k_q in diffusion-controlled reaction is given by

$$k_q = 8RT/3000\eta \tag{1.43}$$

where R is the gas constant and η is the solvent viscosity in compatible units. For typical organic solvents at room temperature, k_q is of the order 10^9 to 10^{10} l mol^{-1} s^{-1}. For viscosities up to 10^7 poise, exothermic triplet–triplet energy transfer is nearly diffusion controlled with a rate inversely proportional to viscosity provided that the donor energy is higher than that of the acceptor by a few kJ mol^{-1}. This requirement is necessary so that energy transfer from the donor to the acceptor is not reversed by simple thermal excitation.

Many experimentally determined energy transfer rates agree very well with those calculated on the basis of this modified Stokes–Einstein relation. However, there are a number of cases of exothermic transfer in non-viscous solvents for which the measured rates are smaller than are the diffusion-controlled rates. In this case, $k_q = \alpha k_{DIF}$ where $\alpha \leqslant 1$, and represents a measure of the efficiency of energy transfer in each encounter of D with A. Wagner and Kochevar (1968) have correlated α with viscosity for a large number of solvents and have suggested that deviations from the diffusion-controlled rate may result from steric hindrance, which prevents a good overlap of the appropriate molecular orbitals, or from restrictive Franck–Condon factors as in cases where the donor and acceptor possess very structured spectra.

In another formulation, α can be related to the rate constant k_{ET} for energy transfer during a collision between D and A by

$$\alpha = \frac{k_{ET}}{k_{ET} + k_{-DIF}} \tag{1.44}$$

where k_{-DIF} refers to the rate constant of diffusion out of the encounter complex. Alpha thus represents the probability that energy transfer will occur during the lifetime of the encounter complex. In cases where α is equal to unity, and the rate constant for energy transfer is large compared to k_{-DIF}, quenching will occur at each collision and the observed Stern–

Volmer rate constant will be equal to k_{DIF}. In its simplest form, k_{DIF} can be approximated by

$$k_{DIF} = 2 \times 10^5 (T/\eta) \qquad (1.45)$$

where T is in K and η is in poise.

1.17 Energy transfer in solid matrices

Energy transfer may also occur in solid solutions under conditions where molecular diffusion is not possible. It may also occur between donor and acceptor combinations which cannot alter their relevant positions because of attachment to a rigid molecule. In this case, the Stern–Volmer formulation which assumed collisional processes for donor and acceptor molecules, is no longer valid and the Perrin equation is used. In this derivation it is assumed that there is an effective quenching sphere around the donor with radius R. If a quencher molecule is within this sphere, the excited state of the donor is deactivated with unit efficiency. However, if the quencher molecule is outside the quenching sphere, it does not quench at all. It is also assumed in this derivation that neither donor nor acceptor can undergo displacements in space during the lifetime of the excited state donor.

The Perrin model (Perrin 1924) predicts the following relationship:

$$\ln(\phi_0/\phi) = VN[\mathrm{Q}] \qquad (1.46)$$

where ϕ_0 and ϕ are the quantum yields for donor emission in the absence and presence of quencher, respectively, V is the volume of the active sphere, N is Avogadro's number and [Q] is the concentration of quencher. A plot of $\ln(\phi_0/\phi)$ versus [Q] yields NV and hence V may be directly evaluated. However, it is customary to express the data in terms of the radius R of the quenching sphere rather than in volume, and R is given by:

$$R = (3V/4\pi)^{1/3} = 6.5[\mathrm{Q}]^{1/3} \qquad (1.47)$$

where R is the radius in Å of the quenching sphere, and [Q] is the concentration of the quencher (in mol liter^{-1}). It is obvious from (1.47) that relatively high concentrations of quencher are required in order to assure that a quenching molecule is within the quenching sphere. For example, at a concentration of 1 mol liter^{-1}, the average distance is 6.5 Å. This drops to 14 Å at 10^{-1} mol liter^{-1} and 30.2 Å at 10^{-2} mol liter^{-1}.

1.18 Energy transfer in polymers

Energy transfer in polymeric systems is a particular interest of this book and will be dealt with in detail in subsequent chapters. However, it is worth noting at this point that the structure of polymers made by conventional polymerization and copolymerization usually results in a polymer chain having chromophores or quenchers A or Q located at regular intervals along its length. The long-chain and conformational properties of the polymer result in the proximity of these chromophores, even in very dilute solution. This proximity can lead to interaction between the chromophores of any energy transfer type, leading to unusual photophysical or photochemical properties. For example, the energy may migrate from one chromophore to another by the process of energy exchange or the energy existing in a chromophore may be quenched by another quencher molecule located in the same chain. These processes of intramolecular energy are of considerable theoretical and practical importance and will be discussed in further detail in chapter 9.

2

Polymer structure and reactivity

2.1 General considerations

The study of kinetics and mechanisms of gas reactions and of small molecules in solution is now highly advanced. Very accurate descriptions of the rates of these simple reactions can be obtained in terms of collision theory and 'molecular dynamics'. However, many biological processes of importance, including photosynthesis, vision and metabolism, occur in systems which involve very large molecules for which the concept of 'billiard ball' collisions between molecules of approximately equivalent size is invalid. The completely random motion of spherical molecules, illustrated in figure 2.1*a* which is characteristic of chemically reacting species in both the gas and the liquid phase, is replaced by a more coordinated motion when a macromolecule is dissolved or swollen in a solvent, as shown in figure 2.1*b*. A still further reduction in independent motion occurs when one considers a solid polymer matrix, illustrated in figure 2.1*c*.

According to the classical theory of thermal reactions, the collisional energy available in an encounter must be sufficient to transfer at least one of the reacting species to some excited state complex from which products are derived. The random thermal motion thus acts as an energy source to drive chemical reactions. The restrictions on motion described in figure 2.1 will not alter the average thermal energy of the system, but may result in a different distribution of energy between colliding partners. For

Figure 2.1. Restrictions on mobility caused by polymer structure. (*a*) Solution; (*b*) polymer solution; (*c*) polymer bulk.

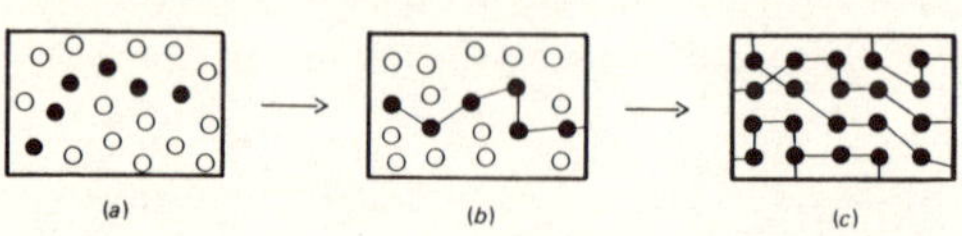

example, it may require more collisional energy to induce an excited state of a large polymer molecule which has many degrees of freedom, than of a small molecule which has fewer alternative modes dissipating the collisional energy. For this reason, many early scientists believed that polymer molecules would be very unreactive.

Another factor which would be expected to affect the reactivity of polymer molecules is the fact that the reactants must be expected to achieve some particular geometric relationship with respect to each other in order to form the excited complex from which reaction occurs. If mobility is severely restricted, this geometry may not be achievable within the lifetime of the excitation and chemists not familar with macromolecular reactions might assume that the restrictions of motion described above would be so great that reaction would not occur at all. In fact, this is not the case and there are now well authenticated examples of photochemical reactions, for example, which are just as efficient in solid polymers as they are in dilute solutions of small molecule analogs. On the other hand, there are indeed reactions which are completely inhibited in the solid phase. The key to the understanding of these effects is to determine how much movement of atoms is required in order to form the geometry of the excited or transition state and to allow some separation of products. Reactions differ in their requirements in these respects and so we can expect to observe different effects when we study chemical reactions in polymeric systems. Perhaps the most powerful tool for studying these relationships involves the use of polymer photochemistry.

2.2 Molecular motion in polymers

We must first ask what types of motions of molecules will be most restricted by a polymeric environment. Polyatomic molecules in the gas phase undergo translation, vibration and rotation. The latter two are quantized, giving rise to spectra; those corresponding to vibration are usually observed in the infrared, whereas those for rotation are normally observed in the microwave region. Fortunately, electronic spectra, when carefully analyzed, also contain information about both the vibrational and rotational motion of small molecules.

Studies of the infrared spectra of polymeric materials both in solution and in the solid phase have shown that, in general, the vibrations of small groups such as the bending and stretching modes of C=O, and stretching of C–H, C–Cl and C≡N, are almost unaffected by the presence of the polymer environment. For this reason, infrared spectra are most valuable in analyzing polymer structures since the location of the bands is usually much the same as in small molecule analogs. The rotational spectra of molecules are more complex, however, since for a very large polymer molecule, rotations of the whole molecule will be very slow indeed compared to those for a small molecule analog and may not occur at all in the solid phase. In solution, while rotations do occur, they are not usually a coordinated rotation of the molecule as a whole, but a sequence of rotations of individual segments. These segments are not necessarily defined by the monomer units from which the polymer is derived, but in hydrocarbon polymers often consist of segments of from 20 to 50 carbon atoms in length. The extent of this segmental motion will, of course, depend on the molecular architecture of the polymer chain, being extensive in flexible macromolecules and much reduced when molecules with stiff chains are considered.

This segmental motion occurs in polymer chains both in dilute solution and in the solid amorphous phase provided the polymer is above its so-called glass transition, T_g. This characteristic temperature for polymers is considered later in §2.7. The major difference between a polymer molecule in dilute solution and the same molecule in an amorphous solid phase is that in dilute solution, there can be a net translational motion of the centre of mass of the polymer, whereas in the solid, the rate of this process is exceedingly small. Even in solution the motion of the centre of mass is very slow compared with the velocity of small molecules of the solvent or of any reactant that it might encounter. For this reason, the rate of chemical reaction in solution is usually determined by the velocity of the small molecule reagents rather than by the translation motion of the polymer, except possibly in the case where two macromolecules are reacting together.

2.3 Shape and size of macromolecules in solution

Let us first consider the rate of reaction of a polymer in dilute solution. Both synthetic and natural biopolymers in solution can usually be categorized as flexible coils, rigid rods or globular particles (figure 2.2). Flexible coils are usually formed by polymers having carbon–carbon single bonds in the backbone, or other single bonds capable of free rotation. Random collision with solvent molecules continually moves segments of the polymer chain so that its shape changes continuously with time but will, on average, take up spheroidal geometry characterized by what is called the *random coil*.

Methods of calculating and measuring the dimensions of this random coil configuration have preoccupied polymer physical chemists for several decades and the theory is now well developed. The random coil dimensions are usually defined in terms of two parameters illustrated in figure 2.3. The root mean square end-to-end distance $\langle \bar{R}^2 \rangle^{1/2}$ can be considered either to be the average value of the end-to-end distance for a single macromolecule averaged over time, or the instantaneous average obtained from a large group of macromolecules of the same molecular weight averaged over space. R_ε is the radius of the spherical volume which is swept out by the polymer during its random segmental movements caused by

Figure 2.2. Shapes of polymers in solution: (*a*) flexible coil; (*b*) rigid rod; (*c*) globular particle.

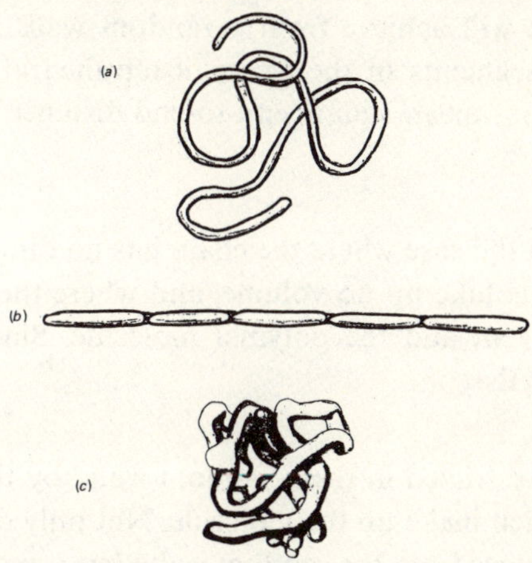

Figure 2.3. Dimensions of a random coil. $\vec{R}$ is the end to end distance, R_ε the radius of equivalent sphere and C the centre of mass.

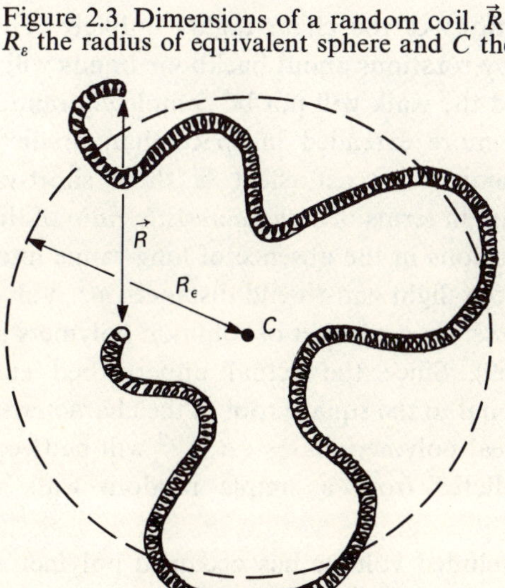

Brownian motion. It is sometimes referred to as the radius of the 'equivalent sphere', which defines the hydrodynamic volume in solution. Another quantity commonly used is the root mean square radius of gyration, $\langle \bar{S}^2 \rangle^{1/2}$, which is the average value of the first moment of all segments of the chain with respect to the centre of mass of the chain. Theory shows that for a completely flexible macromolecule in a random coil, the following simple relationship exists between the root-mean-square radius of gyration and the root-mean-square end-to-end distance:

$$\langle \bar{R}^2 \rangle^{1/2} = 6 \langle \bar{S}^2 \rangle^{1/2} \tag{2.1}$$

It is possible to calculate exactly from theory the dimensions which a flexible chain of any length will achieve from a random walk in three dimensions. If there are *n* segments in the chain of tetrahedral carbon atoms of length *l* then the root-mean-square end-to-end distance is given by:

$$\langle \bar{R}^2 \rangle^{1/2} = n^{1/2} l \tag{2.2}$$

However, this is true only in the case where the chain has no dimensions, i.e., where the chain segments take up no volume, and where there is no interaction between the solvent and the polymer molecule. Since *M* is proportional to *n*, it follows that

$$\langle \bar{R}^2 \rangle^{1/2} \propto M^{1/2} \tag{2.3}$$

Real polymer chains are restricted in their 'random walk' by the finite dimensions of the atoms which make up the molecule. Not only does this exclude certain volumes of space from the random walk, but it introduces steric hindrance (such as the well known 'pentane' interference) which makes certain conformations of the chain highly unlikely. Thus some conformations achieved by rotations about backbone bonds will be more probable than others, and the walk will not be completely random. Real chains will therefore be more extended in space than predicted from simple random walk theory. The net effect of these short-range interactions can be expressed in terms of a *characteristic ratio* of the square of the actual chain dimensions in the absence of long-range interactions to the square of the random flight end-to-end distances nl^2. Values of the characteristic ratio $\langle \bar{R}_0^2 \rangle / nl^2$ for a number of common polymers are given in table 2.1 (Flory 1969). Since the actual unperturbed end-to-end distance will be proportional to the square root of the characteristic ratio, it is clear that for most real polymer chains $\langle \bar{R}_0^2 \rangle^{1/2}$ will be two to three times greater than predicted from a simple random walk in three dimensions.

The problem of the excluded volume has occupied polymer scientists for a number of years. Theoretical solutions to this type of problem have recently been achieved by the use of Monte Carlo simulations of random walks on lattices such as the diamond lattice where it is possible to determine the number of chains which intersect at the same point and thus must be eliminated from the total distribution. Readers interested in further details of these treatments are referred to the book *Statistical Mechanics of Chain Molecules* (Flory 1969).

The results of such calculations show that at sufficiently high values of n (i.e., high chain length) the proportionality between $\langle \bar{R}_0^2 \rangle^{1/2}$ and the square root of the number of segments is retained. However, a further expansion of the coil is predicted. Solvents with strong interactions with the polymer chain segments will also cause an expansion of the coil. Both of these effects can be included in an expansion factor α, defined by

$$\alpha \equiv \langle \bar{R}^2 \rangle^{1/2} / \langle \bar{R}_0^2 \rangle^{1/2} \tag{2.4}$$

While so-called good solvents will tend to expand the macromolecular coil, poor solvents will tend to make it contract below the value of its unperturbed dimensions. Substantial contraction below this value will, however, result in precipitation of the polymer from solution. However, it is usually possible to find certain solvents at a particular temperature, or mixtures of solvents and nonsolvents with the appropriate solution power, to arrange that the compression of the molecular chain by the solvent just balances the expansion of the chain caused by the excluded volume effect. Under these conditions, known as θ or Flory conditions, the coil will achieve the exact dimensions corresponding to the unperturbed coil. Studies in a θ solvent are very useful if one is concerned about relating observed results to theoretical calculations relating to coil dimensions.

It is important to point out that recent results of neutron scattering studies of solid polymers have shown that the original Flory prediction

Table 2.1. *Characteristic ratios for various polymers*

Polymer	$(\langle R_0^2 \rangle / nl^2)^{\frac{1}{2}}$	$\langle R_0^2 \rangle / nl^2$
Polyethylene	2.6	6.7
Atactic polystyrene	3.2	10.0
Isotactic polypropylene	2.4	5.7
Poly(methyl methacrylate)		
Atactic	2.6	6.9
Isotactic	3.0	9.3
Syndiotactic	2.6	7.0
Poly(ethylene oxide)	2.0	4.0
Nylon 66	2.4	5.9

relating to the size of polymer coils in the amorphous solid phase has been confirmed. Accordingly it is generally safe to assume that in the truly amorphous solid polymer, the dimension of each interpenetrating polymer coil will be very close to that in a θ solvent, i.e., that predicted for the random coil configuration.

The most powerful technique for simulating the properties of real polymer chains involves computer-assisted calculations using the 'rotational isomeric state' model. These computations take into account the energy barriers to rotation in extended hydrocarbon chains along with actual bond lengths and angles to generate the profile of a polymer chain by stochastic processes. Figure 2.4 shows the results of such calculations for chains of 500 and 3000 carbon atoms (Flory *et al.* 1982) and represents the projection of the three-dimensional chain on the x–y plane in only one of a multitude of possible configurations. Inspection of large numbers of such figures shows that the actual chain is not spherical, but elongated in a particular direction (depending on the direction chosen for the first bond). It is only when large numbers of such figures are superimposed on each other that one observes a spherically symmetric distribution of segments. This means that the properties of the flexible macromolecule will have spherical symmetry only if observed over times long enough so that thermal motion will allow the chain to adopt many conformations during the time scale of the experiment.

2.4 Motion of flexible macromolecules

When a flexible macromolecule in solution moves in response to a shear field the random coil will move along with solvent molecules trapped within the coil. If the flow is measured at low shear rates, so that the coil is not distorted by the shearing action of the solvent, such as in a capillary viscometer, the polymer will behave as if it were a sphere with an 'effective hydrodynamic volume' which can be estimated from viscosity measurements.

Einstein (1906, 1911) derived the viscosity law which predicts the increase in viscosity of a liquid by the addition of spherical particles:

$$(\eta-\eta_0)/\eta_0=2.5(N_2/V)V_e \qquad (2.5)$$

where N_2/V is the number of spheres (molecules) per unit volume and the effective volume of the individual spherical particle

$$V_e=\frac{4}{3}\pi R_e^3 \qquad (2.6)$$

where R_e is the effective radius of the sphere and η and η_0 are the solution and solvent viscosities, respectively. Replacing N_2/V by $cN/100\ M$, where c is the polymer concentration in g/100 ml, N is Avogadro's number, and M

is the molecular weight of the polymer, yields

$$\frac{1}{c}\left(\frac{\eta-\eta_0}{\eta_0}\right)=0.025\frac{NV_e}{M} \tag{2.7}$$

In measurements of polymer solutions it is found that this law is obeyed only at very low concentrations where the polymer coils do not interact with each other so that the quantity on the left-hand side of (2.7) is replaced by the intrinsic viscosity $[\eta]$, defined by:

$$[\eta]=\lim_{c\to 0}\left[\frac{1}{c}\left(\frac{\eta-\eta_0}{\eta_0}\right)\right] \tag{2.8}$$

Figure 2.4. Computer-generated random configurations of hydrocarbon chain molecules consisting of (*a*) 500 and (*b*) 3000 carbon–carbon bonds. Indicated for comparison with each configuration are spheres representing how much space would be occupied by the atoms of the molecule if the molecules were fully compacted. (From Flory *et al.* 1982.)

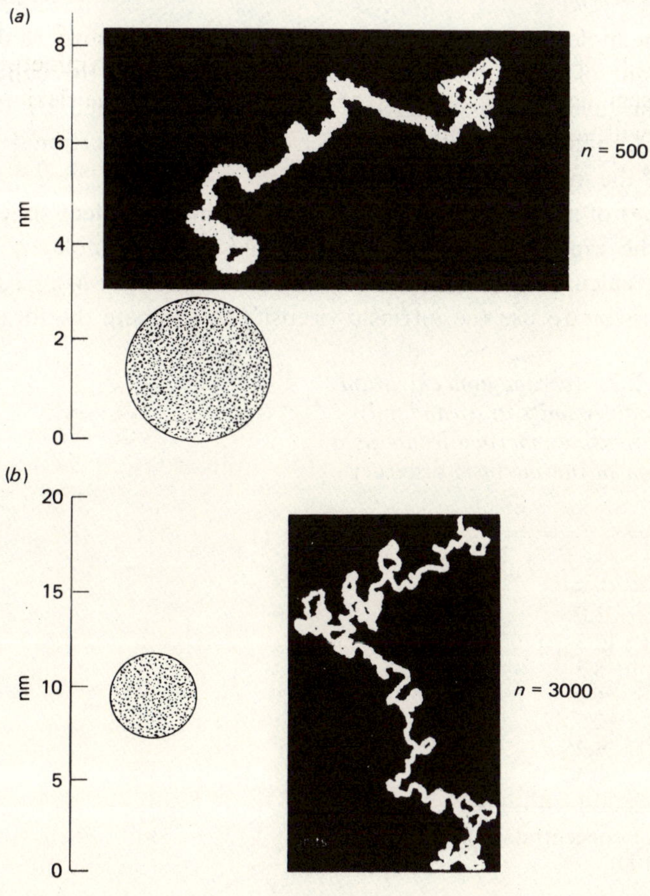

Thus for polymers we have

$$[\eta]=0.025\frac{NV_e}{M} \tag{2.9}$$

By measuring the intrinsic viscosity of a flexible polymer in a suitable solvent one can therefore estimate the 'effective size' of the coil. The segment density in a random coil decreases exponentially with the square of the distance from the centre of mass of the molecule (Flory 1953), so that the polymer is much more 'concentrated' at the centre than on the periphery of the equivalent sphere. However it is useful to use (2.9) to calculate the average concentration of the polymer within the random coil. Rearranging (2.9) it is easy to show that the average polymer concentration c in g/100 ml is given by:

$$c=2.5/[\eta] \tag{2.10}$$

and c' in monomer units (base moles) per liter is

$$c'=25/M_0[\eta] \tag{2.11}$$

where M_0 is the molecular weight of the monomer unit. Obviously as the molecular weight of a polymer increases, the expansion of the coil is greater and the 'average concentration' is lower, as indicated by the inverse proportionality of c with $[\eta]$ which in turn is proportional to some power of the molecular weight.

Typical values of average concentrations within the equivalent sphere occupied by the random coil are given in table 2.2. The molar concentrations are calculated using a monomer molecular weight $M_0=100$.

It is also possible to use the intrinsic viscosity to estimate the mean-

Table 2.2. *Average concentration of monomer units in a randomly coiled flexible macromolecule as a function of the intrinsic viscosity* $[\eta]$

$[\eta]$	$c\,(\text{g dl}^{-1})$	$c'\,(\text{mol l}^{-1})$*
0.1	25.0	2.5
0.2	12.5	1.25
0.3	8.3	0.83
0.5	5.0	0.50
1.0	2.5	0.25
2.0	1.25	0.125
5.0	0.50	0.05

* Molar concentration assuming $M_0=100$.

square end-to-end distance $\langle \bar{R}^2 \rangle$ for the random coil in any solvent from the relation

$$[\eta] = \Phi \frac{\langle \bar{R}^2 \rangle^{3/2}}{M} \tag{2.12}$$

where Φ is a 'universal' constant having the value of 2.3×10^{21} when R is expressed in cm, M in units of molecular weight and $[\eta]$ in dl g^{-1}.

The expansion α of the coil over its unperturbed dimensions is given by:

$$\alpha = \langle \bar{R}^2 \rangle^{1/2} \langle \bar{R}_0^2 \rangle^{1/2} \tag{2.13}$$

and can be measured experimentally by obtaining the intrinsic viscosity of the polymer in a θ solvent and using the relation:

$$\frac{[\eta]}{[\eta]_\theta} = \frac{\langle \bar{R}^2 \rangle^{3/2}}{\langle \bar{R}_0^2 \rangle^{3/2}} = \alpha^3 \tag{2.14}$$

In certain types of photophysical and photochemical experiments, where processes such as quenching are concentration-dependent, it is no longer appropriate to treat the random coil as if it were a sphere with a uniform concentration of chain units corresponding to the average value calculated from (2.10). In this case a better level of approximation can be introduced by assuming a Gaussian distribution of chain elements (monomer units) around the centre of mass (Debye and Bueche 1952).

2.5 Motion of rigid macromolecules

Unlike flexible coil macromolecules, both the rigid rod and the globular particle characteristic of certain viruses and globular proteins (e.g., insulin) are restricted by their molecular architecture from changing either their shape or dimensions in solution. However, like the flexible coil, their relatively large mass compared to the solvent molecules results in relatively low diffusion rates. Reaction with small molecule reagents will thus be controlled more by the rate of diffusion of the small molecule components, and for many reactions one can consider the centre of mass of the polymer will be stationary during the course of a reaction.

However, in all three types of macromolecules, there are difficulties in relating the reaction rate to concentration. In its simplest and most fundamental definition, a rate constant represents the proportionality factor relating an observed rate to the concentrations of reagents. It is usually assumed in rate theories that the mobility of molecules is such that one can consider the distribution of molecules throughout the medium as being completely random. Thus the probability of reaction does not depend on the location of a particular reagent in the medium. However, as a polymer system is diluted, whether it be flexible coil, rigid rod, or globular protein, this is no longer the case. Both the rigid rod and the globular particle actually represent separate phases from which other reagent molecules may be excluded. In most cases, only the surface of these particles is available for reaction, not the total volume. On the other hand, while the random coil molecule at high concentrations is sufficiently flexible that it will fill most of the available volume, allowing small molecular reagents to diffuse freely through the spaces within the coils, it is clear that as the dilution of the polymer is increased and the coils separate from each other, then there will be regions of space containing no polymer molecules and other regions where the concentration of segments is quite high, as shown in figure 2.5. It is also likely that the concentration of any reagent will be different inside and outside the polymer coil because of the differing attractive forces between the polymer and the solvent which surround it. One must, therefore, be very careful in interpreting the significance of rate constants for reactions in polymer systems in cases where segregation of reagents, either by concentration or by osmotic or specific interactions, is possible. The importance of these effects will, of course, be much greater with relatively rapid reactions where the reagents and products do not have time to equilibrate through diffusional processes.

Special considerations are also necessary when the components of a reacting system are present either on two separate macromolecules or at different positions in the same macromolecule. In the first case, the rate of reaction will depend on the molecular diffusion of the centre of mass of the two macromolecules and may also contain factors relating to the reluctance of macromolecular chains to interpenetrate the space occupied by another macromolecule. This is due to the reduction in entropy which occurs when one coil penetrates the volume of space occupied by another. If the two macromolecules involved have different chemical structures, in addition to the entropy factor, there is a strong enthalpy factor, which also reduces the tendency of the coils to interpenetrate. This 'incompatibility' of polymer chains is important except under conditions of the very highest dilution and is often the cause of substantial reductions in the observed rate of macromolecular reactions.

Figure 2.5. Effect of dilution on concentration in randomly coiled macromolecules.

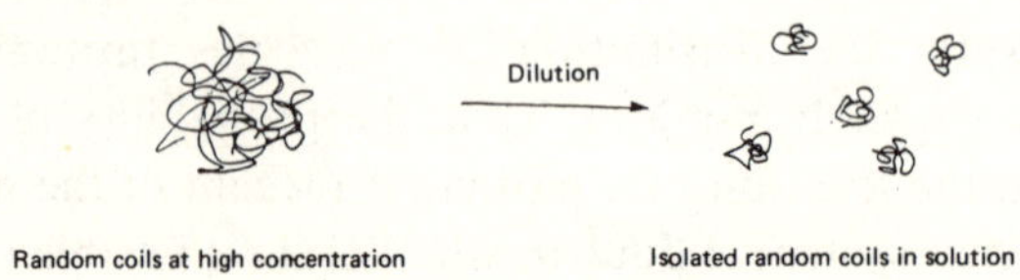

Reactions between groups on the same chain will not involve diffusion of the centre of mass at all, but will depend on the relative rate of conformational change of the random coil as it adopts various configurations in space. This will depend on the separation of the two reagents, the rotational freedom in the chain and the thermal activation provided by collision with solvent molecules.

2.6 Reactions in the solid phase

For reactions in polymer solids, it is helpful first to consider the nature of the amorphous solid phase for macromolecules having some degree of chain flexibility. In this case, we can assume that some approximation to the random coil configuration will be retained in the solid through interpenetration of a large number of polymer chains. If a polymer is cooled slowly to absolute zero, it would be expected that it would achieve its most stable geometry and closest packing. However, because of the large macroscopic viscosity of the solid, it seldom achieves this packing density in finite times unless it undergoes crystallization at a relatively high temperature and hence even at absolute zero the polymer matrix will contain some free volume. 'Free volume' in this context could be defined as space unoccupied by atoms because the maximum packing density cannot be achieved. At absolute zero, of course, this free volume would be relatively small, and because there is no energy to activate the molecules, they will not move to occupy this space. As the temperature is increased towards room temperature, thermal motion will begin and the solid will expand. Additional free volume will be created by this expansion and small scale rotations or oscillations of the atoms will ensue. The change in specific volume of an amorphous polymer as a function of temperature is shown schematically in figure 2.6. As the free volume increases, it becomes possible at particular temperatures for certain types

Figure 2.6. Specific volume and 'free volume' of polymeric material.

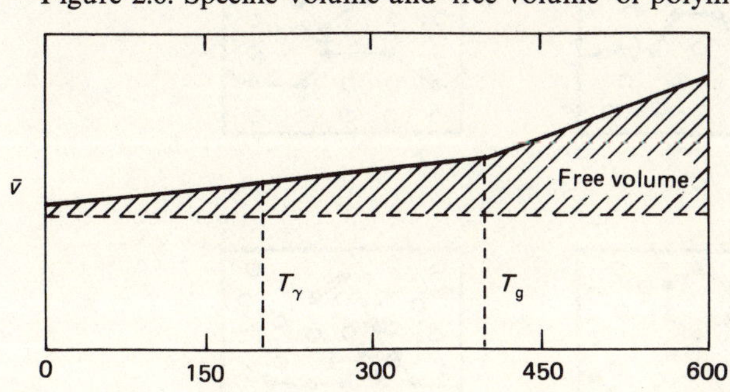

of small-scale motion to be observed in polymer chains. Some of these are illustrated schematically in figure 2.7.

It will be obvious that the temperature at which any type of group motion is first observed will depend on the frequency of the method used to detect the motion. Thus measurements made at high frequencies detect motion in groups at higher temperatures than those involving lower frequencies and therefore greater time intervals. Motion of this kind can be detected by a variety of physical measurements on the polymer matrices including dielectric relaxation, and thermal mechanical deformation. The interested reader is referred to the excellent reviews by Haward (1975) and North (1975) for further details.

As the temperature is raised, typical motions which are observed at characteristic temperatures become evident. The point designated by T_γ in figure 2.6 represents the temperature at which sufficient free volume is available in polystyrene to observe the motion of phenyl rings at a frequency of about 10 Hz. As the temperature is raised still further, additional free volume becomes available and at the point indicated by T_g, a change in the slope of the specific volume curve is observed which is due to the occurrence of motion of very long segments of the polymer chain consisting of units of from 20 to 50 carbon atoms. The types of motion observable below T_g in solid polymers will vary with the chemical nature of the polymer. Specific motions which have been identified in polymer systems include the rotations of phenyl and methyl groups, rotations of carboxyl and carboxylic ester, nitrile, and keto groups, etc. Conformational transformations such as the movement of cyclohexane rings from the chair to boat form also can be observed in

Figure 2.7. Some modes of motion required for reaction. (From Guillet 1977.)

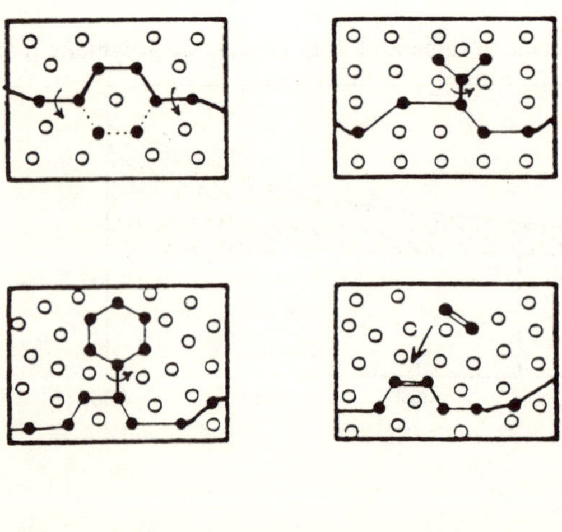

particular temperature regions. Above T_g, the motion of flexible chains becomes very rapid indeed and, in fact, under these conditions, the conformational mobility of polymer chains appears to be nearly as great as if they were dissolved at high dilution in a good solvent.

2.7 Relations between molecular mobility and polymer properties

Polymer molecular movement can involve a large variety of specific motions of atoms or groups of atoms within the molecule. On the smallest scale are the atomic vibrations familiar to the infrared spectroscopist, while on the macroscopic scale, motion is observable in the phenomenon of fluid flow. Between these extremes lie the molecular movements which determine a variety of physical and chemical characteristics, including modulus of elasticity, electrical permittivity and loss, solution or melt viscosity and dynamic specific heat, in addition to chemical reaction rates.

In photochemistry we will be concerned primarily with three types of polymer motion: translational motion of the centre of mass of molecules (figure 2.8*a*), segmental rotation deriving from internal rotation about covalent bonds contained in the backbone of a linear polymer chain (figure 2.8*b*) and group rotation of functional groups attached to the side chain of a polymer backbone and again due to rotation about single bonds (figure 2.8*c*). It is important to stress that although segmental rotation is due to rotation about single bonds, the segments themselves cannot necessarily be identified with particular chemical groups or monomer units. Thus, the segments identified pictorially in figure 2.8 may be composed of a number of monomer units, often as many as 10 or 20, provided that this number is less than the number in the whole chain except for very stiff molecules. In the case of both segmental and group rotations, it is not necessary that a full 360° rotation should occur around the individual bonds. Often it is sufficient that a librational movement is possible, giving rise to the observed mobility.

The first two of these motions, translation and segmental rotation, are generally considered to control the bulk physical properties of a macromolecular system. The hardness or stiffness modulus of a typical polymer will change with temperature as thermal energy is imparted to different motions of the polymer chain. A plot of the logarithm of the modulus as a function of absolute temperature is shown in figure 2.9. At low temperatures, polymers are usually brittle glasses with an elastic modulus close to 10^{10} Nm^{-2} which is characteristic of many solids including inorganic glasses. At sufficiently high temperatures, thermoplastic materials flow as

Figure 2.8. Types of motion in polymer molecules: (*a*) translational; (*b*) segmental rotation; (*c*) group rotation.

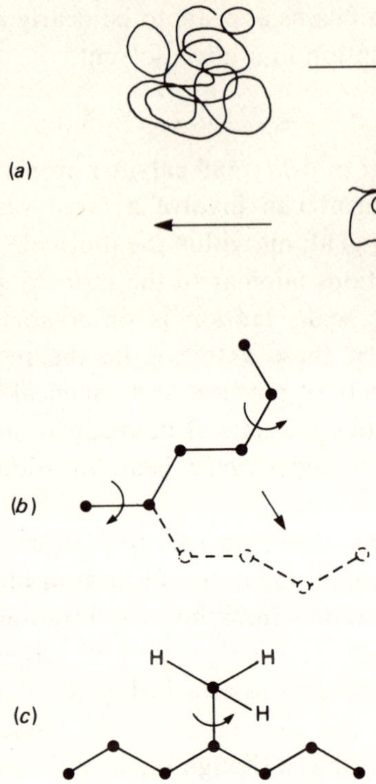

Figure 2.9. Hardness of energy storage modulus (full curve), and energy loss modulus (broken curve) for an ideal polymer. (After North 1975.)

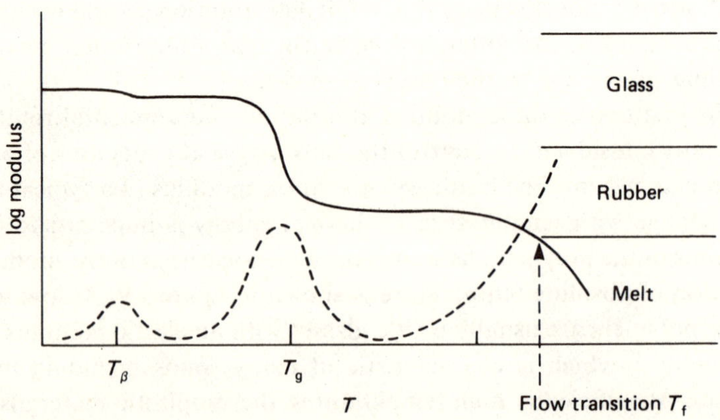

viscous liquid melts, just as do most low molecular weight substances. However, unlike small molecules, linear polymers show an intermediate region, often designated the 'rubbery plateau' region in which, with a relatively low modulus of about 10^6 N m^{-2}, extensive mechanical deformation may occur without excessive flow or creep. A rather ill-defined transition, called the melt flow transition exists between the rubbery and the fluid state, and a second transition called the glass transition, T_g, marks the boundary between the glassy and rubbery behavior. Below the glass transition, neither translational motion nor segmental rotation can occur. However, smaller scale rotations such as functional group rotation, and small scale vibrations of backbone atoms can occur. At and above the glass transition, the combination of thermal energy and increased free volume permits the rather large scale rotations involved in segment motion. This results in a relatively low modulus, but since translational motion cannot occur, deformation is elastic and after stressing, the molecules will return to their previous state unless stressed beyond their elastic limits. At the melt-flow transition, translational motion of polymer molecules becomes possible and the system becomes a viscous liquid. However, it is important to note that the high viscosity of polymeric liquids relates only to the restricted motion of the centre of mass of the polymer molecules. The motion of small molecules in a polymer matrix under these conditions may 'see' a microviscosity many orders of magnitude below that of the observed macroscopic viscosity.

The foregoing emphasizes the fact that it is not possible to discuss the onset of a particular form of molecular motion without including in the discussion the question of the time or frequency of the observation. Thus, for example, dielectric measurements at very high frequency will see very little motion of atoms and molecules, even at relatively high temperatures, while lower frequency studies such as the measurement of creep over periods of days or months will see motions of polymer chains to which dielectric procedures are blind.

The fact that the movement brought about by response to a particular stress occurs at a finite rate means that the modulus of a polymer at a given temperature is both time and frequency dependent. At high frequency, the molecular response will lag in phase behind the perturbation applied to the system, and for this reason, the modulus is composed of real and imaginary parts representing two components differing in phase by 90°. The real part of this complex modulus is the measure of the energy which can be reversibly applied to, and recovered from, the system. It is often called the storage modulus for this reason. It is shown in the full curve of figure 2.9 and corresponds to the observable hardness of the polymer. The imaginary part of the complex modulus is known as the loss modulus and is a measure of the energy dissipated as heat in applying stress to the system. When the temperature is such that the molecules can easily keep up with the applied constraint, or alternatively do not move at all, there is no thermal loss and the loss modulus is consequently zero. However, when the rate of molecular movement is comparable to the rate of change of stress, the molecules absorb heat and exhibit a phase lag in energy dissipation. The loss modulus passes through a maximum called the 'loss peak' at those transition temperatures when the molecular and applied frequencies are of the same magnitude as shown in the broken curve of figure 2.9.

In many organic and most polymer glasses, additional transitions are observed below T_g. It is often tempting to relate these to specific motions of individual subgroups attached to the polymer chain and indeed there is good evidence that this occurs in a number of cases. However, there are well authenticated examples of polymers without side groups, such as polyethylene, which show secondary maxima in a loss modulus and it is clear that this must be due to some motion of smaller segments of the backbone chain. In polyethylene, for example, this has been postulated to be a type of crankshaft rotation, as illustrated in figure 2.10. In other cases, it is clear that a coordinated motion of atoms, both in the backbone and in the side chain groups of a polymer, contribute to secondary loss maxima. Because of the difficulty in identifying the exact process responsible for the secondary maxima, it is conventional to label the transitions in sequence below the glass transition denoted T_g or T_α by T_β, T_γ, T_δ, etc. The β transition is of particular importance to the physical properties of polymer glasses since the range between T_β and T_g often defines the useful plastic range for good physical properties.

The significance of the β transition to photochemical or other chemical processes in polymer glasses is not yet fully understood. However, it is clear that in polymers above the glass transition, the large scale of the

Figure 2.10. The crankshaft mechanisms of (*a*) Shatzki and (*b*) Boyer. (After McCrum, Read and Williams 1967.)

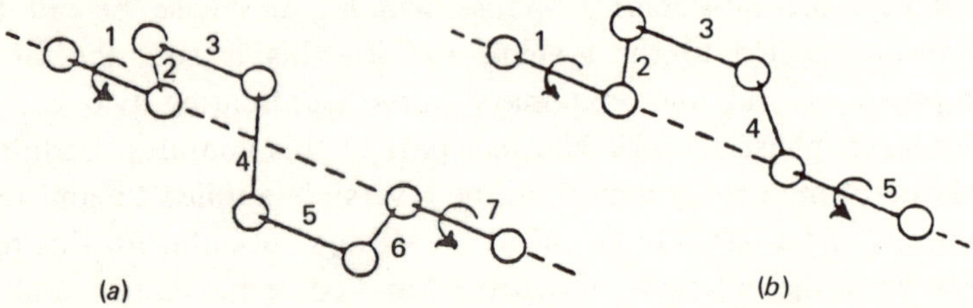

segmental motion available to the polymer suggests that the same degree of conformational freedom is available to the molecule as would be in dilute solution in a typical solvent. This has important consequences in the study of chemical and photochemical reactions in polymeric media.

Experimentally, it has been shown that the smaller the scale of a particular molecular motion, the smaller is the amount of energy required to activate it. Thus, the activation energies associated with the β transition are usually of the order of 8 to 24 kJ mol^{-1}, while those associated with the glass or α transition range from 40 to 120. The lower transitions, such as the γ and δ transitions, will have activation energies of 4 to 8 kJ mol^{-1}. As a consequence of this, the temperature difference observed between the two transitions will depend strongly on the frequency of the observation. For example, figure 2.11 shows a plot of the log of the rate of a particular observation versus $1/T$ for the α, β and γ transitions and it can be seen that the frequency–temperature plots tend to merge at high frequencies and temperatures. This means that the temperature difference between the transitions is greatest for low frequency measurements or long-term observations. Details of the theory and practice of the determination of segment and molecular motion in polymer systems is beyond the scope of this book and the interested reader is referred to the reviews by North (1975) and by Shen and Eisenberg (1970).

2.8 Crystallinity in polymer systems

No discussion of polymer structure–property relationships would be complete without reference to polymer crystallinity. In spite of the very large size of polymer molecules, many of them are able to crystallize to a high degree of perfection. Two types of crystallinity can be identified: those associated with globular proteins in which each of the

Figure 2.11. Arrhenius diagram for α, β, and γ transitions in typical polymer. (After North 1975.)

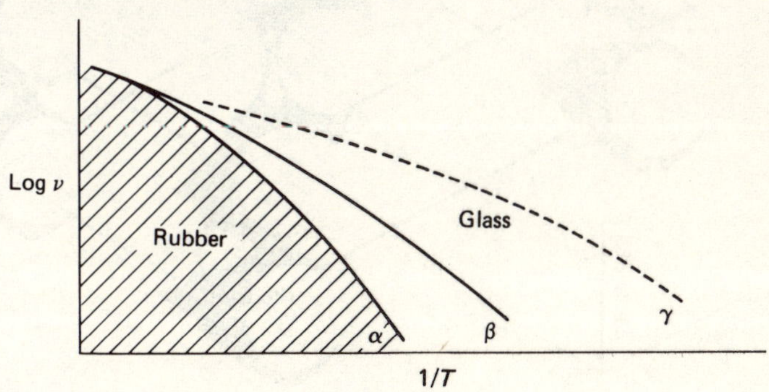

unit cells is associated with a single polymeric molecule, and crystallinity occurring in such polymers as polyethylene, in which segments of polymer chains pass through many unit cells (figure 2.12). The idea that large flexible molecules could form crystalline regions was at first considered to be very unlikely, but the indisputable evidence of crystallinity in cellulose and later in polyethylene showed that even polymers with both stiff and flexible chains could crystallize under appropriate conditions. In fact, the tendency of polyethylene to crystallize is so great that it is very difficult to obtain linear polyethylene in a completely amorphous state at room temperature.

The degree to which a polymer can crystallize will depend to a large extent upon the regularity of its structure. Inclusion of side groups at regular intervals, branches of variable length, and other defects will reduce the crystallinity of a polymer. However, under carefully controlled conditions it is possible to prepare single crystals of such long flexible chains as polyethylene and polypropylene as well as more polar polymers, such as nylon and poly(ethylene terephthalate).

The exact mechanism whereby a long flexible chain could crystallize into a single crystal was not understood until the work of Keller (1957) and Schlesinger and Leeper (1953) who showed that flexible polymers will crystallize by forming folded chains in which the lamellar thickness is of the order of 10 nm. Single crystals can be formed in this way in which

Figure 2.12. Arrangement of chains in the unit cell of polyethylene. (From Billmeyer 1962.)

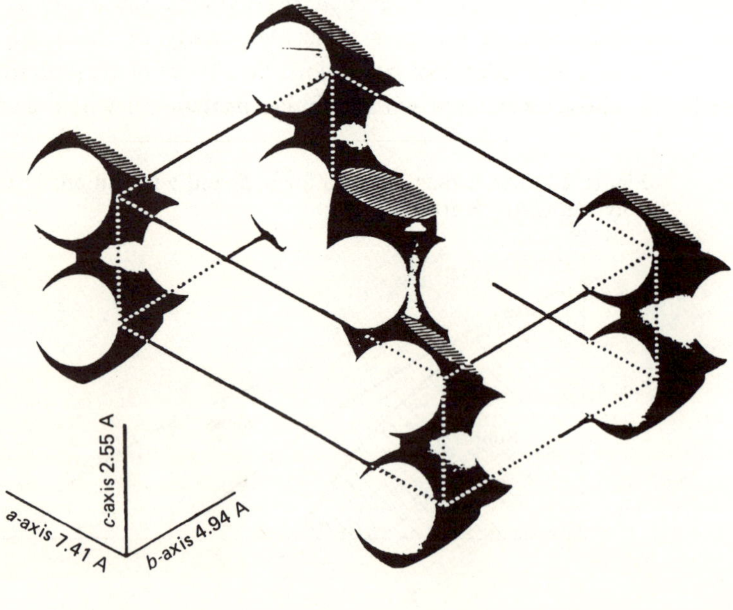

the only amorphous material remains at the surfaces in the crystal. Although single crystals can be prepared from such polymers by crystallization in dilute solution, most polymers as molded specimens will not contain single crystal units. Furthermore, except in very special cases, the crystallinity will not be 100%, that is, the polymer will consist of a mixture of amorphous and crystalline zones, or of a disordered crystalline array composing a single phase but of incomplete regularity. These two concepts of crystallinity in semi-crystalline polymers are illustrated in figure 2.13. These two types of crystalline arrangements are known as the 'fringed micelle' model and the 'para crystalline' model.

Figure 2.13. (*a*) Fringed micelle model of the crystalline-amorphous structure of polymers. (From Bryant 1947.)
(*b*) Lattice model and optical diffraction pattern for a two-dimensional ideal crystal. (From Hoseman and Bonart 1957.)
(*c*) Lattice model and optical diffraction pattern for a two-dimensional paracrystalline fiber. (From Hoseman and Bonart 1957.)

(*a*)
(*b*)
(*c*)

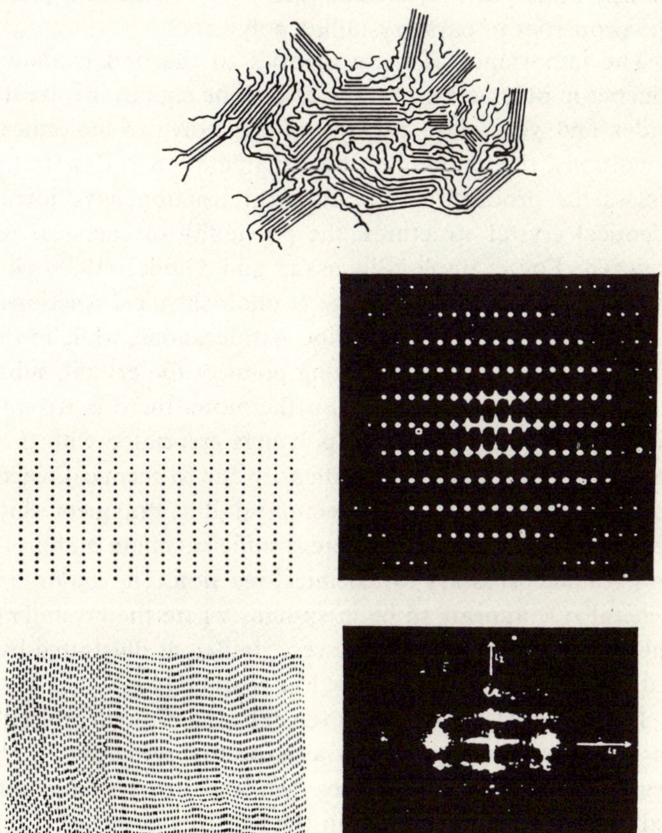

The validity of these two models has some importance in polymer photochemistry. The fringed micelle model, whether it consists of folded or oriented chains, is essentially a two-phase model in which crystalline regions are surrounded by amorphous polymer, except possibly at very high degrees of crystallinity. On the other hand, a para crystalline material is a single homogeneous phase in which the degree of crystalline order is reduced. In a para crystalline polymer, all molecules will have rather similar mobility and will therefore be of equal chemical reactivity. In a semi-crystalline material the amorphous regions will be characterized by a rather high mobility and therefore greater chemical reactivity, whereas the crystalline regions will be characterized by high symmetry and lower free volume with a consequent reduction in chemical reactivity. There remains a great deal of controversy about the actual structure of particular polymers; however, there is strong evidence that polyethylene, polypropylene, nylon and poly(ethylene terephthalate) are semi-crystalline polymers which correspond more closely to the fringed micelle model, while polyacrylonitrile and poly(vinylchloride) may well exhibit the properties of para crystalline polymers.

The importance of these theories to the understanding of chemical kinetics in polymers is that the crystalline regions involve a high degree of order and very close packing of the polymer molecules. Under these conditions, the amount of molecular motion is drastically reduced and unless the products of the chemical reaction have identical or nearly identical crystal structures, the possibility of chemical reaction is also reduced. For example, Slivinskas and Guillet (1973) showed that the classic Norrish type I and type II photochemical reactions did not occur at detectable levels in crystalline 7-tridecanone, while in the molten state a few degrees above the melting point of the crystal, substantial photochemical reaction occurred. Furthermore, there is strong evidence that the photo-oxidation of solid polymers associated with their degradation in outdoor environments is primarily due to chemical reactions occurring in the amorphous phase of semi-crystalline polymers, since it is only in the amorphous phase that there is sufficient chain mobility to provide the free volume necessary for chemical reaction. The major exception to this general rule appears to be in systems where the crystalline structures of the reagents and products are very similar, as illustrated by the polymerization of diacetylenes discussed in chapter 11.

It has also been shown that the diffusion of small molecules like oxygen in semi-crystalline polymers also occurs primarily in the amorphous regions and so reaction of oxygen and other small molecules will be expected to occur primarily in the amorphous phase. There is, however, the possibility that the transition region between the amorphous and crystalline phases of the polymer are under considerable steric strain, particularly if the polymer is below its glass transition. The possibility that this strain can accelerate the rates of chemical reactions in the boundary region must be considered. The details of the morphology of most experimental polymers are often unknown, and even those of very common polymers such as polyethylene and nylon are not completely understood. For a good review of this work, the reader is referred to the books by Wunderlich (1973) and Geil *et al.* (1974).

2.9 Polymer stereoregularity

Polymers made from the addition of monomers of the general structure of $CH_2{=}CHR$ have two possible configurations for each R group in the chain. Figure 2.14 shows possible arrangements of a long-chain polymer in which the tertiary carbon atom is arranged in either *d*

Figure 2.14. Representation of isotactic (it) and syndiotactic (st) polymers with the base units (–CRr–) and (–CH_2–CRr–), respectively. (From Elias 1977.)

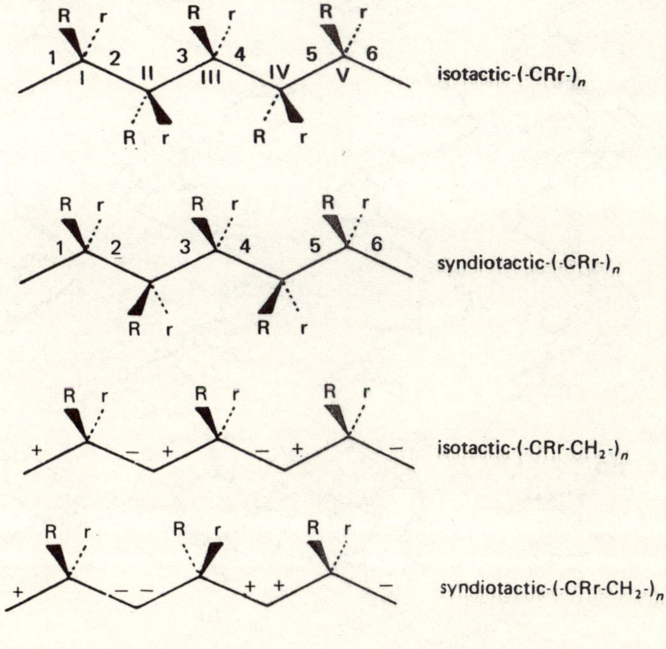

or *l* configuration. Since the regularity of a polymer chain has a strong effect on crystallinity, it is clear that a sequence in which all of the groups are either *l* or *d*, or alternating *d* or *l* in a regular fashion, will have a greater ability to crystallize than one in which the *d* and *l* sequences are random. The nomenclature of such polymers is due to Natta, who also synthesized examples of some of these structures with propylene. Polymers which contain long regular sequences of *l* or *d* units are called *isotactic*, whereas polymers containing regularly alternating sequences of *d* and *l* groups are *syndiotactic*. Polymers which contain random distributions of *d* and *l* configurations are called *atactic*. Because of their irregular structure atactic polymers are usually amorphous, whereas isotactic and syndiotactic polymers are usually capable of crystallization. However, the degree of crystallinity will depend on the thermal history of the polymer and it is possible that a completely syndiotactic or isotactic polymer will remain amorphous under most methods of preparation. For example, isotactic polystyrene is often difficult to obtain in the completely crystalline form. It has been shown experimentally that the isotactic

Figure 2.15. Schematic representation of the different kinds of helix occurring in various isotactic polymers $\text{-(}CH_2\text{–}CHR\text{)-}_n$. (I) 3_1; (II) 7_2; (III, IV) 4_1. (After Natta, Corradini and Bassi 1955.)

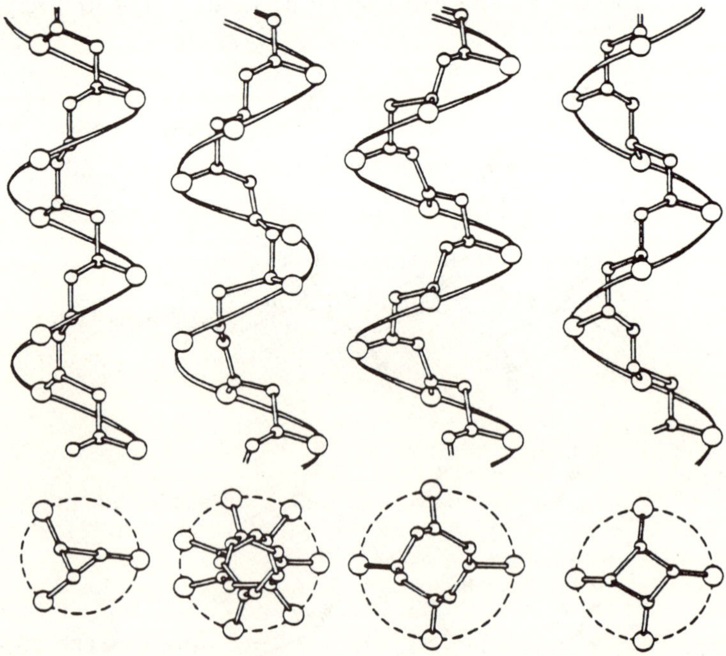

I II III IV

polymers form regular helices in the crystal lattice and that this is primarily responsible for the packing necessary to form a crystal (figure 2.15). There is also some evidence that these helical arrangements can be retained in the melt and in solution above the melting point, and these regular arrangements can affect both the photophysics and the photochemistry of polymer systems, as will be seen in later chapters.

3

Diffusion and permeability in polymers

3.1 Diffusion theory

The kinetics of many photoprocesses in polymeric systems will often depend on the rate of diffusion of one or more molecular species in the reaction medium. The theory of such processes has been well developed for reactions in gases and conventional liquids, however the special nature of polymeric solids and solutions requires further consideration.

The theory of diffusion in isotropic substances is based on the hypothesis that the rate of transfer of a diffusing molecule through unit cross-sectional area is proportional to the concentration gradient measured normal to the section. This is known as Fick's law:

$$F = -D(\partial c/\partial x) \tag{3.1}$$

where F is the flux (i.e., the rate of transfer per unit area of section), c is the concentration of diffusing substance, and x is the space coordinate measured normal to the section. D is known as the diffusion coefficient, and in the simplest case (Fickian diffusion) is independent of c. If F and c are expressed in the same units, D has dimensions of l^2t^{-1} and is usually expressed in units of $cm^2\ s^{-1}$. In the case of three-dimensional diffusion through a volume element where D is constant the fundamental differential equation for diffusion is given by

$$\frac{\partial c}{\partial t} = D\left(\frac{\partial c}{\partial x^2} + \frac{\partial^2 c}{\partial y^2} + \frac{\partial^2 c}{\partial z^2}\right) \tag{3.2}$$

Where D is not independent of c, the most general form of the fundamental equation becomes:

$$\frac{\partial c}{\partial t} = \text{div}(\text{grad } c) \tag{3.3}$$

For diffusion across films or membranes the concentration gradient can often be treated as unidirectional (i.e., along the x axis), in which case (3.2) becomes:

$$\frac{\partial c}{\partial t}=D\left(\frac{\partial^2 c}{\partial x^2}\right) \tag{3.4}$$

This is known as Fick's second law.

For a plane sheet or membrane of thickness l whose surfaces are exposed to constant concentrations c_1 and c_2 of penetrant, a steady state will be reached where the rate of transfer F across all sections is constant and given by:

$$F=-D\left(\frac{\mathrm{d}c}{\mathrm{d}x}\right)=\frac{D(c_1-c_2)}{l} \tag{3.5}$$

In some cases in which a gas or vapor diffuses through a membrane, the surface concentrations may not be known and provided that Henry's law is obeyed, c will be proportional to the partial pressure of the gas, thus

$$c=Sp \tag{3.6}$$

where p is the partial pressure and S is called the 'solubility' or 'solubility coefficient'. The flux under these conditions is given by

$$F=\frac{P(p_1-p_2)}{l} \tag{3.7}$$

Where P is called the permeability, or permeability coefficient, and is given by

$$P=DS \tag{3.8}$$

It is important to recognize that the rates of reactions in thin films will be controlled not by the diffusion constant alone, but also by the solubility. In many polymer films, for example, the diffusion constant for oxygen is quite large, but because of the low solubility, oxidation reactions may be very slow.

Experimental values of P and D for various organic permeants and oxygen in low density polyethylene (table 3.1) illustrate this point. The concentration of the diffusing species, c, is proportional to the ratio P/D (cf. (3.19)). Thus at a given partial pressure, p, of the diffusing species, the concentration of benzene in polyethylene would be $\sim 3\times 10^4$ greater than that of oxygen, for example.

Experimentally, both D and S can be determined by an experiment in which a film is suddenly exposed to a constant partial pressure p_1 of penetrant on one side and the partial pressure p_2 measured on the other. A plot of p_2 versus time will be of the form shown in figure 3.1.

In this experiment there will be a time-lag before any of the diffusing species reaches the other side of the film, after which the increase in p_2

approaches a straight line. The slope of this line gives the steady-state flux F which can be used to calculate the permeability P via (3.7). The intercept L of the straight line portion on the time axis can be shown to give the diffusion constant D from the relation:

$$L = l^2/6D \tag{3.9}$$

This is now the standard procedure for determining diffusion constants and solubilities of small molecules in polymeric films. The reader is referred to the excellent text, *Diffusion in Polymers*, by Crank and Park (1968) for further theoretical and experimental details. More recently, Guillet and co-workers have developed gas chromatographic procedures

Table 3.1. *Diffusion constants* (D) *and permeabilities* (P) *for small molecules in low density polyethylene*

Permeant	$D \times 10^9$ (cm^2 s^{-1})	$P \times 10^{10}$ (cm^3 (STP) cm cm^{-2} s^{-1} cmHg^{-1})
Oxygen	460	2.9
Decane	3.5	~600
Benzene	8.2	5300
Hexane	—	2910
Carbontetrachloride	—	3810
Ethyl alcohol	—	56
Ethyl acetate	—	513

Source: Bandrup and Immergut (1975).

Figure 3.1. Rate of permeation of a penetrant across a thin film.

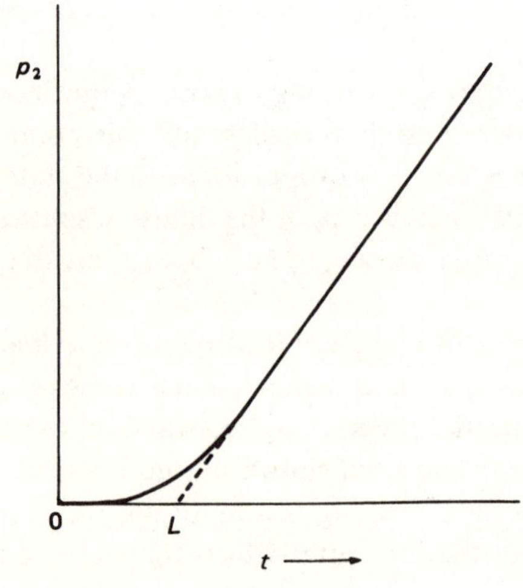

for measuring solubilities and diffusion constants for small molecules in polymers (Gray and Guillet 1973, Lipson and Guillet 1981). In addition, a number of authors have described methods for determining diffusion constants by luminescence quenching experiments (Heskins and Guillet 1970, Nowakowska *et al.* 1976, MacCallum and Rudkin 1978).

3.2 Effect of temperature on diffusion

Since it is generally recognized that diffusion is a thermally activated process involving small scale translational motions of both the diffusant and the medium, the temperature coefficient of diffusion is usually expressed in Arrhenius form:

$$D = D_A \exp(-E_d/RT) \tag{3.10}$$

where D_A and E_d are the pre-exponential factor and the activation energy, respectively. Typical values of the activation energy E_d for the self-diffusion of small solvent molecules in liquids are 8 to 12 kJ mol^{-1}, while diffusion in polymeric solids shows significantly higher values, ranging from 20 to 80 kJ mol^{-1}, but usually about 40 kJ. This is clearly related to the greater energy required to displace large polymer segments to provide 'holes' for the diffusing penetrant.

Table 3.2 shows values of the diffusion constants for helium, oxygen, carbon dioxide and methane in natural rubber, semi-crystalline poly-

Table 3.2. *Diffusion constants and activation energies for gases in polymers at 25 °C*

Polymer	Gas	$D \times 10^7$ cm^2 s^{-1}	D_A	E_d kJ mol^{-1}
Natural rubber	He	216	0.031	18
(cis-polyisoprene) (amorphous)	O_2	15.8	1.94	35
$T_g = -60$ °C	CO_2	11.0	3.7	37
	CH_4	8.9	1.97	36
Polyethylene	He	68	0.13	25
($d = 0.914$ (semicrystalline)	O_2	4.6	4.48	40
$T_g = -40$ °C	CO_2	3.7	1.85	38
	CH_4	1.9	17.2	46
Polystyrene	He	104	0.0019	13
(glassy) $T_g = 80$ °C	O_2	1.1	0.125	35
	CO_2	0.58	0.128	36

ethylene and glassy polystyrene. The values of D decrease both as the polymer chain becomes less flexible and with the size of the penetrant. The diffusion constants for the larger gases are about an order of magnitude lower in glassy polystyrene than in 'rubbery' polyisoprene. As the size of the diffusing molecule increases, both the diffusion constant and the activation energy increase. Thus Gray and Guillet (1973) estimate values of $D=0.35\times10^{-8}$ and $0.82\times10^{-8}\ \mathrm{cm^2\ s^{-1}}$ for *n*-decane and benzene in polyethylene at room temperature, roughly 50 and 20 times smaller than for methane. On the other hand, the diffusion constant for *n*-tetradecane in polyethylene at 130 °C was $1\times10^{-8}\ \mathrm{cm^2\ s^{-1}}$ with an activation energy of 67 kJ mol^{-1}. The significance of the large values of E_d for polymers is that at high temperatures their internal mobility approaches that of small molecule liquids with respect to the permeation of small molecules. Values for gaseous diffusion in ordinary liquids are typically about two orders of magnitude greater than in polymers, e.g., O_2 in cyclohexane, $D=5.3\times10^{-5}\ \mathrm{cm^2\ s^{-1}}$ at 30 °C, as compared with 4.6×10^{-7} in low density polyethylene.

3.3 Processes in dilute solution

Many photoprocesses in the gas phase occur at virtually every collision between reacting species. These include radical–radical recombinations, fluorescence quenching and other 'allowed' energy transfer processes. In solution the rate of such processes may be limited by the rate of mutual diffusion of the two particles. These are often (and somewhat misleadingly) called diffusion-controlled reactions to denote the fact that the rate is not dependent on the reactivity of the species, but on the frequency of encounters of the reagents.

The diffusion constant for a spherical particle is given by Stokes law:

$$D=kT/6\pi\eta r \tag{3.11}$$

where k is the Boltzmann constant, T is the absolute temperature, η is the viscosity of the medium and r is the radius of the particle. Thus the rate of diffusion will be inversely proportional to the viscosity of the medium.

The rate constant for a diffusion-controlled reaction is often estimated from the modified Debye equation (Debye 1942):

$$k_{\mathrm{diff}}=\frac{1}{4}\left(2+\frac{d_1}{d_2}+\frac{d_2}{d_1}\right)\frac{8RT}{3000\eta} \tag{3.12}$$

where k_{diff} is in units of liter mol^{-1} s^{-1} when η is in poise, $R=8.31\times10^7$ erg mol^{-1} deg^{-1} and d_1 and d_2 are the diameters of the reacting species which are assumed to be spherical. Within the approximations of the theory the two species are usually assumed to have similar dimensions, in which case (3.12) becomes:

$$k_{\text{diff}} = \frac{8RT}{3000\eta} = 2.2 \times 10^5 (T/\eta) \tag{3.13}$$

The assumption of equivalent size of the diffusing species is no longer valid if one of the two is a polymer molecule. In this case a modified form of the Smoluchowski relation has been used (Heskins and Guillet 1970):

$$k_{\text{diff}} = \frac{4\pi\rho' N}{1000}(D_1 + D_2) \tag{3.14}$$

where ρ' is the sum of the radii of the colliding groups and D_1 and D_2 are the corresponding diffusion coefficients. In studies of the quenching of naphthalene fluorescence by both polymeric and small molecule ketones this equation gave consistent results using a value of 4 Å for ρ', which is equivalent to assuming that only the C=O group is involved in the collision with naphthalene. If the second reactant is part of a copolymer chain, it is likely that segment diffusion rates will be more than an order of magnitude smaller than the translational diffusion of a small molecule, and in this case (3.14) reduces to

$$k_{\text{diff}} = \frac{4\pi\rho' ND}{1000} \tag{3.15}$$

where D is now the diffusion constant for the small molecule reagent in the solvent.

The Debye relation (3.12) can be used to determine diffusion constants at experimental temperatures if the value of D is known at some other temperature T_1. This is done by measuring the viscosity of the solvent at T_1 and the experimental temperature T_2. The required value of D_2 is given by:

$$D_2 = D_1 (T_2/T_1)(\eta_1/\eta_2) \tag{3.16}$$

where η_1 and η_2 are the solvent viscosities at T_1 K and T_2 K, respectively.

The rate of a diffusion-controlled reaction can be changed significantly by changing the viscosity of the solvent, as (3.13) shows. Table 3.3 gives the viscosity of a number of common solvents and the calculated values of k_{diff} (Calvert and Pitts 1966).

From diffusion theory it is possible to calculate the average distance x that a particle travels in time t from the relation

$$x = (2Dt)^{1/2} \tag{3.17}$$

where x is the average displacement of the particle and D is the diffusion coefficient. For non-viscous fluids, D for a small molecule is usually of the order of 10^{-5} cm^2 s^{-1}, while in a viscous or solid medium, D can be of the order of 10^{-10} cm^2 s^{-1}. Using the above relation, one can easily calculate that in a fluid medium, a particle will diffuse about 15 Å in 1 ns, whereas in a viscous medium, it will be essentially stationary on a nanosecond scale, but will move 50 Å in a millisecond (Turro 1978). These numbers will be useful later in interpreting the results of the kinetics of photochemical reactions in fluid and solid media.

3.4 Processes in solid polymers

Many photophysical and photochemical processes occur readily in polymer systems. As we have seen, the free volume in a polymeric matrix increases with temperature so that diffusion of small molecule reagents becomes possible. For example, oxygen and other small molecules will penetrate plastic matrices at measurable rates at temperatures well below 100 K. Since there is usually no net motion of the centre of mass in a solid polymer it is usual to consider this diffusive motion as occurring by a series of hops as free volume is opened up by segment motion in spaces adjacent to the moving molecule. For example, the

Table 3.3. *Viscosity[a] and diffusion-controlled rate constants for various solvents*

	Viscosity, $\eta \times 10^3$ (poise) at a temperature (°C) of					$k_{\text{diff}} \times 10^{-10}$ liters/mol-s
Compound	0	10	20	30	40	at 20°C)
Isopentane	2.78	2.49	2.25	2.05	—	2.9
n-Pentane	2.79	2.55	2.35	2.16	—	2.8
n-Hexane	3.81	3.62	3.13	2.85	2.62	2.1
n-Heptane	5.26	4.67	4.18	3.77	3.43	1.6
n-Octane	7.13	6.20	5.47	4.87	4.37	1.2
Chloroform	6.99	6.25	5.63	5.10	4.64	1.2
Toluene	7.73	6.70	5.87	5.20	4.65	1.1
Methanol	8.08	6.90	5.93	5.15	4.49	1.1
Benzene	—	7.60	6.49	5.62	4.92	1.0
Carbon tetrachloride	14.03	11.35	9.69	8.42	7.38	0.67
Cyclohexane	—	11.80	9.80	8.26	7.04	0.66
Water	17.93	13.10	10.09	8.00	6.54	0.64
Ethanol	—	—	11.92	—	—	0.55
n-Propyl alcohol	38.6	29.2	22.6	17.7	14.0	0.29
Isopropyl alcohol	45.6	32.5	23.7	17.7	12.9	0.27
Ethylene glycol	—	—	173	—	—	0.038
Glycerol	—	—	10 700	3800	—	0.000 61

[a] Data from *International Critical Tables*, McGraw-Hill Book Co., New York, 1933; Selected values of physical and thermodynamic properties of hydrocarbons and related compounds, *API Project* 44, Carnegie Press, Pittsburgh, 1953.
Source: Calvert and Pitts (1966).

theory of Bueche (1962) relates the diffusion constant D to a jump frequency ϕ and a jump distance δ through

$$D=\tfrac{1}{6}\phi\delta^2 \tag{3.18}$$

It is expected that the jump distance δ will be relatively constant from one polymer to the next, and that variations in D will be caused primarily by differences in the jump frequency ϕ. The frequency will depend on the probability of collecting a sufficient number of 'packets' of free volume to give a hole of volume V^* of sufficient size that the diffusing molecule can move into it. The creation of a hole in a liquid will require the expenditure of a certain quantity of energy E_h and the probability of forming such a hole can be calculated from Boltzmann statistics to be proportional to $\exp(-E_h/RT)$. Application of the theory to polymeric and small molecule liquids has given jump distances δ corresponding to the dimensions of small molecules (i.e., 2–20 Å) and jump frequencies ranging from 10^4 to 10^6 per second. These and other free volume theories are described in more detail in Crank and Park (1968). The important concept, common to all of these theories, is that diffusion occurs in polymers through free volume obtained by minor displacements of side groups or segments of the chain but without net translational movements of the centre of mass of the polymer.

It is important to realize that for amorphous solids above the glass transition temperature, the rates of diffusion of small molecules are only one or two orders of magnitude below those in ordinary liquids. The relatively high mobility of small molecules in polymers above their glass transitions and the almost complete conformational freedom of the polymer chain means that reactions may take place under these conditions nearly as rapidly as in solution. The polymer is thus not a viscous medium in the conventional sense because the diffusing molecule reagents are not affected by motion of the centre of mass but only by segment diffusion. For example, measurements of k_{diff} by Heskins and Guillet (1970) for naphthalene in polyethylene show an effective viscosity according to (3.13) of about 1 poise at 80 °C (i.e., higher than ethylene glycol but lower than glycerin (*cf.* table 3.3). It is clear that the microviscosity of polymeric systems is many orders of magnitude lower than their macroscopic viscosity, which in the case of solid polyethylene would be *ca.* 10^{10} poise. This fact has gone unrecognized in many studies of polymer reactions.

As a polymer chain is crosslinked the coordinated segment motion available for permeation is reduced, with a consequent reduction in the diffusion constant D. For example, table 3.4 shows that high degrees of

Table 3.4. *Effect of crosslinking on diffusion of N_2 in natural rubber*

% Sulfur	$D \times 10^7$ ($cm^2\ s^{-1}$)	E_d ($kJ\ mol^{-1}$)
0	11.0	—
1.7	10.8	33
2.9	8.0	35
7.1	2.7	40
11.3	1.1	45

Source: Barrow and Skirrow (1948).

sulfur crosslinking reduce the diffusion of nitrogen in natural rubber at 25 °C by a factor of ten, and, as expected, the activation energy for diffusion increases because of the greater rigidity imposed on the system by the crosslinks. However, low degrees of crosslinking (0–3% S) show little effect.

It should be pointed out that the simple theories of diffusion in polymer matrices described in the previous sections usually only apply at very low concentrations of the diffusant or penetrant. At higher concentrations, additional effects are observed. For example, below the glass transition temperature, diffusing molecules which are solvents for the polymer will reduce the glass transition in the region immediately surrounding the diffusing front. This will cause substantial deviations from the simple laws of diffusion. For this reason it is advisable to carry out experiments under conditions where Fick's law is obeyed.

3.5 Solubility of small molecules in polymers

Since the flow rate of a gaseous molecule through a polymer depends on both the concentration gradient, which is determined by the solubility S and on the diffusion constant D some discussion of the factors controlling solubility in polymers is appropriate. It is important to note that while polymers are usually soluble in only a very small number of organic solvents, most small molecules are soluble in polymers to some extent. This is the reason for the often remarkable permeability of plastic films to a variety of organic compounds.

For most small molecule gases at pressures up to 1 to 2 atm, Henry's law (3.6) is experimentally obeyed. Therefore, the concentration of gas in the film is directly related to the permeability constant P by:

$$c = (P/D)p \tag{3.19}$$

For glassy polymers, the solution process often becomes more complex and the relationship between *c* and *p* approximates a Langmuir isotherm. There is some evidence that this is due to the effects of true molecular solution complemented by adsorption or solution in cavities or holes formed in the glass as it is cooled below T_g. However, for partial pressures below 1 atm this phenomenon is relatively rare and Henry's law becomes a reasonable approximation even for glassy polymers. The temperature dependence of *S* includes the heat of condensation which is always negative, and the heat of mixing. Although this is usually small and positive, very strong specific interaction between the diffusing molecule and the polymer can sometimes occur. As a result, heats of solution can be either positive or negative. As the size of the gas molecule increases, the heat of condensation increases more rapidly than does the heat of mixing, giving rise to higher solubilities, regardless of the sign of the mixing term. The value of $\log S$ has been shown to be a nearly linear function of both T_b and T_c, the normal boiling point and critical temperature (K) of the diffusant gas (figure 3.2).

Figure 3.2. Solubility coefficients of gases in natural rubber at 25°C *vs* their critical temperatures (T_c) and boiling points (T_b). Units cc STP gas/cm^3/atm. (From Crank and Park 1968.)

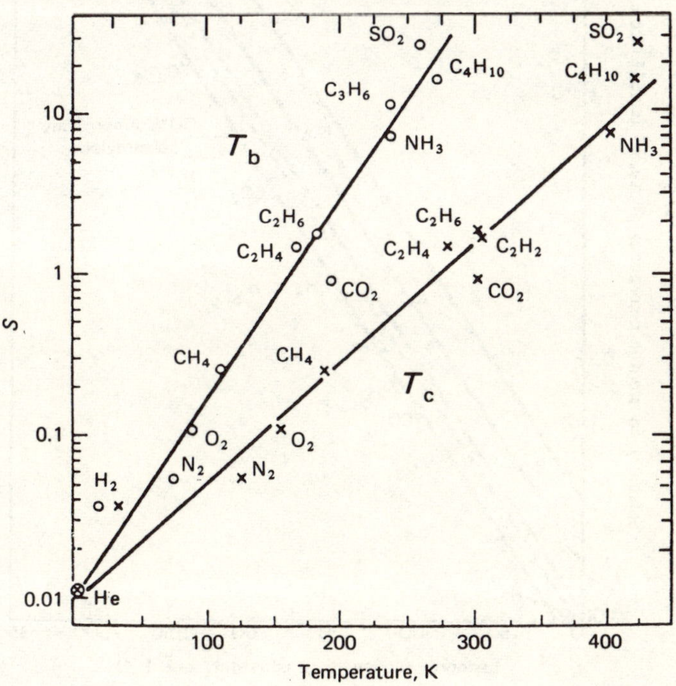

In the absence of other experimental data this figure provides a useful way of estimating solubilities in polymers.

Solubilities can also be related to the magnitude of the van der Waals attractions in the diffusant. Figure 3.3 shows the linear relationship observed between the logarithm of the solubility constant and the Lennard–Jones force constants for a variety of gases in amorphous polyethylene, natural rubber, benzene and heptane.

Since the lines are parallel, the ratios of solubilities are 1:1.5:2:4 for amorphous polyethylene, polyisoprene, benzene and heptane, respectively, for any diffusing molecule. This provides a useful rule of thumb, since if the solubility of a gas or vapor is known in heptane or benzene at

Figure 3.3. Correlation of the solubility coefficients of gases in polyethylene at 25°C with the Lennard–Jones force constants. Data for n–C_7H_{16} (circle) and C_6H_6 (triangle). (From Crank and Park 1968.)

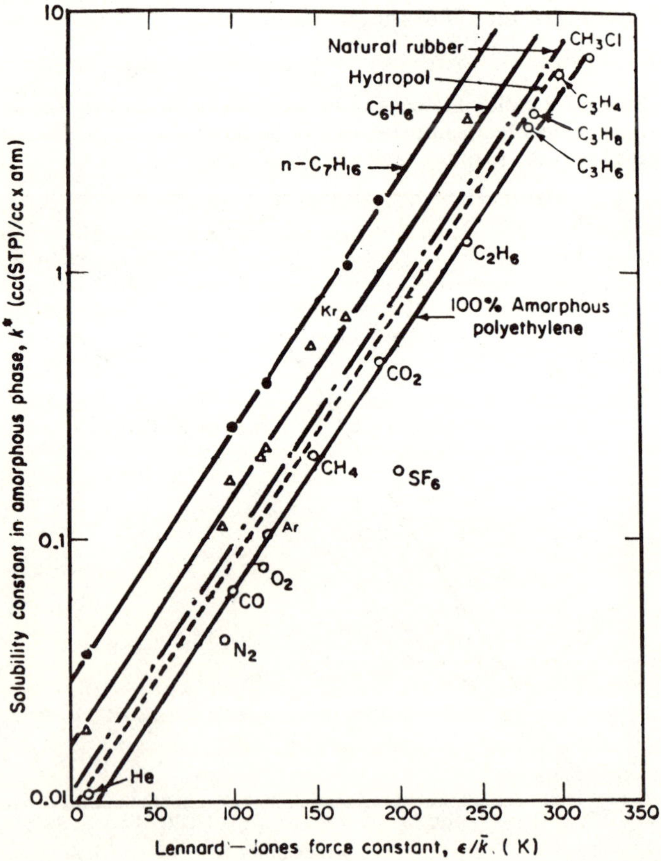

a given temperature, the solubility in polyethylene can be assumed to be, respectively, one-quarter or one-half of this value. It is important to remember, however, that this will be true only for the amorphous phase. Since most polyethylenes are from 40 to 90% crystalline, both the solubility and diffusivity will be modified by the presence of these impermeable regions. When dealing with semi-crystalline polymers D will be reduced by the tortuosity of the diffusion path imposed by the crystallites and S must be calculated on the basis of the fraction of amorphous polymer actually present in the polymer.

In dealing with glassy polymers at high partial pressures of the diffusant, the diffusion constant will no longer be independent of concentration and more complex behavior will be observed. Frisch (1965) has identified a class of diffusion (denoted type B Fickian) which is characterized by non-dilute penetrant polymer concentrations, concentration-dependent diffusion coefficients and energies of activation which are temperature dependent. The application of this theory is beyond the scope of this monograph.

3.6 Determination of permeability in polymers

Although there are extensive tabulations of permeabilities of most of the permanent gases such as H_2, O_2, CO_2, CH_4, N_2, etc. for most common polymers, values of the diffusion constants and solubilities of larger molecules in polymers are not readily available. The standard permeability measurement described in §3.1, although simple in principle, must be carried out carefully in specially constructed apparatus in order to obtain reliable results (Crank and Park 1968).

A novel technique developed in the author's laboratory (Guillet 1973) utilizes a conventional gas chromatograph to obtain these data, and has shown some advantages with respect to the ease with which data can be obtained on large numbers of penetrants in polymers. Since the method is novel and not yet well known, the principles and practice will be briefly summarized in the following sections.

3.6.1 *Determination of solubilities*

The determination of solubilities of penetrants in polymers by gas chromatography is quite simple. The polymer to be studied is coated on a conventional gas chromatography support at concentrations of 5 to 15 wt-%. The support is dried and packed into a column and the retention time is measured for penetrants of interest under conditions where the polymer is above its glass transition and where surface adsorption at the polymer–gas interface can be neglected. A non-interacting marker gas (usually air or methane) is injected with each penetrant (probe) and the time t_N between the peak maxima of the marker and the molecule of interest is determined. Also recorded is the flow rate u using a soap bubble flowmeter, and the pressure drop across the column. A typical GC trace is shown in figure 3.4. Observation of a symmetrical peak is evidence that the partition isotherm is linear and that Henry's law is obeyed.

The net retention volume V_N is a measure of the solubility of the penetrant in the polymer and is given by:

$$V_N = (t_N u)J \tag{3.20}$$

where J is a correction term for pressure changes across the column and is given by:

$$J = \frac{3}{2}\left(\frac{(P_i/P_o)^2 - 1}{(P_i/P_o)^3 - 1}\right) \tag{3.21}$$

where P_i and P_o are the carrier gas pressures at the inlet and outlet of the chromatography column.

If the amount of penetrant is very small, the gas should exhibit ideal behavior and in this case the Henry's law coefficient H_1 will be given by

$$H_1 = \frac{RTw_L}{V_N M_1} = \frac{1}{S} \tag{3.22}$$

where w_L is the mass of polymer in the column and M_1 is the molecular weight of the penetrant. The value of H_1 determined in this way is that corresponding to nearly infinite dilution in the polymer and is usually accurate enough for most work on polymeric solids. It is the reciprocal of

Figure 3.4. Gas chromatographic determination of retention times and peak width.

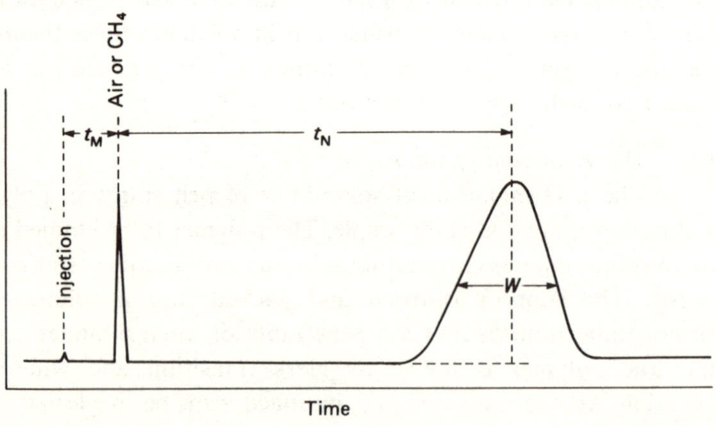

the solubility S defined in (3.6). However, in some cases, greater accuracy can be obtained by correcting for non-ideality in the gas phase and for the solubility of the carrier gas in the polymer phase (Liu and Prausnitz 1976). It is a simple matter to determine the solubility of a large number of penetrants in a polymer by this procedure. Activation energies can also be readily obtained by carrying out the experiment at different column temperatures. Typical values of H_1 for a variety of organic molecules in an ethylene–vinyl acetate copolymer are given in table 3.5. Since this procedure gives a weight fraction Henry's law coefficient the weight fraction c_1 of permeant in the polymer phase in equilibrium with a partial pressure p_1 in the vapor phase will be given by:

$$c_1 = p_1/H_1 = Sp_1 \tag{3.23}$$

Gas chromatographic data are often expressed in terms of the specific retention volumes V_g, which are related to V_N by:

$$V_g = (V_N/w_L)(273/T) \tag{3.24}$$

where T is the column temperature (or temperature at which V_N is measured if different). This is the retention volume per gram of stationary phase reduced to 0 °C and 0 pressure gradient. Many tabulations of V_g values for various absorbing phases are given in the GC literature. Solubilities can be obtained by use of the relations:

$$S = M_1 V_g/273R \tag{3.25}$$

or

$$S = M_1 V_N/RTw_L \tag{3.26}$$

where M_1 is the molecular weight of the permeant. Similarly, solubilities can be calculated from infinite dilution weight fraction activity coefficients $(a_1/w_1)^\infty$ using:

$$S = \frac{1}{p_1^0 (a_1/w_1)^\infty} \tag{3.27}$$

where p_1^0 is the saturation vapor pressure of the permeant vapor at the temperature of the experiment.

3.6.2 *Effect of temperature on solubility*

The simple relation expressed by (3.25) between the solubility of a compound in a polymer and its specific retention volume V_g, also provides an easy determination of the temperature coefficient of the solubility from gas chromatography retention data. It can be readily shown (Guillet 1973) that

$$\frac{\partial \ln S}{\partial(1/T)} = \frac{\partial \ln V_g}{\partial(1/T)} = \frac{\Delta H_s}{R} = \frac{\Delta \bar{H}_v - \Delta \bar{H}_m}{R} \tag{3.28}$$

Table 3.5. *Henry's constants and solubilities for probes in P(E-VAc) at selected experimental temperatures*[a]

Probe	65 °C		75 °C		85 °C	
	H_1 (atm)	S_1 (weight fraction 1)	H_1 (atm)	S_1 (weight fraction 1)	H_1 (atm)	S_1 (weight fraction 1)
1-Chlorobutane	2.29	0.437	3.25	0.308	4.31	0.232
1-Octene	0.832	1.20	1.24	0.807	1.75	0.572
n-Butanol	0.840	1.19	1.32	0.759	2.01	0.498
Chlorobenzene	0.234	4.27	0.360	2.78	0.525	1.91
Chloroform	1.22	0.820	1.78	0.561	2.44	0.410
Methyl propyl ketone	1.37	0.730	1.96	0.510	2.77	0.361
Toluene	0.688	1.45	1.03	0.973	1.44	0.694
n-Decane	0.126	7.94	0.204	4.90	0.313	3.19
n-Hexane	6.34	0.158	8.55	0.117	12.8	0.078
Benzene	1.98	0.505	2.79	0.358	3.78	0.264
n-Octane	0.848	1.18	1.25	0.802	1.77	0.565
n-Pentane	19.7	0.051	24.7	0.040	28.5	0.035

[a] For partial pressure of probe equal to 1 atm.
Source: Lipson and Guillet (1981).

where $\Delta\bar{H}_s$, $\Delta\bar{H}_v$ and $\Delta\bar{H}_m$ are respectively the molar heats of solution, vaporization and mixing of the probe molecule with the polymer.

A simple determination of the variation of the retention time with column temperature allows the evaluation of $\Delta\bar{H}_s$. If the solubility S_1 is known at one temperature T_1, the value at S_2 at any other temperature, T_2 can be calculated from

$$\ln(S_1/S_2)=(\Delta H_s/R)[1/T_1)-(1/T_2)] \quad (3.29)$$

3.6.3 *Theoretical treatments of solubility in polymers*

In theoretical treatments of polymer solubility, solute activity is often expressed as the sum of two terms, a combinatorial entropy and a non-combinatorial free energy of mixing. In the Flory–Huggins approximation one has

$$\ln a_1=\ln\phi_i+[1-(1/r)]\phi_2+\chi\phi_2^2 \quad (3.30)$$

where ϕ_i is the volume fraction of component i, χ is the Flory–Huggins interaction parameter, and r is the ratio of molar volumes of the polymer and solute. Values of the χ parameters have been determined for a variety of solute and solvents by a number of experimental procedures and are tabulated in the literature (*Polymer Handbook*, Bandrup and Immergut 1966, 1975). Neglecting corrections for non-ideality of the gas phase, the relationship between χ and the solubility S is

$$\ln S=\ln(v_2/v_1)-\ln p_1^0-\chi-1 \quad (3.31)$$

where v_2 and v_1 are the specific volumes of the polymer and probe, respectively. Table 3.6 shows typical values of solubilities for a variety of solutes in polyethylene and polystyrene determined by the chromatographic method. The values here are at infinite dilution but the procedure

Table 3.6. *Henry's constants and solubilities for five probes in polystyrene at 183 °C*

Probe	H_1 (atm)	S_1(weight fraction 1)
n-Decane	13.6	0.074
n-Dodecane	5.09	0.196
n-Tetradecane	1.86	0.539
cis-Decalin	3.80	0.263
trans-Decalin	5.02	0.199

[a] For a partial pressure of probe equal to 1 atm: using the data of DiPaola and Guillet.
Source: Lipson and Guillet (1981).

can be extended to much higher concentrations (Braun and Guillet 1976, Brockmeier *et al.* 1972*a*, 1972*b*, 1973).

3.6.4 *Determination of diffusion constants*

In principle, the gas chromatographic procedure described above can also be used to determine the diffusion constants of the penetrating molecules. This is due to the fact that if the experimental conditions are adjusted appropriately, it can be arranged that the rate-controlling step in the movement of the probe molecule through the gas chromatographic column is the rate of diffusion into and out of the polymer stationary phase.

Figure 3.5. Van Deemter curves (eq. 3.32) for various columns: (1) benzene polyethylene, column A at 27°; (2) benzene–natural rubber, column D at 25°; (3) benzene–polyethylene, column B at 25°; (4) benzene–natural rubber, column C at 23°; (5) *n*-nonane–polyethylene, column B at 50°; (6) decane–natural rubber, column E at 40.2°; (7) dodecane–natural rubber, column E at 685°; (8) hexadecane–polyethylene, column A at 200°. (From Gray and Guillet 1973.)

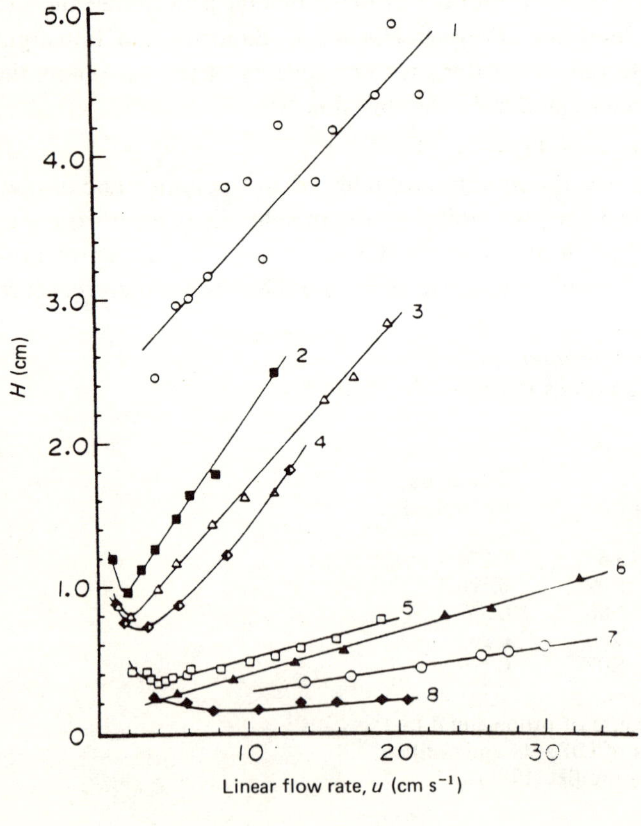

Experimentally this means that the breadth of the observed gas chromatographic peak will be directly related to the diffusion constant by an equation of the form:

$$H = A + B/u + Cu \qquad (3.32)$$

In this equation, H is the height equivalent to a theoretical plate defined by

$$H = \frac{l}{5.54}\left(\frac{W}{t_M + t_N}\right)^2 \qquad (3.33)$$

where W is the peak width at half maximum height, obtained from the chromatogram (figure 3.4), l is the length of the column, t_N is the net retention time for the penetrant molecule and t_M is the retention time for a non-interacting molecule such as nitrogen or methane and essentially represents an estimation of the open space in the chromatographic column. Values of H are determined as a function of the carrier gas velocity u in the column and the slope of this line at large values of u is used to estimate the value of C. The diffusion constant D is obtained from the value of C according to:

$$C = \frac{2}{3}\left(\frac{d_f^2}{D}\right)\left(\frac{t_N}{t_N + t_M}\right)^2 \qquad (3.34)$$

where d_f is the thickness of the polymer on the support. Typical experimental data are shown in figure 3.5 and table 3.7.

Table 3.7. *Diffusion coefficients from gas chromatographic measurements on low-density polyethylene*

Probe	Temp (°C)	Van Deemter C term (s × 10^3)	k	$k/(1+k)^2$	D_l (cm^2 s^{-1} × 10^8)
n-Tetradecane	125	20 ± 7	20.4	0.045	~0.85
	140	13.0 ± 0.6	11.3	0.075	2.2
	150	10.3 ± 0.4	7.83	0.100	3.7
	160	12.5 ± 0.6	5.47	0.131	4.1
	170	8.6 ± 0.5	3.88	0.163	7.4
n-Decane	30	30.0 ± 1.8	34.3	0.028	0.35
	50	24.6 ± 0.7	13.2	0.066	1.03
	60	31.8 ± 0.7	10.0	0.083	1.00
	65	29.9 ± 1.5	8.0	0.099	1.28
	80	38.2 ± 1.0	5.3	0.133	1.34
Benzene	25	114 ± 6	1.50	0.240	0.82

Source: Gray and Guillet (1973).

The major uncertainty in the use of this equation is the determination of the thickness d_f of the film into which the penetrant is moving. In order to have uniform coating thicknesses of known geometry Gray and Guillet (1973) used spherical glass beads as the packing material. The thickness was then calculated from the observed weight of the polymer and the dimensions of the beads. Alternatively, conventional GC supports can be used and the thickness determined by calibration with a penetrant whose diffusion constant is known from other procedures. This method is particularly useful if one wishes to correlate the rates of diffusion of a large number of organic penetrants in polymer systems. Again, it is important to point out that because of the high dilution of the GC system, the diffusion constant D_0 determined by this procedure is that corresponding to very low concentrations of the diffusing penetrant.

4

Determination of Scission and Crosslinking in Polymers

4.1 Polymer molecular weights and distributions

One of the consequences of the interaction of light with polymer molecules is often the breaking or making of a chemical bond. If the chemical bond broken is in the backbone of the polymer chain, a decrease in the molecular weight of the polymer results. Alternatively, if a bond is formed between two polymer molecules, the average molecular weight of the polymer increases. Measurements of polymer molecular weight are then of prime importance in the elucidation of mechanisms of photoreactions of polymer systems.

Polymer chemists have devised a number of important methods for determining the molecular weights of macromolecules and nearly all of them have been applied to studies of polymer photochemistry. These include the measurement of all the colligative properties of solutions, such as osmotic pressure, freezing point depression, boiling point elevation, and vapor pressure depression. Development of the theory of scattering of light by colloidal particles by Debye in the 1930s led the way to the use of light scattering to determine the size of polymer molecules in solution and more recent progress in the development of exclusion chromatography has resulted in the method known as gel permeation chromatography. With the exception of the latter, all of these methods can, in principle, give absolute molecular weights defined on a thermodynamic basis. The most precise of all methods, but one which requires calibration by other standard procedures, is viscometry and it is the preferred method if small changes in molecular weight are to be observed.

Unlike small molecules, pure polymer samples do not usually consist of a single molecular species but of a series of homologues of varying molecular weight. These homologues are nearly identical in chemical composition but differ in the length of the polymer chain. For precise work, it is essential to characterize the distribution of molecular weights in the polymer sample and to fit it to some empirical or theoretical equation. Any distribution of molecular weights can be described in terms of an average over all species present. These averages are usually defined as follows:

$$M_n = \sum_i n_i m_i \Big/ \sum_i n_i = P_n M_0 \tag{4.1}$$

$$M_w = \sum_i n_i m_i^2 \Big/ \sum_i n_i m_i = P_w M_0 \tag{4.2}$$

$$M_z = \sum_i n_i m_i^3 \Big/ \sum_i n_i m_i^2 = P_z M_0 \tag{4.3}$$

where M_n, M_w and M_z are, respectively, the number, weight and z-average molecular weights, n_i is the number of molecules of molecular weight m_i present in the sample, P_n, P_w and P_z are the number, weight and z-average degrees of polymerization, respectively, and M_0 is the molecular weight of the monomer unit.

A convenient measure of the breadth of a particular distribution of molecular weights is given by the dispersity γ defined by

$$\gamma = M_w / M_n$$

Carefully prepared vinyl and condensation polymers will often have dispersities in the range of 1.5 to 2.5, but many polymers can have values of γ exceeding 10, depending on the method of preparation. Polymers with crosslinks, or prepared from trifunctional (or higher) monomers will tend to have very large dispersities, and great care should be taken in these cases to be sure that only methods giving number-average molecular weights are used to evaluate reaction kinetics.

To study the kinetics of a reaction involving either the breaking of a polymer chain backbone, or the crosslinking of two polymer molecules it is only necessary to know the number-average molecular weight. The number of bond breaks or crosslinks can be calculated directly from the change in the number-average molecular weight or degree of polymerization as a function of the reaction time or other conditions, for example, the light absorbed during a photochemical process.

Number-average molecular weights are given directly by measurement of the colligative properties of a solution, i.e., measurements of osmotic pressure, freezing point depression, boiling point elevation or vapor pressure lowering. Of these, the most reliable is the measurement of the osmotic pressure. The osmotic pressure, π, is measured as a function of concentration and the molecular weight calculated from the relationship:

$$\lim_{c\to 0}(\pi/c) = RT/M_n \tag{4.4}$$

Unfortunately, the osmotic pressure is not conveniently measured in kinetic experiments and as a result it is more common to measure a property of the solution which can be determined more rapidly, namely the solution viscosity.

4.2 Viscosity methods

When a macromolecule is dissolved in a suitable solvent, the viscosity of the solution increases by an amount related to the actual size the macromolecular coil takes up in solution. If η is the viscosity of the solution and η_0 the viscosity of the solvent, we can define the relative viscosity η_{rel} as

$$\eta_{\text{rel}} = \eta/\eta_0 \tag{4.5}$$

and the specific viscosity by

$$\eta_{\text{sp}} = (\eta - \eta_0)/\eta_0 \tag{4.6}$$

The quantity which is directly related to the molecular weight is known as the intrinsic viscosity $[\eta]$ which is defined below:

$$[\eta] = \lim_{c\to 0} \eta_{\text{sp}}/c \tag{4.7}$$

In his early studies of macromolecules, Staudinger (Staudinger and Heuer 1930) proposed a simple relationship between the viscosity and molecular weight:

$$[\eta] = KM \tag{4.8}$$

which was subsequently shown to be only approximate in form. Modern theory gives the relationship which is now generally known as the Mark–Houwink equation:

$$[\eta] = KM^a \tag{4.9}$$

in which K and a are both constants depending on the particular polymer and the solvent used. A more correct formulation of this equation is

$$[\eta] = K_v M_v^a \tag{4.10}$$

where M_v is the viscosity-average molecular weight, which is defined by

$$M_v = \left[\sum_i w_i M_i^a\right]^{1/a} \tag{4.11}$$

It is important to realize that the simple relationship given by (4.9) will only be valid for polydisperse macromolecules if the breadth of the distribution, that is the ratio of M_n to M_w to M_z and thus to M_v remains invariant. As we shall see later, this condition is obeyed in relatively few cases and consequently the results obtained by using viscosity-average molecular weights must be carefully interpreted, particularly if information is not available about the distribution of the macromolecules in the sample. On the other hand, (4.10) is valid for all distributions, but requires a knowledge of the exact distribution in each polymer sample to calculate the more useful moments of the distribution, namely M_n, M_w and M_v.

The values of K and a for a large number of polymers have been determined in a variety of solvents and are tabulated in the *Polymer Handbook* (Bandrup and Immergut 1975). If they are not available, they can be obtained experimentally by careful fractionation followed by molecular weight determination; however, this is a tedious and difficult process and modern theory provides a number of approximate methods of estimating the values of these constants from data on other solvents.

A rapid procedure for determining K and a has been reported by Kilp and Guillet (1977). This method involves the use of photoscission of the macromolecule to obtain a series of fractions with well defined distributions, followed by determination of the viscosity and osmotic pressure. Determination of the intrinsic viscosity usually involves measurement at a series of concentrations and extrapolation to zero. This also is a tedius procedure and it is recommended that η be determined by means of the Solomon–Ciuta equation:

$$[\eta]=(\sqrt{2}/c)(\eta_{sp}-\ln \eta_{rel})^{1/2} \tag{4.12}$$

This single point determination of the intrinsic viscosity gives values identical to those determined by extrapolation and is certainly satisfactory for most determinations of polymer reaction kinetics.

4.3 Light scattering

An additional procedure which is of particular relevance to the study of photochemical reactions involves the determination of the molecular weight of the polymer by light scattering. Based on the early work of Debye (1944, 1947), it has been shown that the intensity of light scattered from the polymer molecule in solution at a particular angle is related to its molecular mass by the following equation:

$$H(c/\tau)=1/M+2A_2c+\ldots \tag{4.13}$$

where

$$H=\left(\frac{32\pi^3n^2}{3N\lambda^4}\right)\left(\frac{\mathrm{d}n}{\mathrm{d}c}\right)^2 \tag{4.14}$$

N is Avogadro's number, n is the refractive index, λ is the wavelength of light, c is the concentration in g l^{-1} and the turbidity τ is given by

$$\tau=\left(\frac{32\pi^3}{3\lambda^4}\right)\left(\frac{RT_c}{A_0}\right)\left(n\frac{\mathrm{d}n}{\mathrm{d}c}\right)^2 \Big/ \frac{\mathrm{d}\Pi}{\mathrm{d}c} \tag{4.15}$$

and Π is the osmotic pressure. For very large molecules, it is sometimes necessary to measure the scattering intensities as a function of angle and extrapolate both to zero angle and to zero concentration; however, more modern techniques involving low angle laser scattering allow a single point determination of the molecular weight. These should become more utilized when the cost of the equipment is reduced to make the apparatus more generally available. Further details of the experimental technique for light scattering and the theory related to it are beyond the scope of this text, however, excellent reviews are given in Tanford (1961) and Pecora (1972).

The use of light scattering from a laser beam following a laser pulse has been used by Schnabel and co-workers (Lindenau *et al.* 1977) to determine important characteristics of the photodegradation of a number of polymer molecules. It should be pointed out in this context, however, that the light scattering method determines the weight-average molecular weight which must be converted to number average from a knowledge of the polymer molecular weight distribution, both at the initial stage and during the course of the degradation process in order to obtain the precise data relating to the rates of chemical reactions.

4.4 Determination of rates of bond scission in polymers

Photochemical reactions in polymers frequently result in the scission of a bond at random in the backbone of the polymer chain. Breakage of a single C–C bond reduces the molecular weight of that particular chain by one-half. In early studies on degradation reactions in polymers, it was common practice to plot either the molecular weight or the intrinsic viscosity $[\eta]$ or more properly $[\eta]^{1/a}$ as a function of the reaction time. When this is done, a curve of the form shown in figure 4.1 is typically obtained. The shape of such a curve in which there is a high initial slope followed by a lower value was often taken by early workers to signify the presence of 'weak bonds' in the polymer which were broken in the early stages of the reaction. Then more stable bonds which were resistant to either temperature or radiation were broken at later stages. However, a simple consideration of the stoichiometry of the bond breaking process indicates that plots of this kind are not related to the actual rate of bond breaking in a simple way. The reason for this is that it takes only one bond break to cut the molecular weight in half, regardless of the initial molecular weight of the polymer. Polymers with very large molecular weights will therefore require fewer breaks per unit weight than polymers with low molecular weight in order to reduce their molecular weight by the same amount. For this reason, the only appropriate method of plotting such data is to plot $1/\bar{M}_n$ or the reciprocal of $[\eta]^{1/a}$ as a function of either time or the radiation absorbed, as shown in figure 4.2. If the process involves a constant rate of bond breaking, then a straight line will be observed, the slope of which can be related either to the rate constant or the quantum yield of thermal or photochemical processes, respectively. Either upward or downward curvature of plots of the form of figure 4.2 can then be accurately interpreted in terms of changes in

Figure 4.1. Plot of molecular weight vs. time or light absorbed.

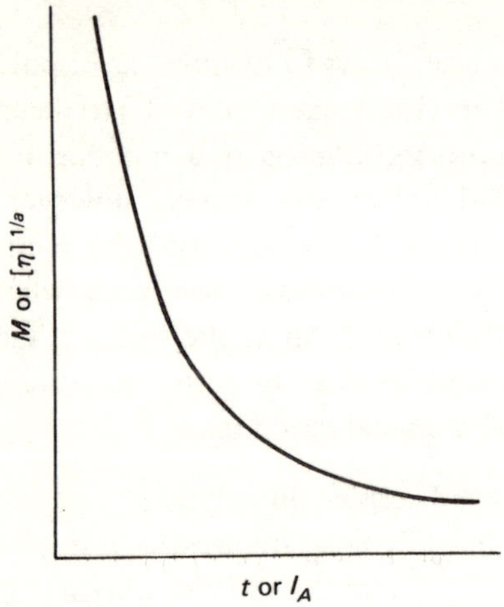

Figure 4.2. Plot of inverse molecular weight vs. time or light absorbed.

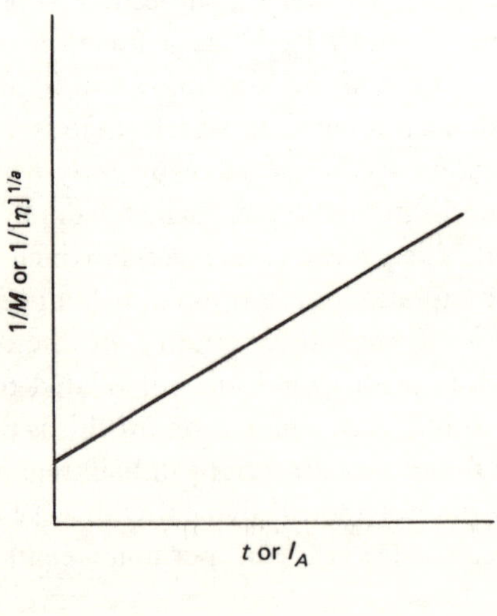

either the mechanism or rate of some of the fundamental processes involved in degradation. Because of the difficulty in interpreting data plotted in the form of figure 4.1, the use of such plots should be discouraged.

It is important to point out, however, that the $1/M$ plot will only be linear if M is the number-average molecular weight M_n. If any other average molecular weight is used, then it is necessary to determine the changes in distribution which occur in the course of the degradation process and convert the average to M_n.

4.5 Theory of random degradation and crosslinking

The theory of polymer degradation has been studied for many years and has been reviewed in a number of papers (Saito 1972, Scott 1974). The mathematical theory of random scission and crosslinking in polymer chains has been developed extensively by Charlesby (1960) and by Saito (1958). Because of its importance, we will summarize briefly the relationships derived for random scission, but for details of the mathematical treatment, the reader is referred to the original paper (Saito 1958).

In Saito's theory, the central assumptions are as follows: (1) that the probability of main chain scission is independent of the position in the polymer chain, (2) that polymers are linear, and (3) that the degree of polymerization is continuous from zero to infinity so that summations can be treated as integrals. For a polymer which is undergoing degradation by random scission, Saito derives the equation:

$$n(P,\tau)=\{n(P,0)+2\tau\int_P^\infty n(l,\tau)\mathrm{d}l+\tau^2\int_P^\infty \mathrm{d}l\int_1^\infty n(m,0)\mathrm{d}m\}\mathrm{e}^{-P\tau} \qquad (4.16)$$

where $n(P,0)$ is the number of polymer molecules having degree of polymerization P after zero irradiation and τ is the probability of scission per monomer unit. $n(P,\tau)$ is the number of polymer molecules having degree of polymerization P after reaction sufficient to give a probability of monomer rupture equal to τ. The initial polymer molecular weight distribution, $n(P,0)$, is normalized so that

$$\int_0^\infty Pn(P,0)\mathrm{d}P=N \qquad (4.17)$$

where N is the total number of monomer units in the system. Using (4.17) one can write the various molecular weight averages as

$$M_n(\tau)=M_0N/\int_0^\infty n(P,\tau)\mathrm{d}P \qquad (4.18)$$

$$M_w(\tau) = (M_0/N)\int_0^\infty P^2 n(P,\tau)\mathrm{d}P \tag{4.19}$$

$$M_v(\tau) = M_0\left[(1/N)\int_0^\infty P^{1+a} n(P,\tau)\mathrm{d}P\right]^{1/a} \tag{4.20}$$

where M_0 is the monomer molecular weight and a is the Mark–Houwink viscosity law exponent, which usually has a value between 0.5 and 1.0.

The molecular weight averages can thus be calculated, as a function of τ, for different initial distributions. A tabulation of commonly used distributions is given by Scott (1974). One such distribution is the Schulz distribution, defined by

$$\mathrm{n(P,0)} = \left(\frac{b+1}{U}\right)^{b+2}\left(\frac{N}{\Gamma(b+2)}\right)P^b \mathrm{e}^{-(P/U)(b+1)} \tag{4.21}$$

where $\Gamma(x)$ is the γ function, U is the initial number-average degree of polymerization and b is a parameter which determines the initial dispersity, γ_0. For the Schulz distribution, we have

$$\gamma_0 = \frac{b+2}{b+1}, \quad M_n^S(0) = M_0 U, \quad M_w^S(0) = M_0 U\left(\frac{b+2}{b+1}\right),$$
$$M_v^S(0) = \frac{M_0 U}{b+1}\left[\frac{\Gamma(b+a+2)}{\Gamma(b+2)}\right]^{1/a} \tag{4.22}$$

The superscript S refers to the Schulz distribution.

Applying Saito's degradation equation to the Schulz distribution, Inokuti (1963) has obtained

$$M_w^S(\tau) = \frac{2M_0}{\tau}\left\{1 - \frac{1}{U\tau}\left[1 - \left(1 + \frac{U\tau}{b+1}\right)^{-(b+1)}\right]\right\} \tag{4.23}$$

also

$$M_n^S(\tau) = \frac{M_n^S(0)}{(1+U\tau)} \tag{4.24}$$

One can also calculate an exact expression for $M_v^S(\tau)$ (Nemzek and Guillet 1977). The Schulz distribution is broad for negative b values and becomes monodisperse as b becomes large positively. When b is equal to zero, one obtains the so-called 'Flory distribution', sometimes called the most probable random distribution, which has a dispersity of 2.0 and is defined by

$$n(P,0) = (N/U^2)\mathrm{e}^{-P/U} \tag{4.25}$$

The Flory distribution is the usual distribution of molecular weights attained by free radical polymerization under conditions of constant monomer and catalyst concentrations, and when termination is by disproportionation. It has also been shown, both theoretically and experimentally, that any distribution undergoing a random scission process will approach a Flory distribution relatively rapidly. If τ is the probability of breaking a particular bond in a polymer chain, we can define a quantity $\xi \equiv U\tau$ where U is the initial number-average degree of polymerization and ξ is the number of chain scissions per original number-average molecule. Stoichiometric considerations therefore define ξ as

$$\xi = \frac{M_n(0)}{M_n(\tau)} - 1 \tag{4.26}$$

When discussing the shapes of molecular weight distributions, it is usual to plot weight fractions as a function of molecular weight. The weight fraction is given by

$$W(P,\tau) = (P/N)n(P,\tau) \tag{4.27}$$

and its integral over all P is unity.

Molecular weight distributions are now commonly determined by gel permeation chromatography (GPC). In an idealized GPC trace, the elution volume V is proportional to the logarithm of the molecular weight M for all values of P. The integral of the GPC trace height with respect to V is proportional to the injected weight W of the polymer. The functional form of $G(P)$, the GPC trace, can be shown to be proportional to $P^2n(P,\tau)$ (Nemzek and Guillet 1977). For this reason, GPC traces accentuate the high molecular weight portions of distributions and are not identical with weight fraction plots. The difference between the two plots for two Schulz distributions, each having $b = 10$ and $\gamma_0 = 1.09$ is shown in figure 4.3. It is seen that the GPC trace emphasizes the high

Figure 4.3. Sum of two undegraded Schulz distributions, each with $b = 10$ and $\gamma_0 = 1.09$: $U_1 = 10^3$, $U_2 = 10^4$, $N_1 = N_2$, $\gamma_{0\,\text{total}} = 3.30$. (From Nemzek and Guillet 1977.)

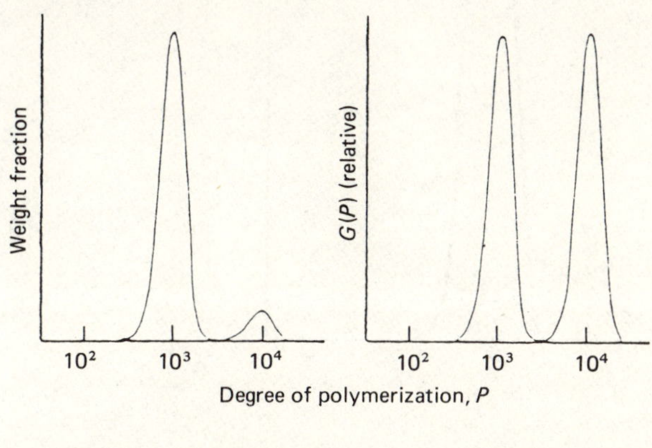

molecular weight material and further that while the maxima in weight fraction curves occur at $P=10^3$ and $P=10^4$ the maxima in the GPC curves occur at $P=1.09\times10^3$ and 1.09×10^4. In general, for a single Schulz distribution, the maximum in the weight fraction curve occurs at degree of polymerization U, the number-average degree of polymerization, while the maximum in the GPC curve occurs at $P=\gamma_0 U$, the dispersity times the number-average degree of polymerization.

4.6 Theoretical calculations of changes in molecular weight distribution

Using (4.18)–(4.24) it is possible to calculate the changes in molecular weight distribution (MWD) during random scission. When a narrow distribution undergoes random scission, the peak maximum is

Figure 4.4. Random scission of a polymer with narrow molecular weight distribution. I starting distribution, II scissioned distribution. (From Kells and Guillet 1969.)

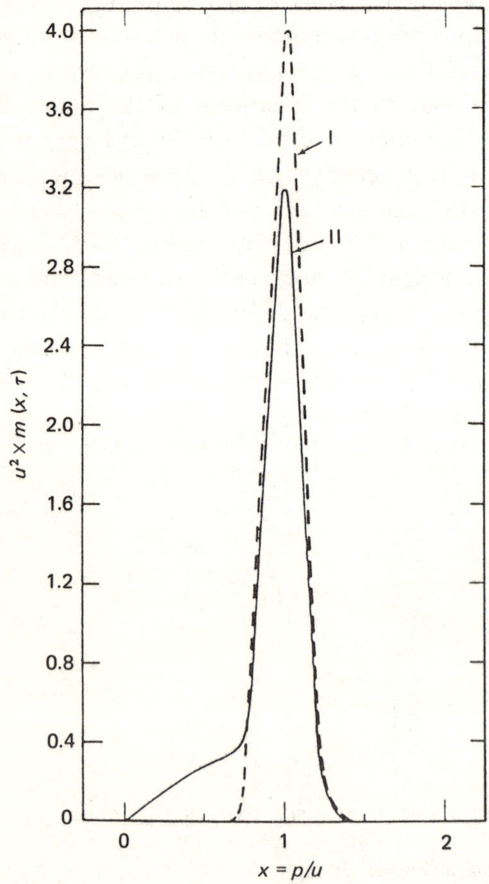

usually unchanged and material is observed developing at lower molecular weights, as shown in figure 4.4 (Kells and Guillet 1969). On the other hand, if a Flory distribution is allowed to undergo random scission, the changes are shown in figure 4.5 (Kells and Guillet 1969).

When random crosslinking occurs, Saito's treatment shows that $n(P,\tau)$ satisfies the following integro-differential equation:

$$\frac{\partial n(P,\tau)}{\partial \tau}+2Pn(P,\tau)=\int_0^P l(P-l)n(l,\tau)n(P-lm\tau)\mathrm{d}l \qquad (4.28)$$

This equation can be solved on a digital computer using the method of finite differences (Kells and Guillet 1969). Figure 4.6 shows the random crosslinking of an initial narrow molecular weight distribution and here it is obvious that one can see distinct species relating to molecules which have one, two and three crosslinks, since in each case the molecular weight increases by equal increments.

On the other hand, figure 4.7 shows the random crosslinking of a starting Flory distribution which is remarkably similar in its general shape to the distribution observed for random scission. Note that both in random scission and in random crosslinking, the peak maximum moves to lower molecular weights. The reduction in peak area in the latter case

Figure 4.5. Random scission of a Flory (normal) distribution. I starting distribution, II scissioned distribution. (From Kells and Guillet 1969.)

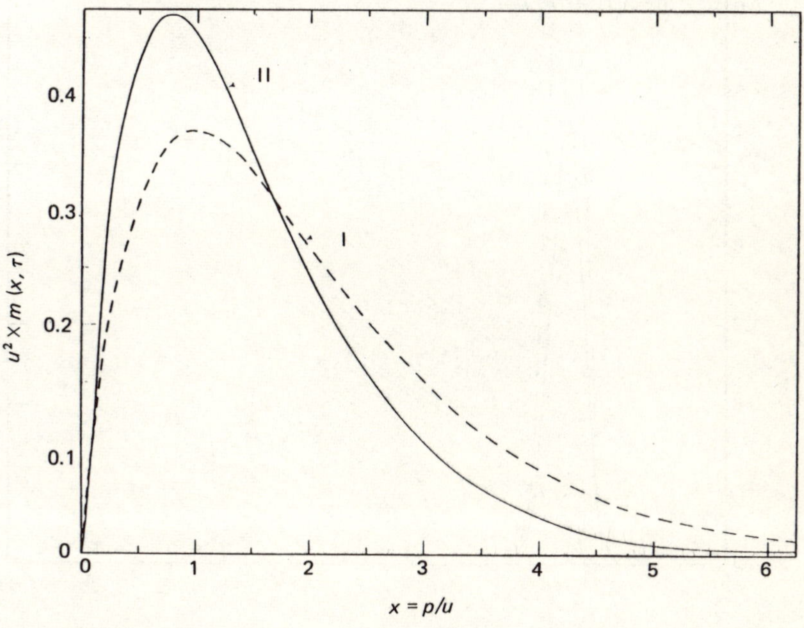

is due to the fact that a large portion of the distribution has been extended outside the scale of the graph. For this reason, the interpretation of whether either crosslinking or random scission is occurring, even in relatively normal distributions such as the Flory distribution, is open to considerable error in interpretation. It has not been generally recognized that random crosslinking of a Flory distribution moves the peak maximum to lower degrees of polymerization. Inspection of molecular weight distribution curves is therefore not a reliable way of determining whether either random scission or crosslinking is occurring.

As might be expected, the combination of random scission and random

Figure 4.6. Random crosslinking of narrow molecular weight distribution. I starting distribution, II crosslinked distribution. (From Kells and Guillet 1969.)

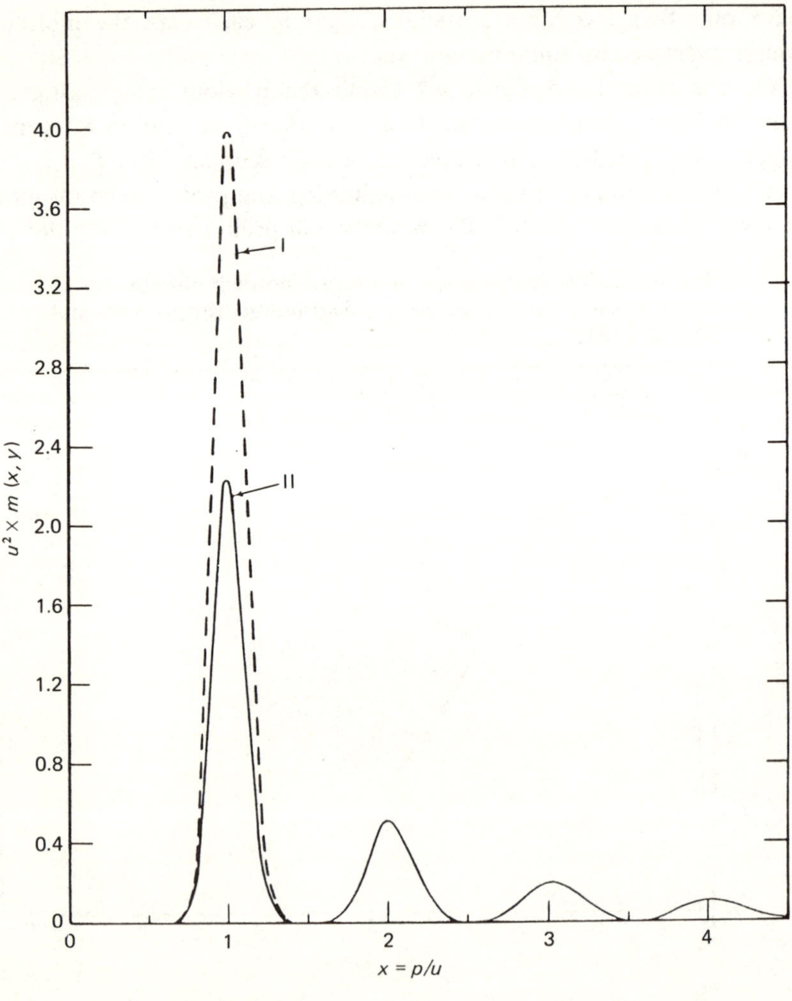

crosslinking occurring in the same sample is even more complex. Only in the case of monodisperse samples are the results readily observable and easily interpreted. Figure 4.8*a* shows calculations of random scission and crosslinking of a polymer system with an initial narrow distribution which may be compared with an actual distribution determined on the analytical ultracentrifuge for a sample of polystyrene irradiated by cobalt-60 (figure 4.8*b*). The ultracentrifuge appears to be one of the better techniques available for detecting random crosslinking in monodisperse polymer systems.

4.7 Viscometric determination of chain scission

When viscometry is used to determine the rate of chain scission in a polymer undergoing photochemical reaction, the parameter of interest is usually the quantum yield of chain scission which is defined by

$$\phi_S = \frac{S_n}{I_{abs}} = \left(\frac{w}{I_{abs}}\right)\left(\frac{1}{M_n(0)}\right)\left(\frac{M_n(0)}{M_n(\tau)} - 1\right) \tag{4.29}$$

where S_n is the total number of moles of chain breaks, I_{abs} is the number of moles of photons absorbed, w is the total weight of the polymer in the solution and $M_n(0)$ and $M_n(\tau)$ are the initial number-average molecular weight and the molecular weight after reaction, such that the probability of rupture in a given monomer unit is equal to τ. Amerik and Guillet

Figure 4.7. Random crosslinking of normal molecular weight distribution. I starting distribution, II crosslinked distribution. (From Kells and Guillet 1969.)

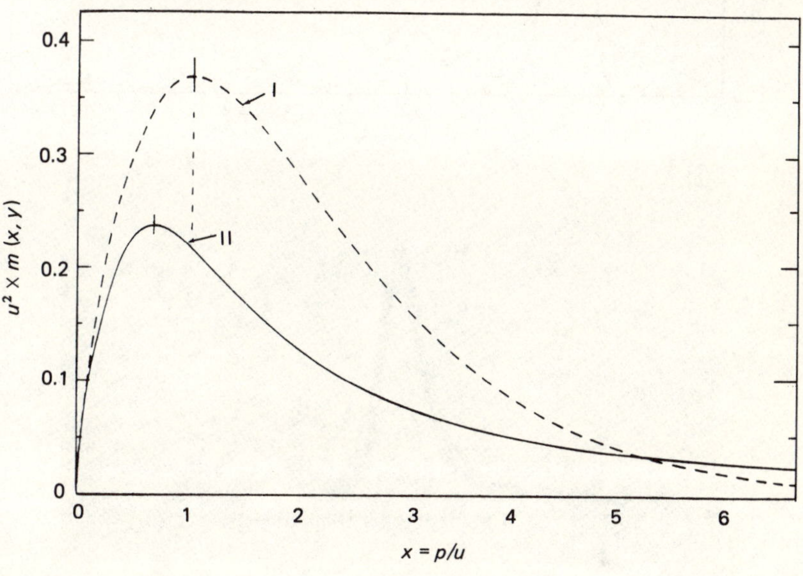

Figure 4.8. (*a*) Random scission and crosslinking of narrow molecular weight distribution. I starting distribution, II crosslinked and scissioned distribution. (From Kells and Guillet 1969.) (*b*) Schlieren photograph of material irradiated to 25.1 Mrad. (From Kells *et al.* 1968.)

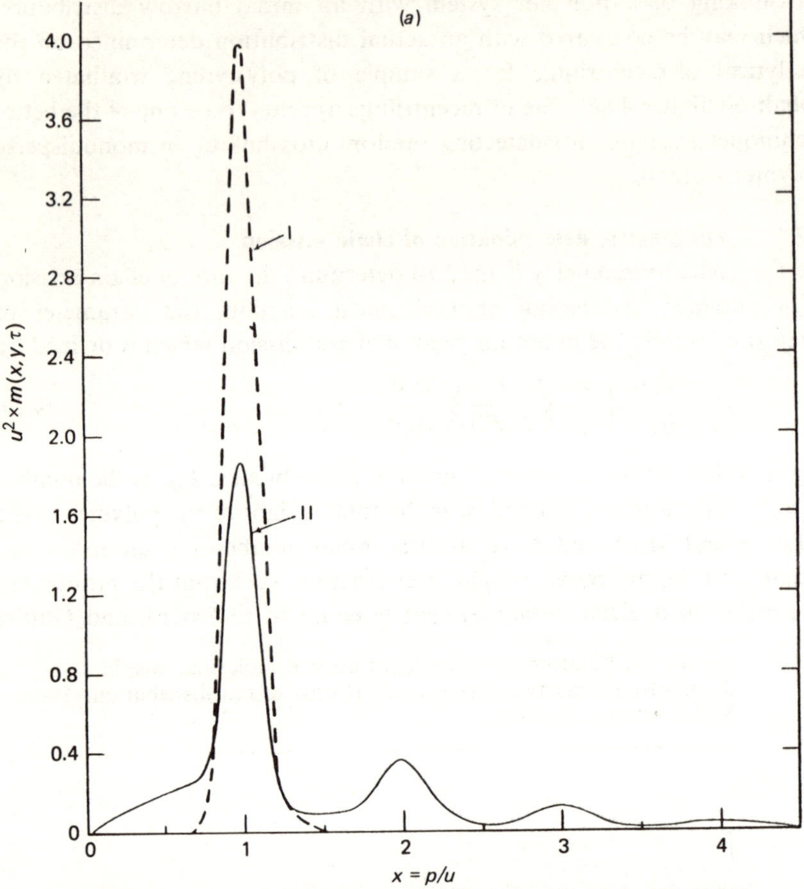

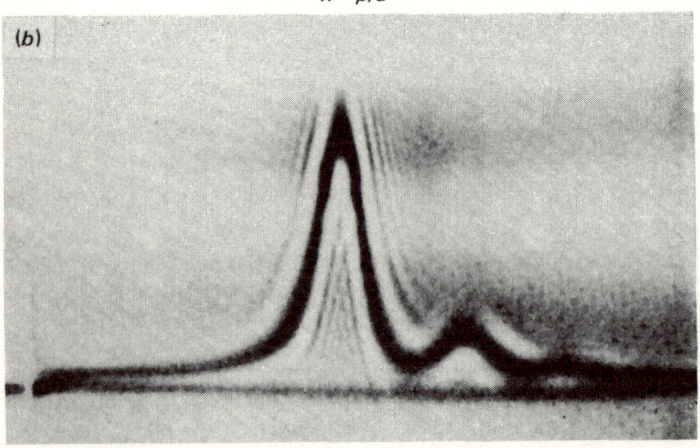

(1971) showed that this exact relationship can be approximated by the relation

$$S_W = \left(\frac{2w}{M_w(0)}\right)\left(\frac{M_w(0)}{M_w(\tau)} - 1\right) \tag{4.30}$$

where $M_w(0)$ and $M_w(\tau)$ are, respectively, weight-average molecular weights which were assumed to be equal to viscosity-average molecular weights ($M_v(0)$, $M_v(\tau)$). More recently, it has been shown by Nemzek and Guillet (1977) that a more precise relationship is given by

$$S_v = \frac{w[\Gamma(a+2)]^{1/a}}{M_v(0)}\left(\frac{M_v(0)}{M_v(\tau)} - 1\right) \tag{4.31}$$

where Γ is the gamma function and a is the exponent in the Mark–Houwink equation. This equation can be used to estimate the quantum yield of chain scission with an accuracy of better than $\pm 20\%$ provided, first, that the initial dispersity of the sample, γ_0, is between about 1.5 and 2.5, secondly that the distribution is of a form that can be approximated by either a Schulz or sum of Flory distributions, and thirdly that the initial degree of polymerization U is greater than about 100. If accuracy greater than $\pm 20\%$ is required and particularly if information is desired in the very early stages of the reaction, it is necessary to determine the initial molecular weight distribution, fit it to either a Schulz or sum of Flory distributions and calculatc a corrcction factor of R_v as a function of ξ by procedures described by Nemzek and Guillet. As the number of chain scissions per molecule ξ increases, S_v becomes an increasingly better approximation to S_n since the distribution approaches the Flory distribution.

4.8 Crosslinking in polymers

The development of a general and precise method for the experimental determination of crosslinking in polymers remains one of the major unsolved problems in polymer science. However, in certain cases, for example with monodisperse polymers, ultracentrifuge measurements provide a direct method since it is possible to resolve the polymer into monomers, dimers and trimers caused by forming zero, one and two crosslinks per molecule, respectively. Figure 4.8*b* (Kells *et al.* 1968) shows an example of such a determination for the crosslinking of monodisperse polystyrene by γ-radiation.

Low degrees of crosslinking lead to the formation of polymer molecules containing long branches. In principle, a branched polymer (figure 4.9) can be characterized by the number of tri- or tetrafunctional crosslink junctions. If it is possible to determine the number of end

Figure 4.9. Types of branching in polymers: (*a*) trifunctional random branching; (*b*) tetrafunctional random branching; (*c*) polyfunctional branching ('star molecule'); (*d*) trifunctional regular branching ('comb molecule').

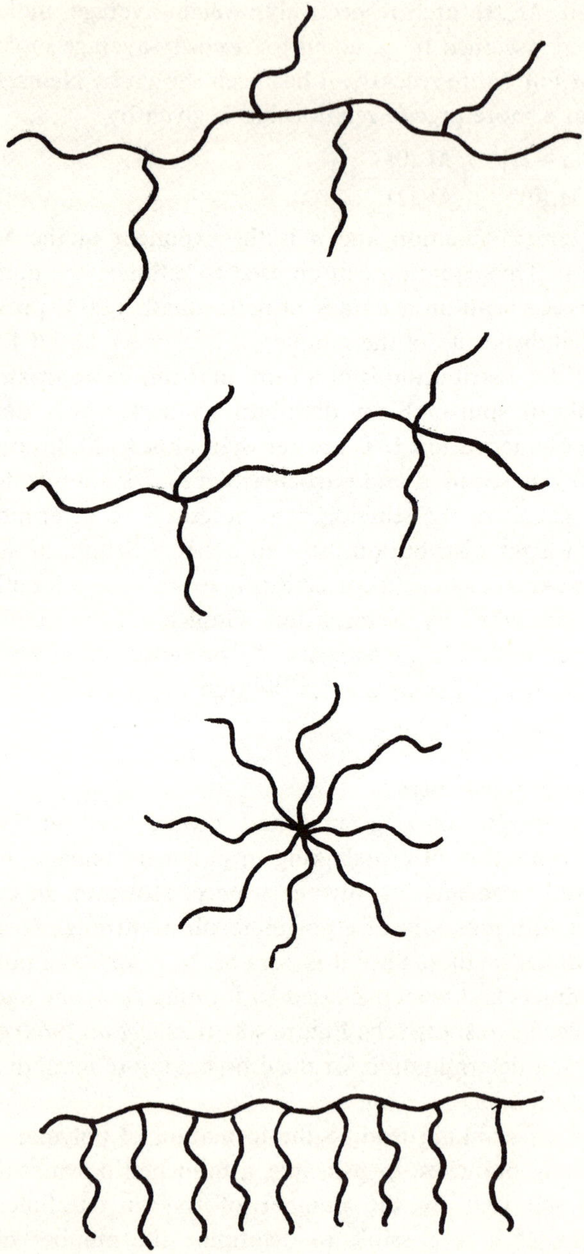

groups N_e in such a molecule, then the number of branches N_b will be equal to $(N_e - 2)$. In general, end group determination is difficult and recourse is made to the determination of solution properties. In general, branched molecules such as those depicted schematically in figure 4.9 will be unable to expand as much in solution as a flexible linear molecule of the same molecular weight. Therefore the volume of the 'equivalent sphere' will be reduced, thus decreasing the intrinsic viscosity and increasing the segment density in the sphere. Since a branched molecule has many ends, it is no longer appropriate to use the end-to-end distance $\langle \bar{R}^2 \rangle$ to characterize the coil dimensions, but the radius of gyration $\langle \bar{S}^2 \rangle$ does retain its significance. The reduction in coil dimensions is usually expressed by a factor g defined by

$$g = \langle \bar{S}^2 \rangle_{\text{branched}} / \langle \bar{S}^2 \rangle_{\text{linear}} \tag{4.32}$$

Values of g can be calculated statistically for various degrees and types of branching (Zimm and Stockmayer 1949). A number of theories have been developed relating g to experimental quantities such as the intrinsic viscosity of branched and unbranched polymers of the same molecular weight. A commonly used relationship is (Zimm and Kilb 1959)

$$[\eta]_{\text{branched}} / [\eta]_{\text{linear}} = g^{1/2} \tag{4.33}$$

However, there is some question as to the power of g to be used in this expression. For example, it was shown that $g^{1.5}$ gave values more consistent with kinetic and other evidence for branching in low density polyethylene (Guillet 1963). For 'star' molecules with relatively low functionality, however, (4.31) seems to give a satisfactory agreement with experiment. This and other procedures for determining branching have been reviewed by Graessley (1968).

4.9 Measurement of crosslinking in insoluble polymers

A linear polymer undergoing random crosslinking, or both scission and crosslinking, provided that the rate of the latter is four times as great, will ultimately become insoluble. Under these circumstances, solution methods of molecular characterization become impossible and resort must be made to determination of the properties of the crosslinked polymer in the solid phase or of the gel formed after the polymer is allowed to reach equilibrium in a swelling solvent.

The most common method of determining the degree of crosslinking in a polymer is by measurement of the gel–sol fraction. The theory of the method has been derived by Charlesby and Pinner (1959).

If we consider an aggregate of polymer molecules all of the same molecular weight undergoing a random crosslinking process, then, in principle, after each polymer molecule has been linked to the next by a single crosslink the molecular weight should become infinite. The molecule will become insoluble. The detailed theory given in Charlesby (1960) shows that no matter what the initial molecular weight distribution, three-dimensional network formation first begins to occur at a crosslinking density corresponding to one crosslinked unit per weight-average molecule. If we define q_0 as the proportion of monomer units crosslinked per unit radiation dose, then the requirement of one crosslink per weight-average molecule gives rise to the following expression for the gel dose:

$$r_{gel} = M_0/M_w q_0 = 1/P_w q_0 \tag{4.34}$$

where M_0 is the molecular weight of the monomer unit, M_w is the weight-average molecular weight, and P_w is the average degree of polymerization. If we define ϕ_x, the quantum yield for crosslinking by

$$\phi_x = \frac{\text{number of crosslinks}}{\text{number of quanta absorbed}} \tag{4.35}$$

then it can be simply related to the gel dose by recognizing that there are two crosslinked monomer units involved in each crosslink between polymer chains, and so

$$\phi_x = 1/2r_{gel} = \tfrac{1}{2}P_w q_0 \tag{4.36}$$

In principle, the use of this equation should permit the determination of the quantum yield for crosslinking by determining the amount of radiation required to form the first noticeable gelation in the polymer. However, experimentally this is difficult because small amounts of gel are almost impossible to detect. The usual procedure is to continue the irradiation and crosslinking, measuring the amount of soluble material (sol) as a function of the dose. The sol fraction S_x will, in general, depend on the molecular weight distribution of the starting polymer, as well as on the dose rate.

Theoretical calculations have been carried out by Charlesby, and are shown graphically in figure 4.10. The sol fraction S_x for any initial starting distribution can be derived by application of an equation originally derived by Flory (1953) for any primary molecular weight distribution $f(y)$

$$S_x = \left[1 \Big/ \sum y f(y)\right] \sum y f(y) [1 - q(1 - S_x]^y \tag{4.37}$$

If the starting distribution is known accurately, then this equation can be used and integrated to provide a more precise determination of the crosslink density q (the proportion of monomer units which are crosslinked). However, in most early work, the assumption was made that the initial distribution could be approximated by the most probable Flory distribution, in which case, as Charlesby and Pinner (1959) have shown, the crosslinking coefficient δ can be determined from

$$S_x + S_x^{1/2} = 2/\delta \tag{4.38}$$

where δ is the number of crosslinks per weight-average molecule. Such a plot is shown for various distributions in figure 4.11. In this plot, $S_x + S_x^{1/2}$ is plotted as a function of r/r_{gel}. The experimental observation of a straight line relationship going through the origin is usually considered to be evidence of the validity of the assumption of an initial random distribution. However, in many systems, the relationships derived above will not hold because they assume that crosslinking is occurring without chain scission. In most systems there will be additional chain scission processes and in this case the Charlesby–Pinner plot will be that shown in figure 4.12 where the straight line is displaced from the origin. Extrapolation of the experimental data to $1/\delta = 0$ will give the ratio τ_0/q_0 for the ratio of degradation to crosslinking.

Recently, Reiser and Edgerton (1979) have made an interesting extension of this experiment to study crosslinking in thin films. Because of the importance of thin film measurements, the treatment is reproduced here.

Figure 4.10. Decrease in soluble fraction when δ exceeds 1: (*a*) initially uniform distribution; (*b*) initially random distribution; (*c*) initially pseudo-random distribution. When δ is less than unity the polymer is completely soluble in all cases. (From Charlesby 1960.)

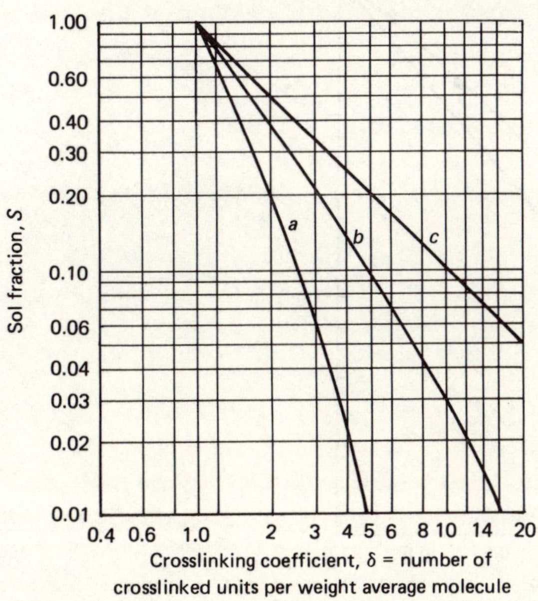

In their experiments, thin films of crosslinking polymers were coated on quartz slides exposed through the back of the slide through a device which gave a range of different intensities. After exposure, the polymer film was washed with a solvent and after drying the thickness of the gel

Figure 4.11. Theoretical curves relating soluble fraction S and crosslinking coefficient δ for an initially uniform, random or pseudo-random distribution. (From Charlesby 1960.)

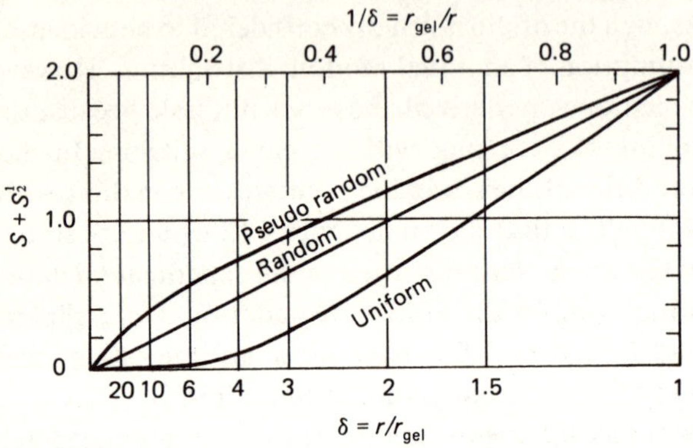

Figure 4.12. Charlesby–Pinner plot of the data for atactic (open circles) and isotactic (solid circles) polypropylene oxide. (From Roberts *et al.* 1971.)

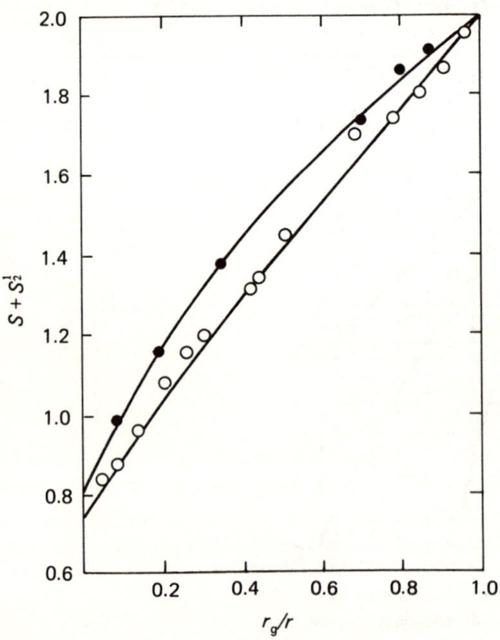

layer was measured with a microinterferometer. A plot of the gel volume h against exposure is called the gel curve of the photopolymer. A typical gel curve is shown in figure 4.13. The gel point exposure E_g can be related in a simple way to the quantum yield of intermolecular crosslink formation.

Consider a unit area of irradiated film of thickness d. If ρ is the specific gravity of the material and M_0 is the molecular weight of the monomer unit, then the number of moles of monomer units in a unit area of the film is equal to $d\rho/M_0$, of which a fraction q takes part in intermolecular crosslinks. The relation between exposure E and crosslinks q is then given by

$$EA = (d\rho/M_0)q \tag{4.39}$$

where E is the incident radiation density (exposure) and A is the fraction absorbed by reactant groups. The gel point exposure E_g will be given by

$$E_g = d\rho/AM_w \tag{4.40}$$

The quantum yield for crosslinking ϕ_x is then given by

$$\phi_x = d\rho/E_g AM_w \tag{4.41}$$

For a thin film where only a small fraction of the light is absorbed in transmission through the film, A can be approximated by

$$A = 2.303\varepsilon M_0 d \tag{4.42}$$

Figure 4.13. Gel curve h and derived exposure function Y of poly(vinyl cinnamate) exposed to monochromatic radiation of 300 nm. (From Reiser and Edgerton 1979.)

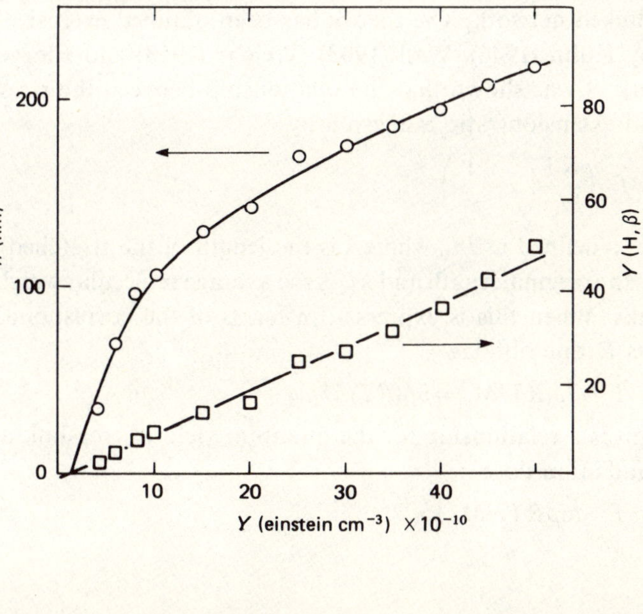

where ε is the molar extinction coefficient of the reacting chromophore at the wavelength of radiation and m_0 is its concentration in the solid film. Combining (4.41) and (4.42) gives the final equation for the quantum yield of crosslinking.

$$\phi_x = (1/E_g)(\rho/2.303\varepsilon M_0) \tag{4.43}$$

If the crosslinking process involves the removal of a chromophore or a change in absorbance of the film, this can be related to the quantum yield of chromophore removal by

$$\frac{\mathrm{d}D}{\mathrm{d}t} = \Delta\varepsilon I_0 A\left(\frac{D - D_\infty}{D}\right)\phi \times 10^3 \tag{4.44}$$

where I_0 is the incident energy flux in einstein cm^{-2} s^{-1}, A is the fraction absorbed, $\mathrm{d}D/\mathrm{d}t$ is the rate of change of optical density D at the wavelength of the chromophore and $\Delta\varepsilon$ is the difference of the molar extinction coefficient of the reagent and the product. Reiser and Egerton integrated the Flory equation (4.37) for the experimentally observed log-normal distribution and used this to correct the observed sol fraction for the distribution of the starting polymer molecular weight. Experimental data for a sample of poly-(vinylcinnamate) are shown in figure 4.13. The linear relationship observed after this procedure is a justification for the theoretical procedures used to evaluate the molecular weight distribution.

4.10 Crosslinking from elastic modulus measurements

An additional method used for determining the quantum yield of crosslinking can be derived from measurements of the elastic modulus of a crosslinked network. The theory has been outlined extensively in early work by Kuhn (1946), Wall (1943), Treloar (1958) and Flory (1953). In this work, it was shown that the relationship between the stress per unit area and extension ratio α is given by

$$f = \frac{\rho RT}{M_c}\left(\alpha - \frac{1}{\alpha^2}\right) \tag{4.45}$$

where α is defined as l/l_0 where l is the length of the stretched elastomer and l_0 is the original length and M_c is the average molecular weight between crosslinks. When this is expressed in terms of the corresponding elastic modulus E, one obtains

$$E = 3\rho(RT/M_c) = 3\rho(RT/M_0)q \tag{4.46}$$

which gives a relationship for the quantum yield of crosslinking in terms of the radiation dose r

$$E = (6\rho RTr/M_w)\phi_x \tag{4.47}$$

and

$$E = 6.24 \times 10^{-6} \rho R T r G_x \tag{4.48}$$

for the G value for crosslinking, G_x. It should be noted that these equations can only be applied at low degrees of extension, where the effect of molecular entanglement can be ignored. This method is difficult to apply with small samples and has not been used extensively in photochemical studies. It is important to recognize, however, that these relationships predict that the modulus of a crosslinking material should increase with the radiation dose, at a rate proportional to the quantum yield or G_x value for crosslinking.

4.11 Crosslinking from swelling measurements

A further method of determining the crosslink density in synthetic polymers is by measurement of the equilibrium swelling. When a crosslinked polymer is exposed to a solvent which would normally dissolve it when it was uncrosslinked, the polymer will swell until the restrictions imposed by the crosslinks prevent further expansion. At equilibrium, the free energy of mixing will contain terms due to the heat of mixing of the polymer with the solvent, and to the elastic component contributed by the swollen network and its resistance to further expansion. The theory on the conditions for equilibrium was derived first by Flory and Rehner (1943):

$$\ln(1 - x_2) + x_2 + \chi x_2^2 + (\rho_2 V_1/M_c)[x_2^{1/3} - (x_2/2)] = 0 \tag{4.49}$$

where x_2 is the mole fraction of polymer at equilibrium, χ is the Flory–Huggins interaction parameter which describes the interactions between the swelling solvent and the polymer segment, ρ_2 is the density of the polymer, M_c is the average molecular weight between crosslinks and V_1 is the molar volume of the solvent. If the degree of swelling is large, so that the mole fraction x_2 is small (4.49) can be simplified to

$$(0.5 - \chi)x_2^2 = (\rho_2 V_1/M_c)x_2^{1/3} \tag{4.50}$$

Commonly, the mole fraction of the polymer is replaced by V, the volume ratio of the swollen to the unswollen polymer, and on rearrangement

$$V^{5/3} = (0.5 - \chi)(M_c/\rho_2 V_1) \tag{4.51}$$

can be expressed in terms of the crosslinked density q in

$$V^{5/3} = (0.5 - \chi)(M_0/\rho_2 V_1 q) \tag{4.52}$$

from which the quantum yield of crosslinking can be obtained from

$$V^{5/3} = \frac{(0.5 - \chi)v_2}{2\phi_x V_1 I_A} \tag{4.53}$$

where v_2 is the volume of the polymer exposed to radiation and I_A is the number of einsteins absorbed, and the G value of crosslinking G_x

$$V^{5/3} = \frac{0.48 \times 10^6 (0.5 - \chi)}{\rho_2 V_1 G_x r} \tag{4.54}$$

It should be remembered that all the above relations assume that the G_x value and the quantum yield ϕ_x are independent of irradiation dose. The Flory–Huggins χ parameter necessary for applying these relations is tabulated in the *Polymer Handbook* (Bandrup and Immergut 1966, 1975) for a number of polymer–solvent systems. Alternatively, it can be determined directly by the procedures described in §3.6.3.

5

Photoprocesses in solid polymer matrices

5.1 Introduction

The clarity and macroscopic rigidity of many common polymers such as polystyrene (PS) and poly(methyl methacrylate) (PMMA) led workers to assume that these matrices were comparable to those of organic and inorganic glasses at low temperature (usually 77 K). However, experiments in which small molecules were dispersed in solid polymer matrices showed that at 300 K they have much greater free volume than do conventional organic glasses and permit significant amounts of molecular motion, particularly of small side groups, although they do provide substantial restraints on diffusion, particularly at low concentrations of the probe molecule. Studies of this kind have provided an important body of knowledge, both about the structure of organic polymer glasses and about photochemical and photophysical processes restricted by a polymeric environment.

Hydrocarbon glasses have been used as matrices to study the spectroscopic properties of small molecules at low temperatures (77 K) since the pioneering studies of G. N. Lewis and his co-workers. Lewis and Lipkin (1942) developed a series of compositions of *n*-pentyl ether or of isopentane and alcohol (known as EPA) which were excellent solvents and easily formed clear glasses at liquid nitrogen temperatures. Later, a number of investigators attempted to do similar studies in glassy polymer matrices such as PS and PMMA at room temperature. For example, Geacintov *et al.* (1968) used a variety of polymer matrices for spectroscopic studies of phosphorescence from 1,2,5,6-di-benzanthracene (I) and found, in all cases, that the emission was quenched by oxygen at room temperature. This is perhaps the earliest evidence that in glassy matrices of polymers, substantial diffusion can occur within the lifetime of the excited state. We showed in §3.3 that small molecules can diffuse up to

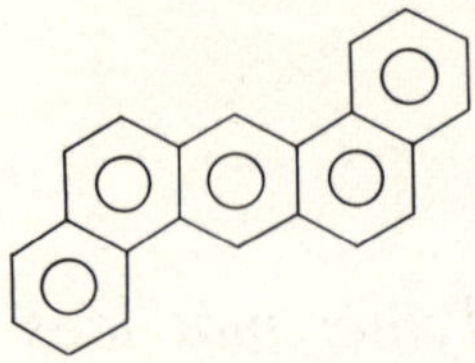

I

5×10^4 Å s^{-1} in a polymeric medium, so we see that if the phosphorescence lifetime is of the order of seconds it can be effectively quenched in the viscous polymer matrix. Clearly this is the case quite often, in marked contrast to rigid glasses, such as the borates, which are impermeable to oxygen even at room temperature (Joussot-Dubien and Lesclaux 1973). Since all glassy polymers will have substantial amounts of free volume at room temperature they can be considered rigid only on very short time scales and for relatively large scale motions of either the host or the guest molecule.

Recent studies in a number of laboratories have helped to delineate the role that polymer matrices can play in polymer photochemistry and photophysics (Farid *et al.* 1979). It now appears that the major factors controlling the possibility of chemical reaction, and the rates of both physical and chemical processes in such polymeric glasses will be the total amount of free volume available within the glass at a particular temperature and the size of the group or group motion relative to the 'packets' of free volume available in the polymer matrix. Glassy polymers are often considered to be amorphous solids containing sub-microscopic 'holes' with a distribution of sizes defined by the length of a moving polymer segment and a jump frequency (§3.4).

In some cases there are examples of specific interactions between the small molecule solute and the polymer which can also affect the rates, although this factor is usually smaller.

5.2 Photoisomerization in polymer glasses

One of the simplest photochemical processes which can be observed in polymer matrices is the photoisomerization of the double bond in olefins such as stilbene. Gegiou *et al.* (1968) showed that the quantum yield of cis–trans isomerization of stilbene decreased with increasing viscosity of the medium.

They found too that a polymer matrix such as polyisobutylene did not inhibit the photoisomerization as long as the temperature was above the glass transition T_g of the polymer where the isomerization followed first-order kinetics, but below T_g the kinetics were so complex as to suggest concurrence of a multiplicity of different first-order processes.

Postulating that an increase in free volume is necessary for the transition from trans- to cis-stilbene

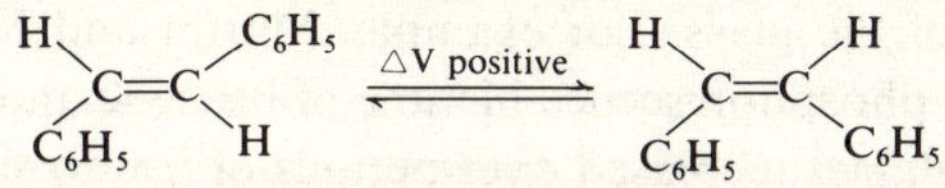

(5.1)

and that restrictions on the rotation of the phenyl groups in solid matrices reduced the quantum yield for the process, they then suggested that the stilbene molecules were located in a range of different guest sites having differing amounts of free volume, a situation equivalent to having a spectrum of microscopic viscosities, each determining the rate constant for the observed process. Similar effects have been reported by other workers for simple isomerizations in glassy polymer matrices and it seems to be a general phenomenon.

5.3 Effect of molecular diffusion

Similar considerations affect the rate of diffusing species in a polymeric system. For example, Heppel (1965) showed that the translational diffusion coefficient of anthracene in solution was unaltered by the presence of polyisobutylene, even when the system was concentrated enough to form a gel. However, in hydrogen-bonded systems, such as glycerol, poly(vinyl alcohol) (PVA) or gelatin gels, a substantial reduction in diffusion was observed. In this case, it seems likely that the hydrogen bonding reduces the mobility of the side groups in the polymer whose coordinated motion is essential in order to provide free volume packets into which the diffusing molecule can move. Buettner (1964) obtained similar results on dry gelatin coatings containing α-naphthol as a guest molecule. He observed the intensity of phosphorescence after flashing the naphthol-containing matrix and showed that the oxygen could be removed from the regions surrounding the naphthol by repeated flashing. Because of hydrogen bonding between the collagen molecules in gelatin, the diffusion rate of oxygen is substantially lower than through polymeric esters such as PMMA which have very little interchain association by hydrogen bonding.

A further result of the reduction in diffusion rates in the hydrogen-bonding polymers such as gelatin and PVA is that no self-quenching of the triplet is observed for either 2-naphthol or erythrosin indicating that these two molecules cannot diffuse a sufficient distance under these conditions for interaction to occur.

5.4 Thermal history

It is generally observed that the phosphorescence lifetime of various guests in polymer or small-molecule glasses will depend on the time since preparation of the glass. For example, Martin and Kalantar (1968) observed that the phosphorescence lifetime of benzene and similar chromophores in EPA glasses increased over periods of five to six hours after the preparation of the glass. If, however, the glasses were allowed to relax over a period of hours before the measurements were made, no change was observed. The effect was found to occur only in glasses whose glass transition was within 30° of the temperature of measurement (77 K). In general, because of the large macroscopic viscosity of polymeric glasses, they will never be at thermodynamic equilibrium below the glass transition. Thus the photophysics of the guest molecule may well depend upon the temperature at which the glass is made. An example of this effect is shown in §7.3.4 where the concentration of excimer sites in PS and poly(vinyl naphthalene) dissolved in PS depends upon the temperature at which the film is coated and dried. After drying, the measurements can be made at considerably lower temperatures without affecting the conformation of the polymer and the resulting concentration of sites. It is possible that the polymer conformation is locked-in under these conditions. If the temperature of observation is significantly below the glass transition, it is unlikely that very large changes in the local site conformation will occur over even quite long periods of time.

5.5 Solid phase transitions

Following the original observations of Charlesby and Partridge (1963) and of Bousted (1970), Somersall *et al.* (1974) observed that the phosphorescence of polymers containing keto and pendant aromatic groups changed as a function of temperature. Discontinuities were observed in the plot of the logarithm of the emission intensity as a function of $1/T$ at temperatures corresponding to the onset of subgroup motion in polymers. Since the phosphorescence in this case is being quenched by oxygen, this is further evidence that the movement of oxygen and other small molecules through polymer matrices is facilitated by the small scale rotations and vibrations of groups pendant to the polymer main chain. As these motions are slowed down on approaching absolute zero, diffusion constants also reach negligible values.

Further information about the role of the matrix in photochemistry can be obtained by studying the formation of excimers and exciplexes in media of varying rigidity and viscosity. The formation of these 'excited state dimers' can be affected by the nature of the medium. The photophysics is dealt with in detail in chapter 7.

5.6 Exciplex emission

Exciplex emission from a chromophore is characterized as being broad, structureless and red-shifted to the corresponding monomer fluorescence. The extent of the red shift depends on the distance between the two components of the complex. It is also characterized by strong dependence on the polarity of the medium. Martic *et al.* (1977) studied exciplexes of anthracene and N,N-dimethyl paratoluidine in toluene and PS. Typical spectra are shown in figure 5.1. At 30° the maximum of the exciplex band in toluene is at 515 nm, whereas in PS, the band is shifted by 25 nm to 490. When PS is copolymerized with 4-N,N-dimethyl amino styrene, and used as the matrix, the exciplex emission is at 480 nm. Furthermore, the maxima of the exciplex emission are temperature dependent and in toluene the maximum shifts to shorter wavelengths as the temperature is increased, while in the polymers the opposite is the case. The maxima approach a common value at the glass transition of PS (figure 5.2). If 20% *n*-butyl benzene is included in the PS matrix to lower its glass transition temperature, the maximum of the exciplex follows the same pattern as in PS by shifting to longer wavelengths with increasing temperature, finally reaching the value measured in toluene at temperatures close to the glass transition of the mixture. Further increases in temperature of the mixed media lead to a shift to shorter wavelengths

Figure 5.1. Emission spectra (uncorrected) of anthracene in the presence of N,N-dimethyl-p-toluidine in toluene and in polystyrene. (From Farid *et al.* 1979.)

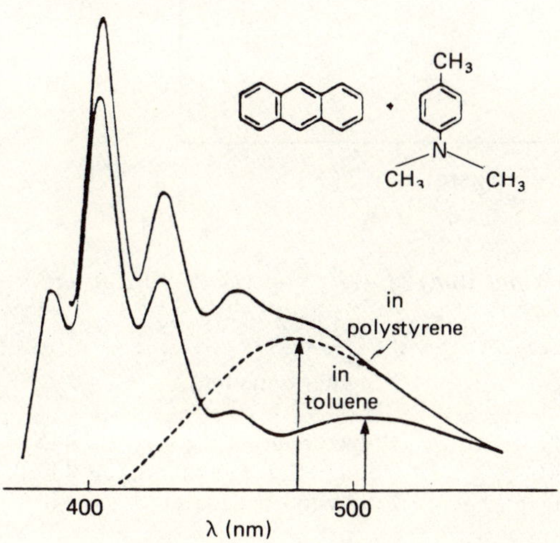

similar to that measured in toluene. It seems likely that this shift to shorter wavelength with decreasing temperature in polymer matrices at temperatures below T_g is due to the greater difficulty in formation of the exciplex geometry which results either in an improper orientation or a larger separation of the exciplex components compared to those in fluid media. Similar results were obtained with pyrene exciplexes.

Farid *et al.* (1979) also report similar results on the formation of exciplexes from small molecules such as methyl 4(1-pyrenyl)butyrate (structure II) in different fluid and polymeric media. The results are shown in table 5.1. Since excimers have negligible dipole moments, their emission

Figure 5.2. Temperature effect on λ_{max} of the exciplex anthracene/N,N-dimethyl-p-toluidine in toluene (full line), polystyrene (dotted line), and polystyrene (80%) + butylbenzene (20%) (broken line). (From Farid *et al.* 1979.)

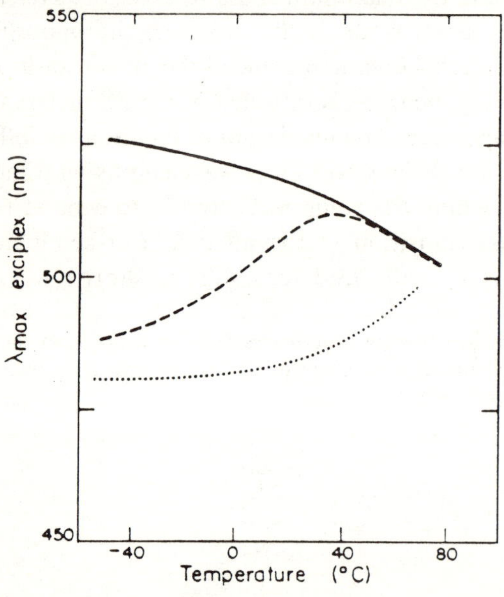

Table 5.1. *Excimer maxima (nm) of 4(1-pyrenyl)butyrate in different fluid and polymeric media*

Fluid medium	λ_{max}	Polymeric medium	λ_{max}	$\Delta\lambda_{max}$
Toluene	480	Polystyrene	445	35
Methyl isobutyrate	480	Poly(methyl methacrylate)	440	40
Methyl benzoate	485	Poly(vinyl benzoate)	440	45

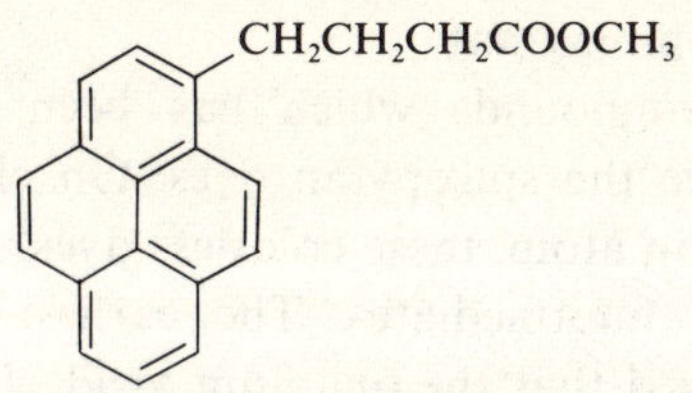

II

spectra should be relatively independent of polarizability. A similar difference between small molecule and polymeric solvents is observed as with exciplexes which suggest that most, if not all, of the shift between exciplex emissions in PS and toluene are the result of destabilization of the complex caused by improper orientation in the polymer matrix. Apparently polarity is less important in these polymer matrices than the restrictions on mobility.

Certain bichromophoric molecules such as compounds III and IV have been shown to form exciplexes intramolecularly when dissolved in fluid

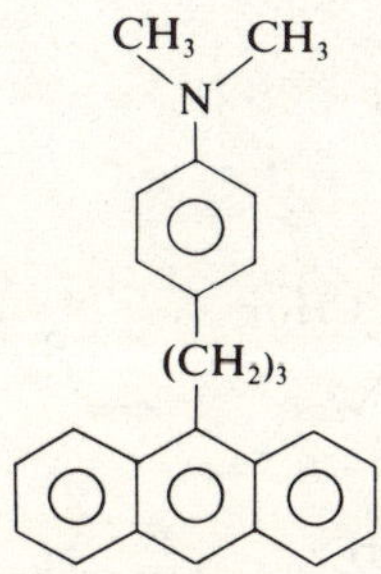

III

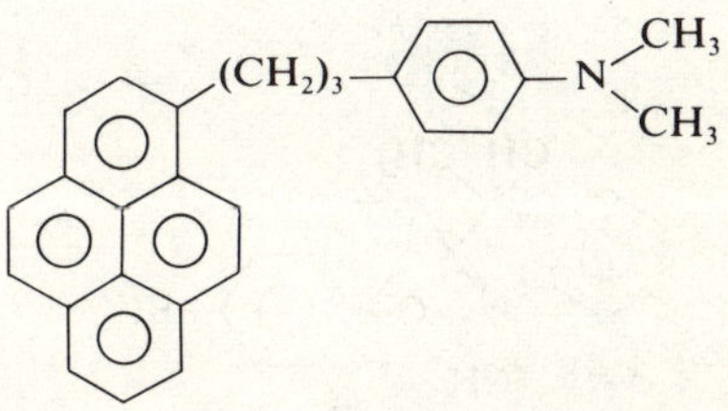

IV

solution. However, neither of these compounds showed exciplex emission when dissolved in PS. The reason for this behavior is presumably that the most favored configuration of the molecule is one in which the two chromophores are rather far apart and there is not sufficient mobility in the matrix to permit two chromophores to approach each other during the lifetime of the excited state. Farid *et al.* (1979) estimate that the rate of folding in such molecules in the solid matrix must be less than $10^6\ s^{-1}$, or at least two orders of magnitude slower in the polymer matrix than in the fluid. The authors suggest that in polymer matrices below T_g, bimolecular interactions are significantly weaker than those in the fluid media, probably due to restrictions on orientation forced by the polymer chain and the poorer solvation which occurs in the polymeric medium. At temperatures above T_g, the interactions in polymers, both solvation and orientational, appear to be the same as those in fluid media, both for exciplexes and excimers.

5.7 Photochromic behavior of spiropyrans

An important class of compounds which has been studied extensively in polymer matrices are the spiropyran dyes. On changing conformation about the spiro carbon atom, these colorless dyes undergo a transformation to colored dye intermediates. The earliest studies (Hirschberg and Fisher 1954) showed that the quantum yield of photoisomerization was decreased in polymer films to the same degree as in EPA glasses at low temperature. Gardlund (1968) studied the thermal bleaching of 6-chloro-8-nitro and 6-nitro-8-methoxyl trimethyl-indoline-benzopyrylospiropyran (V). This dye can be converted to the colored form (VII) by irradiation and revert to the colorless form V via ring closure of intermediate form VI.

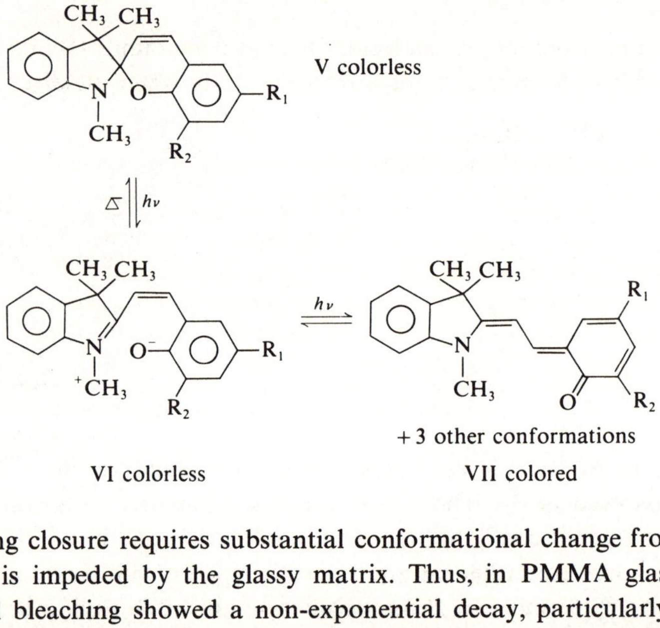

VI colorless VII colored

(5.2)

This ring closure requires substantial conformational change from form V and is impeded by the glassy matrix. Thus, in PMMA glasses the thermal bleaching showed a non-exponential decay, particularly in the early stages, while after long times, first-order decay was approached. On the other hand, in solution, the dark reaction was much faster and obeyed first-order kinetics because the rapid conformational changes allowed facile ring closure from structure VI to structure V. The complex kinetics observed was postulated to arise from the fact that two of the colored forms of VII decay with rate constants differing by at least a factor of ten.

Gardlund and Laverty (1969) later studied a similar series of dyes in which the ester group was changed from methyl to *n*-butyl and cyclohexyl substituents. It was shown that the larger rate constant was independent of the nature of the matrix, whereas that for the slower process increased with the length of the alkyl chain. The activation energy for the bleaching process ranged from 40 to 60 kJ/mole and in all cases the rate constants were essentially identical with those observed in fluid medium above the glass transition temperature.

Smets (1972) has used the spiropyran dyes to probe the molecular mobility of polymer chains in solution. In this case, the dyes were attached chemically to the polymer chains and this work is further discussed in §13.2. Smets has also emphasized the importance of the value of T_g for isomerization reactions in a given polymer matrix. He showed that a photochromic molecule dissolved in poly(isobutyl methacrylate) ($T_g = 45\,°C$) decolored at a rate two to three times faster than in PMMA ($T_g = 100\,°C$). He also showed that at 65 °C, in PMMA, the rate was 40% greater than at 15 °C, which he attributed to the accelerating effect of operating at temperatures near the secondary transition temperature T_β of the polymer.

Studies by Kryszewski *et al.* (1973) of benzospiropyran in amorphous acrylate polymers also showed that the thermal bleaching followed first-order kinetics above the glass transition temperature but, below T_g, required a more complex mechanism postulated to involve the average free volume and segmental diffusion. They pointed out that in polymer matrices with long flexible side groups, such as poly(vinyl butyrate), the decolorization was first order, even below the glass transition temperature.

In an interesting extension of this work, Kryszewski and Nadolsky (1975) studied the electrochromism of indolinobenzospirans in polymer matrices. They studied the effect of a static electric field on the position of the absorption bands of the colored spiran stereoisomers, which were expected to vary depending on the polarity and mobility of the polymer matrices. In the colorless form, the indolinobenzospirans do not show charge separation but on irradiation they form a dipolar species. The electric field imposed after exposure to light in the PMMA matrix produced spectral shifts corresponding to the changes in the thermal equilibrium. These changes were attributed to the increased stabilization of these stereoisomers having larger dipole moments. However, in less polar matrices such as PS and polycarbonate, only small shifts in the absorption spectrum were apparent.

Smets (1975*b*) has reported that the size of the photochromic unit itself influences the rate of the thermal fading. Photochromes of varying size were studied in the same matrix. The size of the photochrome and the effect of orientation by stretching were shown to be the primary factors influencing the rate of thermal return of para-xylene-bisphotochromes. In an interesting extension of these experiments, anthracene photodimer held in a rigid polymer matrix was photochemically cleaved and the rate of recombination observed. It was shown that recombination took place at faster rates than expected because of the fact that the polymer retained the monomers in adjacent positions suitable for recombination to the dimer. The difference in the recombination rate was found to depend upon the experimental temperature and its relationship to T_g of the matrix and also whether or not the films were annealed prior to making the measurement.

5.8 Cis-trans isomerization of azo compounds

Aromatic azo compounds undergo photochemical cis-trans isomerization which is thermally reversible and this reaction has been studied extensively in polymeric glasses. For example, Kamogawa *et al.* (1968) studied the kinetics of 4-methyl-4′-dimethylamino azobenzene isomerization in several polymeric matrices. Upon irradiation, a steady-state concentration of cis (VIII) and trans (IX) isomers is achieved.

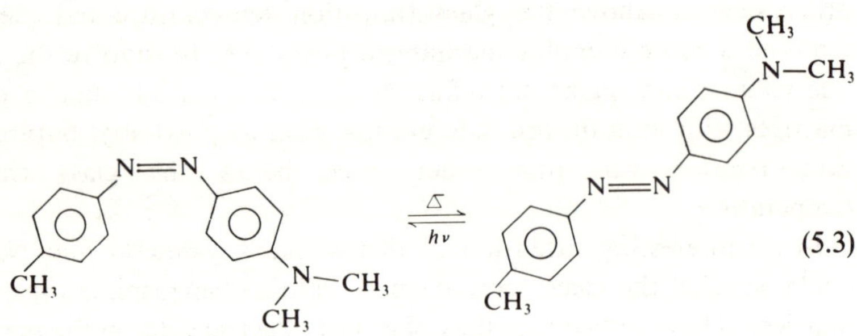

(5.3)

VIII cis IX trans

Light favors the formation of the cis from the trans, whereas at thermal equilibrium the concentration of trans relative to cis is higher. Kamogawa *et al.* reported the recovery times of the thermal reaction at which 50% return of the cis to the trans isomer occurred. These are shown in table 5.2. It is obvious that the PS matrix reduces the rate of

Table 5.2. *Recovery times from cis (VIII) to trans (IX)*

Matrix	Recovery time (min)
Polystyrene	30
Poly(butylacrylate-co-methyl methacrylate)	10
Poly(methacrylamide-co-butylacrylate-co-methyl methacrylate)	1
Solution	8

recovery but in some matrices, particularly those with low glass transitions, an actual increase in the recovery time is observed. Several additional cases were reported in which the recovery time was faster in the polymer matrix than in solution, which led the authors to infer that the isomerization-induced strain in the matrix assisted the thermal return.

In an elegant sequel to this work, Priest and Sifain (1971) studied the cis and trans isomerization of 2,2′-azonaphthalene (X) and 4-ethoxy azobenzene (XI). They used mixtures of PS and *n*-butyl benzene as glass-

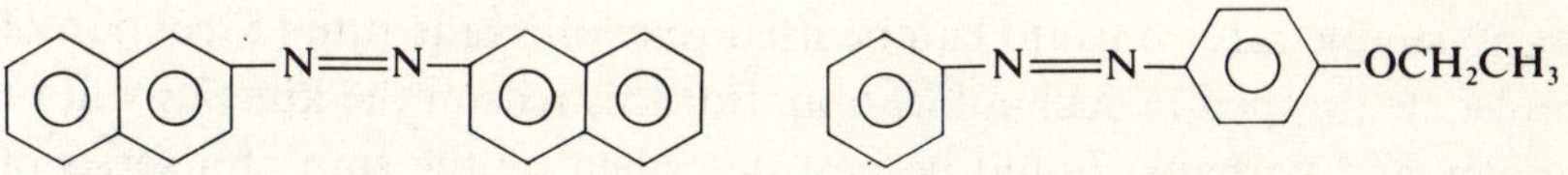

X XI

forming compositions which gave a working temperature range for T_g from that of PS, (104 °C), to that of *n*-butyl benzene (−136 °C). After dark equilibration at 45 °C both compounds existed solely in the trans form but, on exposure to light, the cis isomer formed to steady-state values which depended on the rigidity of the medium. The data are summarized in table 5.3, which shows that the steady-state concentration of cis isomer ranges from 45 % in PS to 78 % in *n*-butyl benzene. As in previous studies, the isomerization rate from the cis back to trans is slower and does not obey first-order kinetics if the experimental temperature is below the glass transition of the matrix. However, the activation energy, which is of the order of 85 kJ mol^{-1}, appears to be independent of whether the dye is in fluid solution or in a solid PS matrix. The authors also observe that the photochemical conversion of the cis isomer was lower in annealed samples of PS than in quenched samples. This is presumably because of the lower free volume available in the annealed specimens. Morawetz and

Table 5.3. *Photochemical conversion of X and dark isomerization rate in* n-*butyl benzene-styrene matrices at 45 °C*

n-Butyl benzene, %	T_g (°C)	% cis at steady state	Isomerization rate × 10^2, (min^{-1})	E_A (kJ mol^{-1})
0	109	45	0.80	89.5
14	75	52	1.97	
33	28	58	2.26	
44	1	58	2.60	
75	−75	73	2.67	
100	−136	78	2.33	94.5

co-workers have also studied cis-trans isomerization of azo compounds attached to polymer chains in order to probe microviscosity effects in solution and in the solid phase (Tabek and Morawetz 1970).

The non-exponentiality of recovery of the azo compounds in polymeric matrices appears to be a real effect and cannot be simply explained by the postulate that differing isomeric forms have different rates of isomerization as was used to explain similar results with the benzospirans. For this reason it now appears likely that the best explanation in both cases is that the matrix contains sites with differing amounts of free volume and that no simple rate constant can be used to express the rate of reaction of a molecule trapped in such a location. For this reason the kinetics will be complex and perhaps should be best described by the sum of a series of rate constants rather than by a single rate constant.

5.9 Molecular motion below T_g

A more direct study of the free volume available in polymer matrices has been made by North and co-workers, who studied the dielectric relaxation of small molecules suspended in a variety of polymer matrices. For example, Lawrie and North (1973) showed that in PS matrices the inversion of motion of the central ring of xanthene (XII) becomes possible at temperatures below the recognized T_β and T_g of the polymer. On the other hand, movement of the oxazole ring in 2,5-diphenyl oxazole (XIII) becomes observable at temperatures corresponding to the onset of the β motion in the polymer chain, but dipole reorientation of diphenyl ether cannot be observed below the main glass transition temperature. These authors also studied the bleaching of the colored isomer of 6-nitro-1,3,3-trimethylindolino benzopyrylospiran (XIV) and also concluded that the fact that simple first-order kinetics was not observed could be best explained in terms of a distribution of differing environments in the glassy matrix surrounding the individual chromophores.

In an interesting extension of his work on polymer photochemistry,

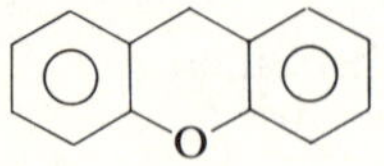

XII

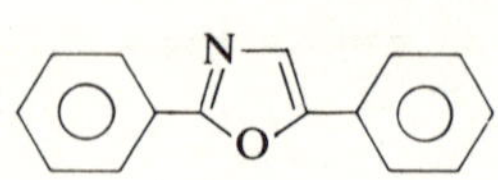

XIII

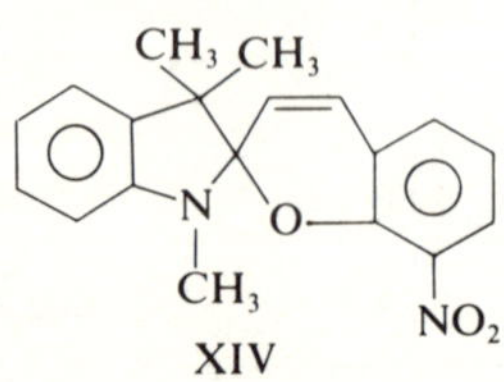

XIV

Smets (Smets and Matsumoto 1976) studied the reversible photocyclization of α-(p-dimethylamino)phenyl *N-m*-nitrophenyl nitrone (XV) to the corresponding oxaziridine (XVI) in solution and in rigid polymer

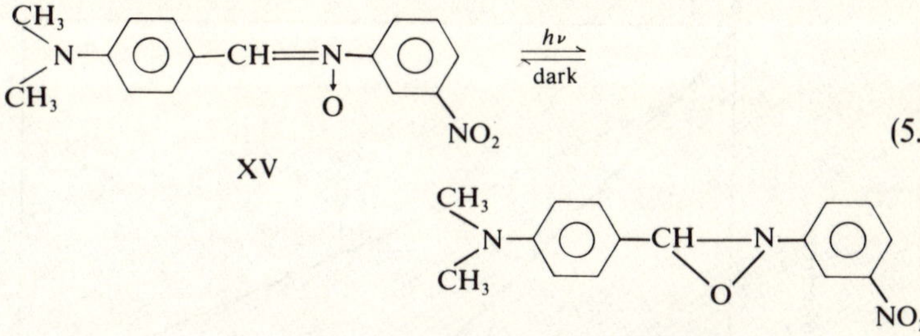

(5.4)

matrices. Irradiation with light of greater than 380 nm causes a clean transformation of XV to XVI in solution, with a quantum yield of about 0.06 which appears independent of the nature of the solvent and irradiation time. However, in PMMA and PS films, the photocyclization showed initial quantum yields of 0.1 and 0.23, respectively, which decrease with irradiation time. Compound XVI reverts to XV in solution in the dark via a first-order reaction but the rate constants are dependent on the solvent.

As with other similar compounds, there is no simple first-order rate law for the thermal return in either polymer matrix. However, the extent of recovery of XV in PMMA films is greater than in ethyl acetate solution. Typical first-order plots for the thermal conversion of XVI to XV in various solvents are shown in figure 5.3. Corresponding curves for the thermal conversion in the two polymer matrices are shown in figure 5.4. The authors explain that the higher quantum yield for the formation of the oxaziradine in the polymeric media is due to the stabilization of the excited state by the rigid matrix. However, it is also possible that the excited state is quenched by oxygen and that this quenching is inhibited by the reduced mobility of oxygen in the stryene and methyl methacrylate matrices. The solvent effect on the return reaction of XVI to XV could also be explained by the stabilization of the nitrone intermediate by media of high dielectric constant. Solvents having hydrogen bonding properties such as ethanol, chloroform and possibly dichloromethane assist the developing negative charge on the nitrone by the formation of hydrogen bonds.

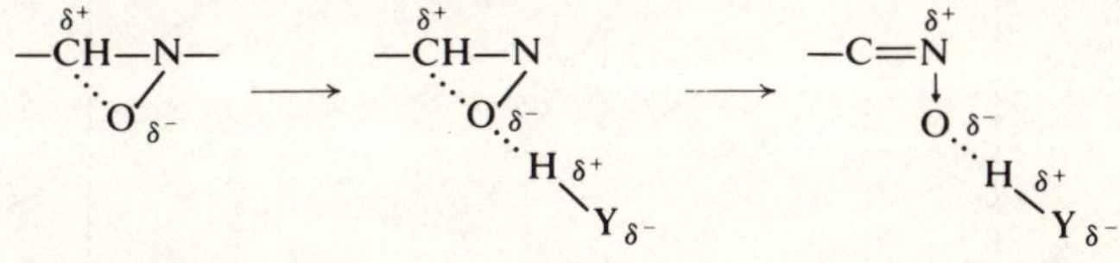

(5.5)

Figure 5.4. First-order plots of the thermal reaction of the oxaziridine into nitrone at 25°C: (1) in polystyrene, (2) in poly(methyl methacrylate). (From Smets and Matsumoto 1976.)

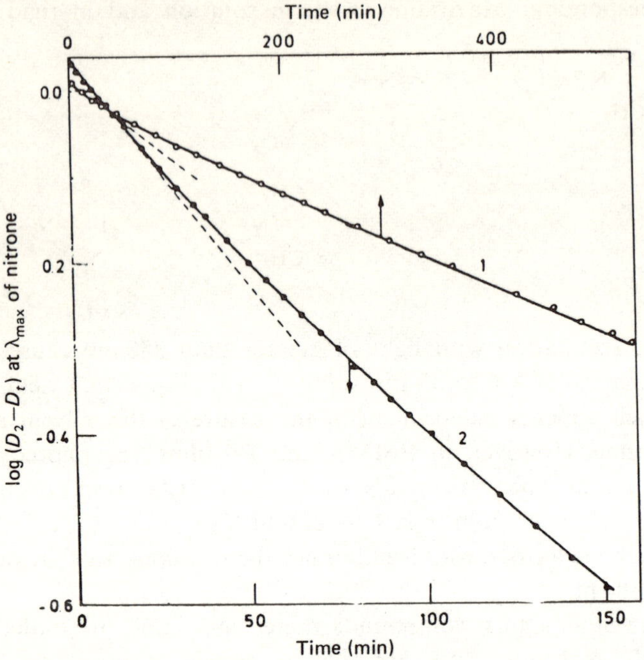

Figure 5.3. First-order plots of the thermal conversion of the oxaziridine into nitrone at 25°C in different solvents: (1) benzene, (2) ethyl acetate, (3) acetonitrile, (4) methylene chloride, (5) chloroform. (From Smets and Matsumoto 1976.)

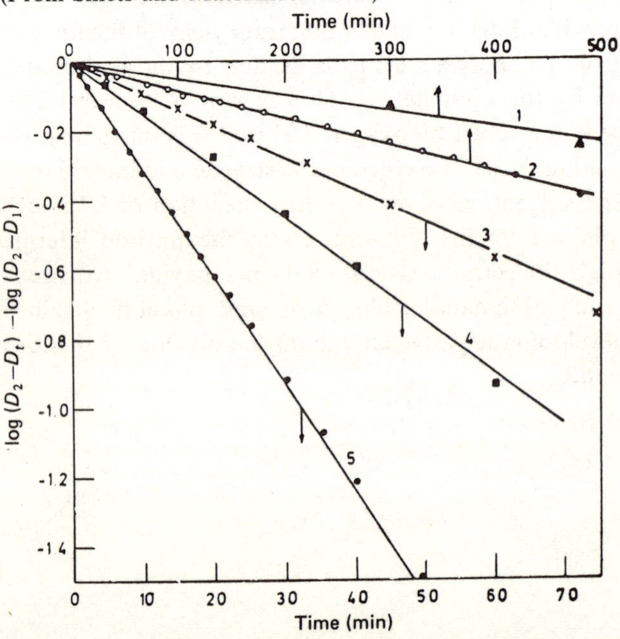

5.10 Latent image formation in glasses

Smets *et al.* (1975) have also described organic photosensitive systems used in the formation of latent images in polymer matrices. These systems, which have some applications for photographic purposes involve a polymer matrix, an activator, usually a polyhalogen compound such as carbon tetrabromide or bromoform, and dye precursors such as styrylic bases, cyanine bases, diphenyl and triphenyl methane leuco derivatives or the corresponding bases. In the systems studied by Smets *et al.*, the three components were poly(vinyl carbazole) (PVCz), carbon tetrabromide and di-β-naphthospiropyrane (DNSP, XVII). In this case PVCz not only acts as a polymeric matrix but also participates actively in the reaction mechanism. Smets *et al.* showed that the photosensitive component is a charge-transfer complex (CT) resulting from the interaction of PVCz in carbon tetrabromide (CT_1). On irradiation in near-uv light, hydrogen bromide is produced by a second reaction:

$$PVCz + CBr_4 \overset{K_1}{\rightleftharpoons} CT_1 \tag{5.6}$$

$$CT_1 \xrightarrow{h\nu} HBr \tag{5.7}$$

and the hydrogen bromide reacts with the dye precursor (DNSP, XVII) to produce the strongly blue colored compound di-β-naphthopyrylium bromide ($DNSPH^+Br^-$, XVIII).

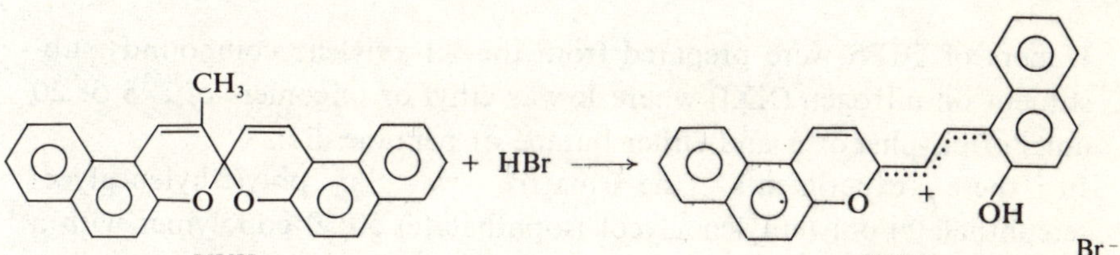

XVII DNSP — XVIII Blue

(5.8)

The blue colored salt (XVIII) also forms a charge-transfer complex with carbon tetrabromide but this absorbs in the visible region and hence a second exposure to visible light will develop an image only at the position of the original sites by the sequence of reactions shown below:

$$DNSPH^+Br^- + CBr_4 \overset{K_3}{\rightleftharpoons} CT_2 \xrightarrow{h\nu} CT_2^* \tag{5.9}$$

$$CT_2^* \xrightarrow{PNVC} HBr \tag{5.10}$$

$$DNSP + HBr \rightarrow DNSPH^+Br^- \tag{5.11}$$

This represents the optical development of an image which can be fixed by removing the carbon tetrabromide activator by heating the film at an elevated temperature for a short period of time after development. In this case, it was shown that the PVCz matrix actually participated in the development of both images through the formation of charge transfer complexes and also by forming a rigid matrix to control the photochemistry. On the other hand, PMMA used as a solid matrix was not effective.

Smets and co-workers also studied the relationship between the size of the chromophore dissolved in the matrix and its photochemical reactivity. They synthesized a number of derivatives of 1-benzyl 6-nitro(3′3′-dimethylspiro[2H-1-benzopyran-2.2′ indoline]) (DIPS, XIX). This compound photochemically isomerizes by the opening of the pyrane ring followed by cis-trans isomerization about the C_3-C_4 carbon bond to form a charge separated colored species.

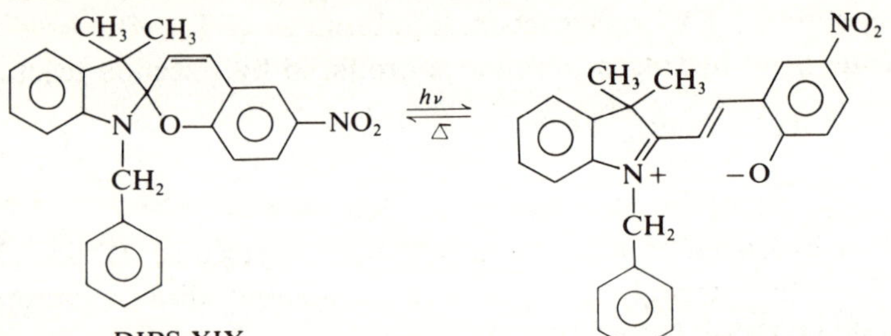

DIPS XIX

XX (5.12)

Dimers of DIPS were prepared from the 1,1-xylylene compounds substituent on nitrogen (XXI) where R was ethyl or oligomers of 2, 8 or 20 units of bis-phenol A and either butane or pentane diol.
In these experiments, the matrix was a poly(ethyleneglycol terephthalate)/poly(ethyleneglycol isophthalate) 50:50 copolymer with a

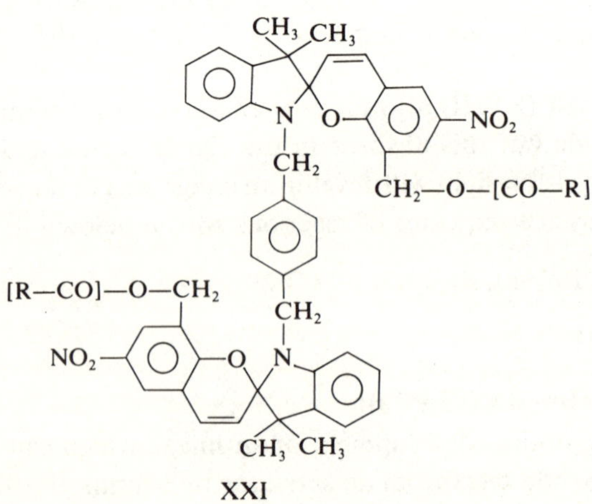

XXI

glass transition temperature of around 46 °C. The thermal fading reaction was much reduced when the size of oligomer was increased. Similar studies in PS matrices have been reported by Vainshtok and Rakhimbaeva (1976).

Gailite and Freimanis (1973) showed that electron transfer occurred both thermally and photochemically between tetrathio tetracene (XXII)

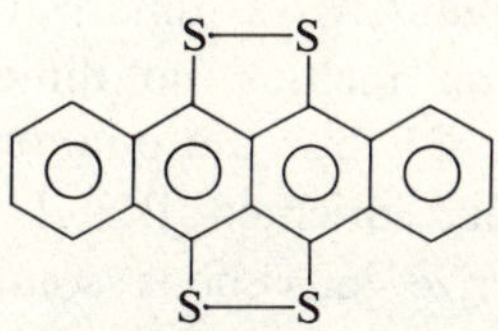

XXII

and various electron acceptors in poly(vinyl acetate) matrices. The photoreactions were said to be second-order with an activation energy of about 78 kJ mol^{-1}.

Smets *et al.* (1975) have reported studies on the photodissociation of the anthracene, 2-butylcarboxylate dimer and in poly(benzpinacol) in poly(butyl methacrylate) and PMMA matrices. They showed that the rates and reversibility were controlled by chain segment mobility and free volume distribution and that the effects were more pronounced when the photosensitive group was incorporated directly in the polymer by a chemical bond rather than when dissolved in the matrix.

5.11 Effect of orientation

An important advantage of polymeric solids as compared to EPA or borate glasses is that they can be readily oriented by stretching. This characteristic has been used to study the optical properties of oriented systems of small molecules in a variety of polymer matrices. For example, Bott and Kurucsev (1975) studied the direction of transition moments in small molecules by means of dichroic spectra in stretched films of poly(vinyl alcohol) and polyethylene (PE). As probe molecules they used naphthalene, anthracene, quiniline and acridine. They concluded that an orientation distribution of the molecules in the films was not entirely determined by molecular shape and that in general the values of the three orientation parameters for each molecule should be determined separately for each solute-polymer matrix system in order to find the transition moment direction from dichroic spectra. In a similar study reported by Aviv *et al.* (1976) the photodimerization of tetraphenylbutatriene (XXIII) was studied in a nematic liquid crystal system

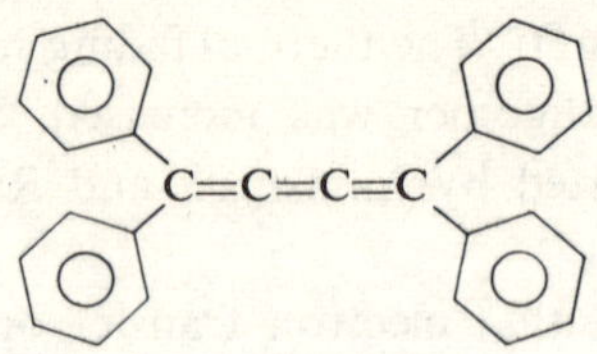

XXIII

and in oriented PE. It was shown that quantitative dimerization was achieved in the cholesteric phase while on melting no dimerization occurred. Dimerization also took place in PE but the quantum yield increased by a factor of three in stretched oriented PE. They also observed that XXIII did not photodimerize in viscous solutions of paraffin oil but did photodimerize in the amorphous phase of PE. The authors proposed that in the polymer media, both stretched and unstretched, the photodimerization is facilitated by extending the lifetime of the excited state. However, it seems more likely that the observed effect is due to preferential orientation on the crystalline phase of PE and in the liquid crystals. It seems clear that the potential for studying the effect of orientation on chemical reactions using polymer matrices is considerable and much work remains to be done in this area.

5.12 Free-radical reactions in glassy solids

In spite of much early work on free radical reactions in low temperature glasses, the extension of these studies to polymer matrices has been relatively recent. Some authors have suggested that photolysis of free radical-forming species in polymer matrices should be inhibited because of an increased 'cage' recombination reaction. However, there is little evidence to support this contention and, indeed, it rests upon improper assumptions about the rigidity of polymer glasses. In fact, it appears that the 'internal viscosity' as seen by a dissociating pair of radicals is little greater than that in simple solution and there is some recent evidence that free radical processes can, in fact, be facilitated in hydrocarbon glasses in polymer matrices (Li and Guillet 1980) due to a reduction in the competing processes. Further evidence against this contention is based on studies of the photo-Fries reaction which is believed to take place via a 'caged radical' intermediate and which appears to occur with equal efficiency in polymers below their glass transition temperature (Li and Guillet 1977) and in fluid solution. Vasilenko *et al.* (1976) have reported quantum yields for photochemical reactions of alkyl and peroxy radicals in PE, deuterated PE, polypropylene and poly(tetrafluoro ethylene). Quantum yields for the alkyl radicals were of the order of 0.1, while for peroxy radicals, quantum yields ranged from 0.01 to 1, depending on the isotopic composition of the polymer and the temperature. Since these workers did not report similar quantum yields in fluid solution, it is not possible to compare the effect of molecular mobility in this case. Similarly, Guseva *et al.* (1973) reported the photolysis of benzoylperoxide in cellulose acetate and PMMA. They also observed rapid decomposition of the peroxide but did not report quantum yields or efficiencies. Further quantitative studies in this area would clearly be helpful.

6

Fluorescence

6.1 Introduction

Because of the relatively short lifetime of fluorescence, which is usually of the order of 10^{-7} to 10^{-10} s, this emission can be used to probe for a variety of interesting characteristics of both large and small molecules. Much of the early work in polymer fluorescence was done by Yasunori Nishijima and his group at Kyoto University, and several extensive reviews have been published (Nishijima 1970, 1973*a*, 1973*b*, Somersall and Guillet 1975, Morawetz 1979, Holden and Guillet 1980) giving details of experimental procedures and the types of information which can be obtained. With the recent development of fast pulsed laser techniques, fluorescence methods have become even more powerful and represent an expanding field of polymer photophysics and photochemistry.

6.2 Fluorescence lifetimes

The process of fluorescence, defined briefly in §1.11.1, involves emission from an excited singlet state, usually the first, to the singlet ground state. The radiative lifetime $^1\tau_0$ of the excited singlet state is defined as the reciprocal of the allowedness of the radiative transfer between S_1 and S_0 and is given by:

$$^1\tau_0 = 1/k_F \qquad (6.1)$$

where k_F is the rate constant for fluorescence emission. This radiative lifetime corresponds to the theoretical lifetime of the excited singlet state which would be expected if fluorescence emission were the only process for the transition from S_1 to S_0. According to the theory of spontaneous emission originally developed by Einstein (1916, 1917), the probability of fluorescence emission is directly related to the transition probability of light absorption. However, the actual fluorescence lifetime τ is generally shorter than the radiative lifetime ${}^1\tau_0$ because the excited molecule may also lose excitation by a variety of radiationless processes, as indicated in the Jablonski diagram (figure 1.2).

The actual fluorescence lifetime is therefore given by:

$$\tau = 1/(k_F + \sum_i k_i) \tag{6.2}$$

where k_i are the rate constants of the various processes resulting in depletion of the excited singlet state. In the photostationary state, the quantum yield of fluorescence q_F is given by:

$$q_F = \tau/{}^1\tau_0 = k_F/(k_F + \sum_i k_i) \tag{6.3}$$

Equation (6.3) can be used to determine the radiative lifetime ${}^1\tau_0$ using experimentally determined values of q_F and τ. From the Einstein relationship, the radiative lifetime ${}^1\tau_0$ will be inversely proportional to the maximum molar extinction coefficient for the absorption band corresponding to the $S_1 \rightarrow S_0$ transition for each compound. Nishijima (1973*a*) reports an approximate relationship:

$${}^1\tau_0 \doteq 10^{-4}/\varepsilon_{max} \tag{6.4}$$

where ${}^1\tau_0$ is in seconds and ε_{max} is in liter mol^{-1} cm^{-1}. This gives an approximate relationship for estimating the value of the radiative lifetime from the molar extinction coefficient. Thus, for an aromatic compound having an ε_{max} of 10^4, ${}^1\tau_0$ should be of the order of 10^{-8} s.

It is now known that one of the structural requirements for an aromatic molecule to be highly fluorescent is a rigid planar configuration. Diphenyl polyenes, for example, such as structure I, show low fluores-

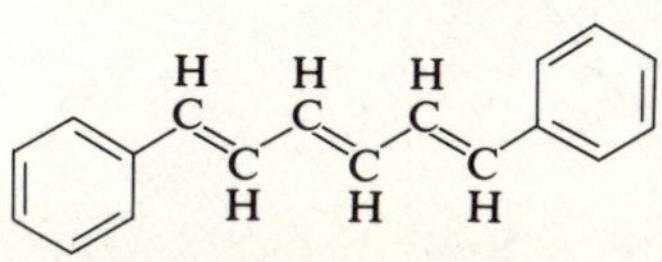

I Diphenyl hexatriene

cence intensity in dilute solution but become highly fluorescent at low temperatures or in viscous media. The low intensity in fluid solution has been attributed to internal quenching due to the internal rotation of the molecule during the lifetime of the excitation. The quantum yield of fluorescence then can be expressed in the form:

$$q_F^0/q_F = 1 + B(\tau^0/\rho) \tag{6.5}$$

where q_F^0 is the limiting quantum yield of fluorescence when internal rotation is inhibited and τ^0 is the limiting lifetime under these conditions, B is a proportionality constant and ρ is the relaxation time for internal rotation. For free rotation in an ordinary solvent $1/\rho$ is generally considered to be proportional to T/η where T is the absolute temperature and η is the viscosity of the medium. The intensity of the fluorescence I_F is then given by:

$$1/I_F = 1/I_F^0 + c(T/\eta) \tag{6.6}$$

where c is a proportionality constant.

Such a plot is shown in figure 6.1 for the fluorescence intensity of a dilute solution of diphenyl hexatriene in ethylene glycol. Consideration of (6.3) and (6.6) shows that there should also be a linear relationship between the fluorescence lifetime and the intensity of fluorescence, which has also been shown experimentally. There are now some well established experimental techniques for measuring the lifetime of fluorescence emission and hence this procedure can be used to determine internal rotation of molecules containing chromophore groups. Also, the effects of external viscosity can be probed in this manner for a solvent or matrix.

6.3 Polarization of fluorescence

In general, fluorescing groups will show some optical anisotropy upon light absorption as well as in fluorescence emission. Using a classical oscillator model, each oscillation for every electronic transition

Figure 6.1. Dependence of the reciprocal of fluorescence intensity ($1/I_F$) on T/η of the medium for a dilute solution of diphenylhexatriene in ethylene glycol. (From Nishijima 1973 *b*.)

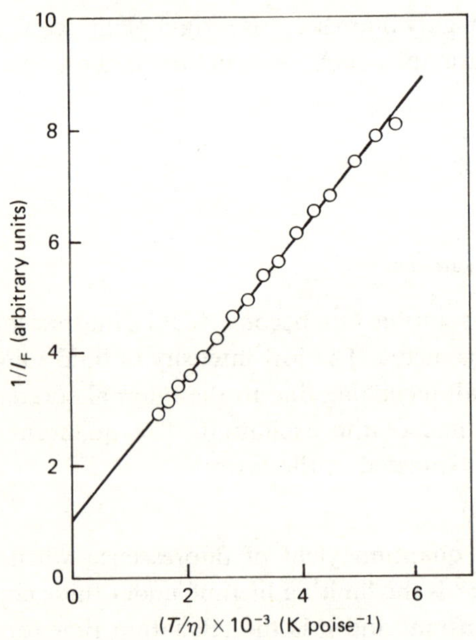

can be assigned along three mutually perpendicular axes which are fixed to the geometrical structure. For highly anisotropic molecules, however, the oscillator axes can be simplified and the molecule can be regarded as a single linear oscillator. Furthermore, when both the light absorption and fluorescence emission belong to the same electronic transition, namely $S_1 \rightarrow S_0$, the absorption oscillator A and the emission oscillator F can be assumed to coincide with a single molecular axis M. The optical coordinate system usually used is shown in figure 6.2. In this diagram, the linearly polarized exciting light propagates along the X axis with its polarized optical vector P_1 parallel to the Z axis. If we assume that the molecule is anisotropic, the probability of light absorption A is given by:

$$A = \kappa \cos^2\alpha \tag{6.7}$$

where κ is a proportionality constant. If the molecule is completely anisotropic, the optical vector of the fluorescence light emitted from the photoselectively excited molecule is parallel to M. When the intensity of fluorescence is measured along the Y axis through an analyzer, the intensity of the polarized component at any angle ξ from P_1 can be expressed as:

$$I_\xi = \kappa' \cos^2\alpha \cos^2\beta \tag{6.8}$$

The polarization characteristics of fluorescence emitted from the system can be obtained by integrating this relation over all molecules in the system.

Figure 6.2. Optical coordinate system (*X, Y, Z*) for characterization of the polarization of fluorescence. (From Nishijima 1973 *b*.)

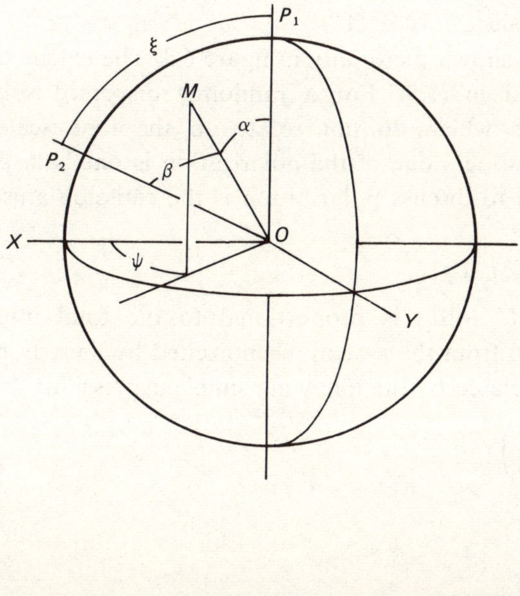

In this treatment, it is assumed that the fluorescent molecules are completely anisotropic and that the oscillators for light absorption and fluorescence both coincide with a single molecular axis. If the molecules are not completely anisotropic, the extent of polarization will be less than that predicted from theory. However, even for completely anisotropic molecules, if the emission oscillator does not coincide with the absorption oscillator, the extent of polarization is lessened. This is often the case when the excitation emission of fluorescence belongs to different transitions, e.g., if the excitation is the $S_0 \rightarrow S_2$ transition and the emission is $S_1 \rightarrow S_0$. Another factor which changes the polarization characteristics of the emission is the depolarization caused by rotations of the molecule or chromophore during the lifetime of the excitation. In this case, the direction of the emission oscillator makes an angle with the absorption oscillator dependent upon the rotational Brownian motion of the molecule.

A further possible cause of depolarization is energy transfer from the absorbing molecule to another chromophore in the system. If the energy transfer process does not require complete alignment of the molecular axes of the donor and acceptor, then energy transfer will generally tend to randomize the emission properties of the radiation (Chapter 9). If the molecular axes of the emitting molecules are mutually aligned, as in stretched crystalline polymers, the polarization of fluorescence will be characteristically oriented also.

If the fluorescent groups are completely anisotropic and the molecular axes do not alter during the lifetime of excitation, a random dispersion of fluorescent molecules will show the following angular distribution of the polarized component of the fluorescence intensity:

$$I_\xi = (\kappa'/15)(\sin^2\xi + 3\cos^2\xi) \tag{6.9}$$

This distribution is shown pictorially in figure 6.3. The extent of polarization P is defined in (1.5). For a randomly dispersed solution of fluorescent emitters which do not rotate on the time scale of the experiment, the limiting value of the polarization is one-half. A similar quantity often used to discuss polarization is the emission anisotropy r which is defined by:

$$r = (I_\parallel - I_\perp)/(I_\parallel + 2I_\perp) \tag{6.10}$$

The denominator $(I_\parallel + 2I_\perp)$ is proportional to the total intensity of fluorescence emitted from the system when excited by linearly polarized light. P and r are related by the following simple expression:

$$\frac{1}{r} = \frac{3}{2}\left(\frac{1}{P} - \frac{1}{3}\right) \tag{6.11}$$

6.4 Rotational depolarization

The spatial orientation of the molecular axes of the excited molecule is photoselected at the moment of light absorption. However, because of the finite time between light absorption and emission of fluorescence, the angular distribution of the molecular axes for emission may be changed by rotation of the molecule before emission occurs. The extent of this rotational depolarization is determined by the ratio of the lifetime of the excitation τ and the rotational relaxation time of the molecule ρ. The relationship between the rotational relaxation time and the observed depolarization P is given by:

$$P_0/P = 1 + (3 - P_0)\tau/\rho \qquad (6.12)$$

where P_0 is the limiting value of P, i.e., where no rotational relaxation occurs and τ and ρ have their usual connotations. In terms of the emission anisotropy r this equation becomes:

$$r_0/r = 1 + 3\tau/\rho \qquad (6.13)$$

For spherical particles, the rotational relaxation time is given by:

$$1/\rho = kT/3V\eta \qquad (6.14)$$

Figure 6.3. Angular distribution of the polarized component of fluorescene (I) for random distribution of completely anisotropic fluorescent molecules. (From Nishijima 1973 *b*.)

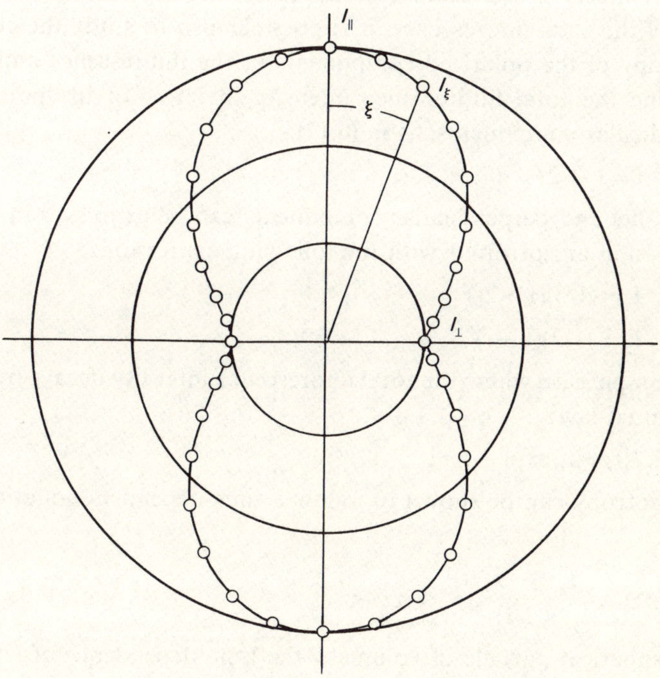

where V is the volume of the particle, T and η are the absolute temperature and viscosity of the medium, and k is the Boltzmann constant. Substitution of this into (6.12) and (6.13) gives the relationship between the polarization of fluorescence and the viscosity of the medium.

$$P_0/P = 1 + (3-\rho)(k\tau/3V)(T/\eta) \tag{6.15}$$

$$r_0/r = 1 + (k\tau/V)(T/\eta) \tag{6.16}$$

These equations have been shown to be valid for solutions of 9,10-diphenyl anthracene in ethylene glycol and in cyclohexanol. In this case, the fluorescence lifetime does not depend on the temperature or the viscosity of the medium. However, in flexible molecules, such as diphenyl hexatriene, the linear relationships predicted by (6.15) and (6.16) are not obeyed because the lifetime of the excited state is changing as a result of the viscosity dependence of the rate constant for internal conversion. In this latter case, it is necessary to plot $1/P$ versus (T/η) times τ, the measured lifetime of fluorescence. When this is done, linear behavior is then observed.

6.5 Transient measurements

The recent development of optical sources having very fast flashes of light from pulsed lamps or lasers has made possible the study of transient effects in fluorescence decay. In addition to studying the rate of decay of the total fluorescence, it is possible also to study the change in anisotropy of the polarized components of the fluorescence emission. If we define the total fluorescence intensity in terms of the parallel and perpendicular components, as in (6.17):

$$I = I_{\parallel} + 2I_{\perp} \tag{6.17}$$

the parallel and perpendicular components can be expressed in terms of the emission anisotropy r with the following expressions:

$$I_{\parallel} = (1/3)(1 + 2r) \tag{6.18}$$

$$I_{\perp} = (1/3)(1 - r) \tag{6.19}$$

In the simple case where the total fluorescence intensity decays by a single exponential law:

$$I(t) = I_0 \exp[-t/\tau] \tag{6.20}$$

the anisotropy can be shown to follow a time-dependent equation of the form:

$$r(t) = \frac{I_{\parallel}(t) - I_{\perp}(t)}{I(t)} \tag{6.21}$$

For a spherical particle of volume V the time dependence of r has been developed by Perrin (1926, 1929, 1934, 1936) and is shown to follow the equation:

$$r(t)=r_0\exp[-3t/\rho] \tag{6.22}$$

where r_0 is the emission anisotropy at $t=0$ corresponding to the limiting value of r for steady-state measurements where no rotation is possible, and ρ is the rotational relaxation time, as defined previously. Further theoretical developments have shown that direct measurements of time dependence of polarization give information on the heterogeneity of the relaxation processes and also the anisotropic rotation of particles. On the other hand, steady-state measurements give only mean values of the relaxation times for these cases.

6.6 Rotational relaxation times in polymer systems

Large macromolecules do not generally assume spherical shapes in solution and also, because of their large size, have relatively low values of rotational diffusion coefficients. Fluorescence depolarization represents an excellent way of determining this important characteristic of the mobility of polymer molecules. The rotation diffusion coefficient Θ can be defined by:

$$\Theta=kT/6V\eta=kT/\xi \tag{6.23}$$

This equation, originally devised by Einstein, relates only to spherical particles of volume V, and also defines ξ the hydrodynamic frictional coefficient. The rotational relaxation time ρ is simply related to the rotational diffusion coefficient Θ by:

$$1/\rho=2\Theta \tag{6.24}$$

It is sometimes useful to approximate anisotropic particles by a spheroid, or ellipsoid of revolution, with two main axes, a and b, where a is the axis of rotational symmetry. In such an anisotropic particle, there are actually three rotational diffusion coefficients corresponding to rotation about each of the three mutually perpendicular axes. For spheroidal particles, the rotational relaxation time ρ_a with respect to the axis of symmetry is given by:

$$1/\rho_a=2\Theta_b \tag{6.25}$$

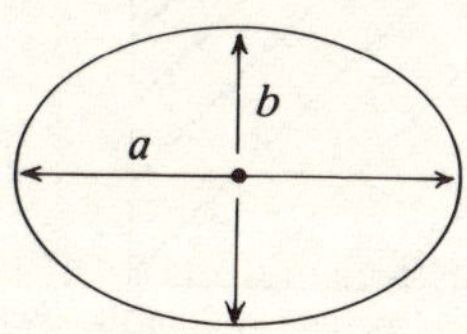

$\sigma=a/b=$axial ratio

where Θ_b is the rotational diffusion coefficient about the axis perpendicular to the main axis, and represents the tumbling motion of the molecule, i.e., moving end-over-end.

It is useful to compare this motion with the relaxation time of a totally spherical molecule of similar size. Values for the rotational relaxation time ρ as a function of T/η for the medium can be calculated from (6.14) and are shown in figure 6.4 for particles of various radii. From this expression it can be calculated that for a spherical molecule of radius 5 Å in water at room temperature, the rotational relaxation time would be about 4×10^{-10} s, whereas the same molecule in glycerol would have a value of 6×10^{-7} s, three orders of magnitude slower. This would be about the same rotational relaxation time as for a rigid spherical macromolecule of radius 50 Å in water.

Figure 6.4. The rotational relaxation time (ρ_0) for spherical particles of various radii (r), plotted as a function of T/η of the medium, according to (6.14). (From Nishijima 1973 *b*.)

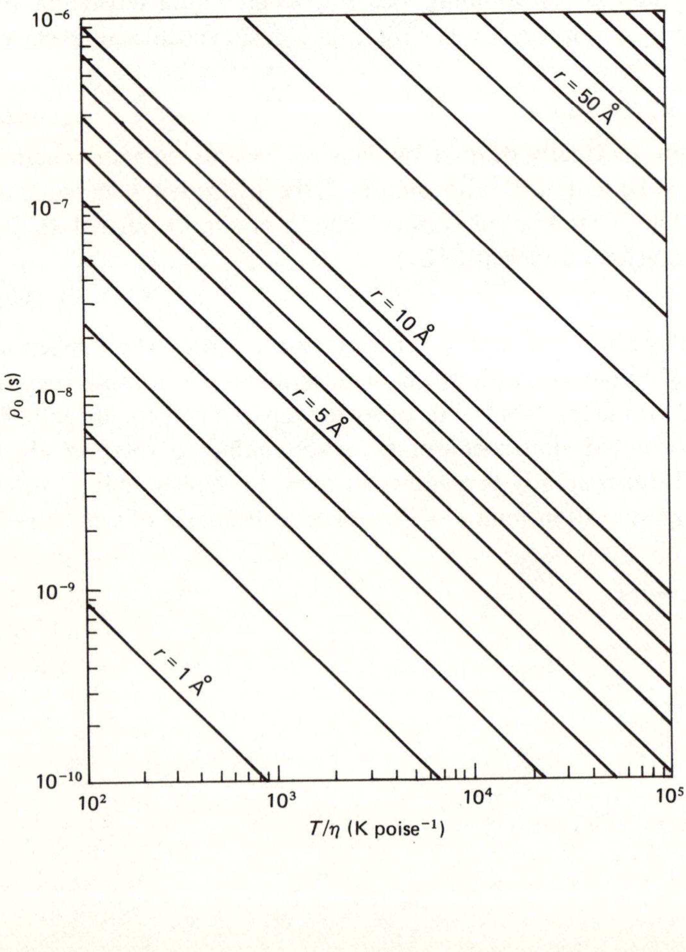

Solutions to the hydrodynamic equations have permitted the calculation of rotational diffusion coefficients for spheroidal particles and figure 6.5 shows the ratio of ρ_a to ρ_0 as a function of the axial ratio σ of the spheroid. As can be seen from figure 6.5 the rotational diffusion coefficient for this tumbling motion may change by several orders of magnitude if the spheroid is very elongated.

Obviously, measurements of this particular property of the polymer will give information about the shape and size of the macromolecule in solution. Applications of this technique have been pioneered by biochemists interested in the size and shape of protein molecules in solution (Weber 1952, 1953). A variety of fluorescent compounds such as fluorescein and 5-dimethylamino naphthalene 1-sulfonyl chloride isocyanate (dansyl chloride) have been used as fluorescence probes which are reacted with functional groups present in the protein molecules (Steiner and Edelhoch 1962).

6.7 Determination of chain dynamics by polarized fluorescence

The procedures described in §6.6 are suitable for the determination of rotational relaxation times for spheroidal particles, such as

Figure 6.5. The rotational relaxation time for a spheroidal particle (ρ_a) relative to the rotational relaxation time for a spherical particle of the same volume (ρ_0), plotted as a function of the axial ratio (σ) of the spheroid. (From Nishijima 1973 *b*.)

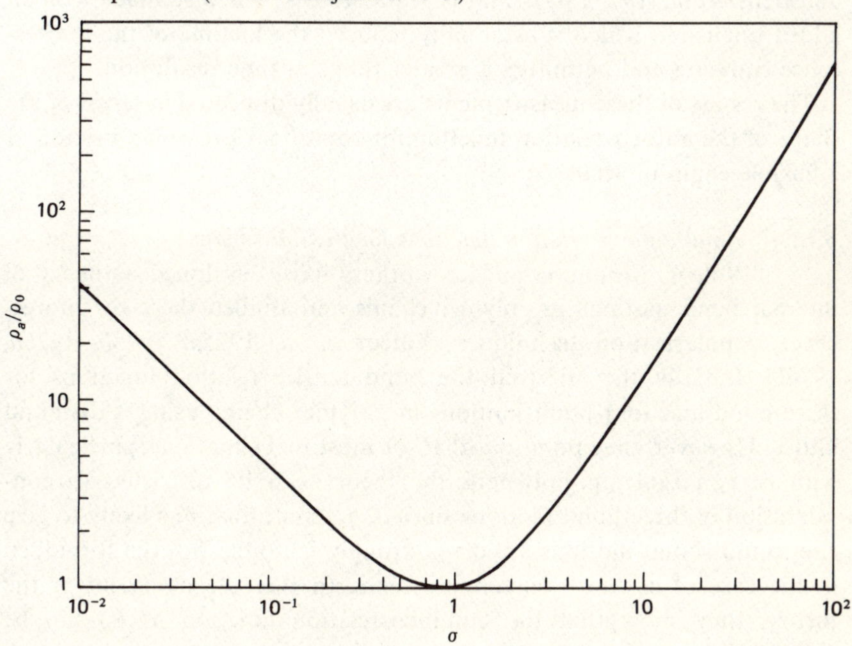

globular proteins, whose shape does not change with time. However flexible macromolecules in solution will not in general undergo coordinated rotations of the entire molecule, but will execute rapid diffusive motions of small segments of the chain, the so-called 'micro-Brownian' motion. The fluorescence of probes attached to segments of such polymers will be depolarized as a result of this motion, and hence provides an important route to the determination and understanding of the dynamics of flexible chains both in solution and in the solid phase. The micro-Brownian motion of flexible polymer chains in dilute solution was first studied by attaching fluorescent groups to the polymer chains (Nishijima 1962). In this way, Nishijima and Mito (1967) showed that the relaxation time of micro-Brownian motion in dilute solutions for many polymers is of the order of 10^{-9} s, but increases with increasing concentration up to $\sim 10^{-7}$ s as the internal structure of the polymer develops in highly concentrated solutions.

In these early studies, fluorescence polarization was measured by steady state procedures and used to calculate relaxation times using some variation of the Perrin equation (6.22). However, transient methods have been developed involving either pulsed laser sources or other fast excitation which permit more information to be obtained. Experimental procedures and the theoretical basis for their interpretation have been described by Yardley and Specht (1976), Valeur and Monnerie (1976*a*) and Bentz *et al.* (1975). In the latter work studies were also made with an added quencher, which substantially reduced the lifetime of the fluorescence emission and permitted a greater range of time resolution.

The results of these measurements are usually discussed in terms of the shape of the autocorrelation function for rotational Brownian motion of a flexible chain molecule.

6.7.1 *Bond autocorrelation functions for flexible chains*

Valeur, Monnerie and co-workers have developed a theory of internal bond motions in polymer chains and studied these by fluorescence depolarization techniques (Valeur *et al.* 1975*a*, 1975*b*, 1976*a*, 1976*b*). Initially they derived the bond autocorrelation functions for three-bond and four-bond motions in polymer chains using a diamond lattice. However, they point out that for most real situations, particularly with carbon skeleton polymers, the theory can be restricted to consideration of three-bond motions since $g^{\pm}g^{\mp}$ conformations likely to give rise to four-bond motions are often strongly inhibited or even forbidden in the case of carbon skeleton polymers. In their development of the theory, they show that the bond correlation function $M_2(t)$ can be expressed in the form:

$$M_2(t)=\exp[\gamma^2 Wt]\text{erfc}(\gamma^2 Wt)^{1/2} \tag{6.26}$$

In this equation, γ is a parameter characterizing the conformational structure of the chain and W is the jump frequency for rotations about these bonds. In this model, the relaxation time ρ will be equal to

$$\rho=1/\gamma^2 W \tag{6.27}$$

Thus the autocorrelation function becomes:

$$M_2(t)=\exp[t/\rho]\text{erfc}(t/\rho)^{1/2} \tag{6.28}$$

where erfc is the error function complement. At long times, this expression reduces to a law in square root of t of the form:

$$M_2(t)\underset{t\to\infty}{\doteq}(\pi t/\rho)^{-1/2} \tag{6.29}$$

It is useful here to introduce the concept of the 'independent kinetic segment' which is the shortest chain segment for which the motion of the central bond is independent of all bond motions outside the segment. In this concept, the segment length will depend on the time considered for the propagation of the perturbation relative to the elementary jump rate W and the conformational parameter γ of the chain. Since for every experimental technique there is a corresponding characteristic observation time, the independent kinetic segment defined in this way depends on the technique employed.

Based on their theoretical studies of these effects, Valeur *et al.* 1975*b*) showed that the effect of a chain end, either fixed or free, attentuates very rapidly and except for measurements at very long times, the effect is negligible for bonds more than 10 away from the chain end.

In real systems, it is likely that the motions cannot be characterized by a single relaxation time ρ but by a distribution of relaxation times ρ_i. In this case the autocorrelation function will be the sum of contributions from different modes given by

$$M_2(t)=\sum_i A_i \exp[-t/\rho_i] \tag{6.30}$$

where A_i represents the amplitude of each relaxation time characterized by its time constant ρ_i. The variation of A_i function of ρ_i represents the distribution of relaxation times $G(\rho)$. If this distribution function can be assumed to be continuous, then $M(t)$ is given by:

$$M_2(t)=\int_0^\infty G(\rho)\exp[-t/\rho]\text{d}\rho \tag{6.31}$$

Equation (6.28) is valid only for a chain which is strictly confined to motions on an ideal tetrahedral lattice. In real chains, both valence angles and internal rotation angles can vary from the values corresponding to the ideal lattice angles. There is also some uncertainty with respect to the angles of internal rotation due to thermal agitation. The combined effect of these deviations has been shown to result in an exponential contribution so that the final autocorrelation function can be written in the form:

$$M_2(t)=\exp[-t/\beta]\exp[t/\rho]\,\mathrm{erfc}(t/\rho)^{1/2} \tag{6.32}$$

where β is a parameter of the relaxation time which reflects the effective discrepancies with respect to an ideal tetrahedral lattice, and is determined by analyzing fluorescence decay curves at long times.

The emission anisotropy $r(t)$ is defined by analogy to (6.10) as

$$r(t)=\frac{I_{\parallel}(t)-I_{\perp}(t)}{I_{\parallel}(t)+2I_{\perp}(t)}\equiv\frac{D(t)}{I(t)} \tag{6.33}$$

where the term in the denominator represents the total intensity $I(t)$ of fluorescence which usually follows an exponential decay of the form:

$$I(t)=I_0\exp[-t/\tau] \tag{6.34}$$

where τ is the lifetime of the fluorophore. The initial value of the emission anisotropy r_0 depends on the angle α between the absorption and emission transition moments according to (6.35):

$$r_0=\frac{1}{5}(3\cos^2\alpha-1) \tag{6.35}$$

and the variation in emission anisotropy can be expressed as:

$$r(t)=r_0M_2(t)=r_0\left\langle\frac{3\cos^2\theta(t)-1}{2}\right\rangle. \tag{6.36}$$

where $\theta(t)$ is the angle for which the emission transition moment turns between time 0 and time t. Combining (6.36) and (6.32) we obtain:

$$r(t)=r_0\exp[-t/\rho]\exp[t/\rho]\mathrm{erfc}(t/\rho)^{1/2} \tag{6.37}$$

6.7.2 *Experimental determination of relaxation times for polymers*

Valeur *et al.* (1976*a*) studied the emission anisotropy of a polystyrene of molecular weight 57 000 prepared containing an anthracene in the centre of the chain with the structure

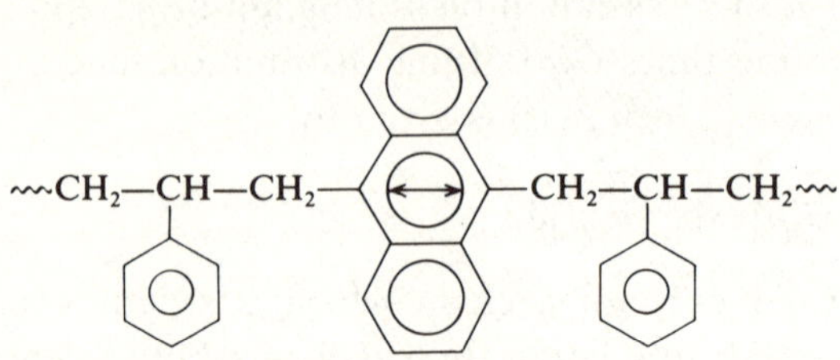

II PS-A

This was prepared by anionic polymerization and termination using 9,10-bisbromomethyl anthracene (Valeur *et al.* 1974). The polymer was studied in mixtures of ethyl acetate and glyceryl tripropionate which permitted the authors to study the internal rotation of the chain in solvents of similar chemical composition but of differing viscosity. The results are shown in figure 6.6. The agreement between the experimental results and theoretical curves is reasonably good: however, some systematic deviations are observed at short times, particularly for very low viscosities, which leaves some doubt as to the validity of the model (Holden 1982).

To confirm that the term $\exp[-t/\beta]$ is not due to the overall rotational diffusion of a large molecule, the authors calculated the rotational diffusion constants from the Kirkwood–Riseman relation (Riseman and Kirkwood 1949) for the rotational diffusion of a polymer coil

$$\tau_D = \frac{2M\eta_0[\eta]}{3RT} \tag{6.38}$$

where M is the molecular weight of the polymer, η_0 the viscosity of the solvent and $[\eta]$ is the intrinsic viscosity of the polymer in the same solvent. This gives a value of the rotational relaxation time τ_D for the complete molecule of 120 ns, whereas the value of β is of the order of

Figure 6.6. Variation in anisotropy versus time for PS-A polystyrene in various ethylacetate-tripropionin mixtures. The theoretical curves are drawn as bold lines. For clarity, each curve has been displaced by 1.5 ns with respect to the preceding ones. Viscosities (cp): 0.43 100% ethylacetate (curve 1); 0.89 (curve 2); 1.50 (curve 3); 2.88 (curve 4); 4.35 (curve 5); 7.47 100% tripropionin (curve 6). (From Valeur and Monnerie 1976 *a*.)

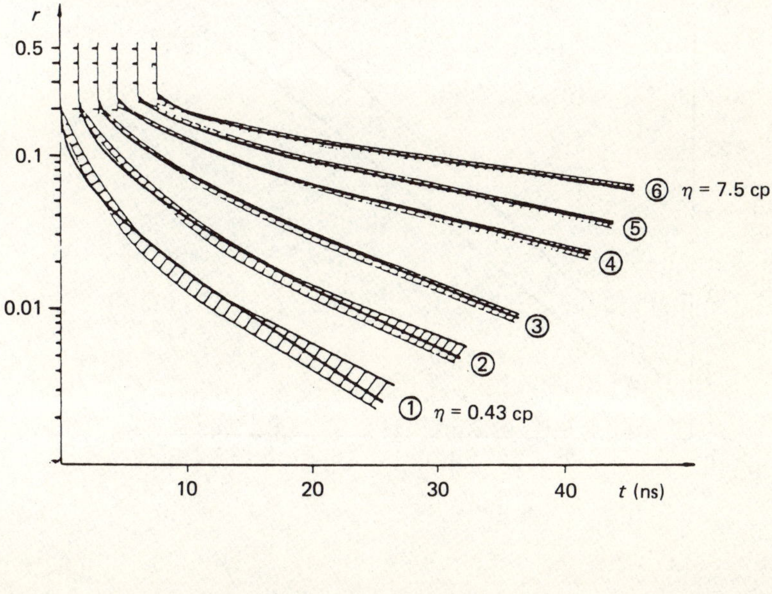

10 ns. The authors also showed that β, the relaxation time reflecting the effective discrepancies with respect to an ideal tetrahedral lattice, increases with the viscosity of the solvent by a relation of the form:

$$\beta = a + b\eta_0 \tag{6.39}$$

If a more general approach is used, in which the system is described in terms of a distribution of relaxation times $r(t)$, one can define an average relaxation time by:

$$\langle\sigma\rangle \equiv (1/r_0)\int_0^\infty r(t)\mathrm{d}t \tag{6.40}$$

A plot of the mean relaxation time as a function of viscosity is shown in figure 6.7. A similar relationship for polystyrene with 9,10-diphenyl anthracene attached to the side chain was obtained by Biddle and Nordstrom (1970).

Valeur *et al.* also studied a statistical copolymer of polystyrene with 9-styryl-10-phenyl anthracene (PS-DPA) which had been previously studied by Wahl *et al.* (1970). This polymer (structure III) permits an estimation of the effect of the direction of the transition moment to the relaxation observed in the system. Since the motion of this chormophore is not so directly constrained to the motion of the backbone of the polymer, the relaxation times were shorter than in the previous case and the decay of

Figure 6.7. Plot of mean relaxation time versus viscosity for PS-A (curve *a*) and PS-DPA (curve *b*). (Adapted from Valeur and Monnerie 1976 *a*.)

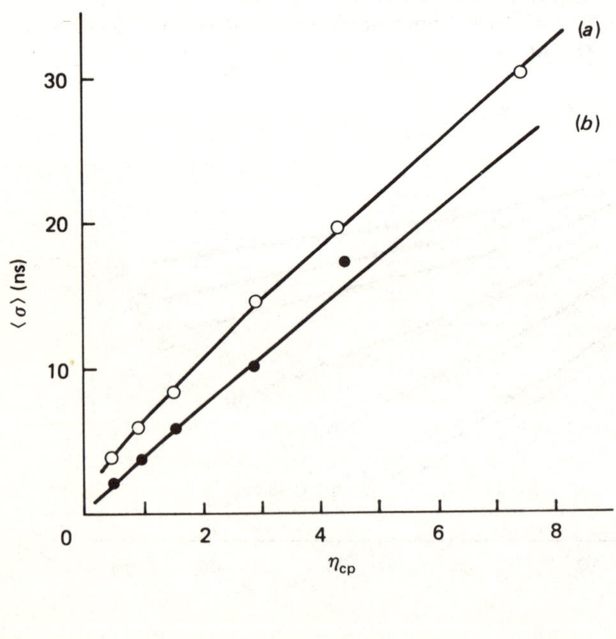

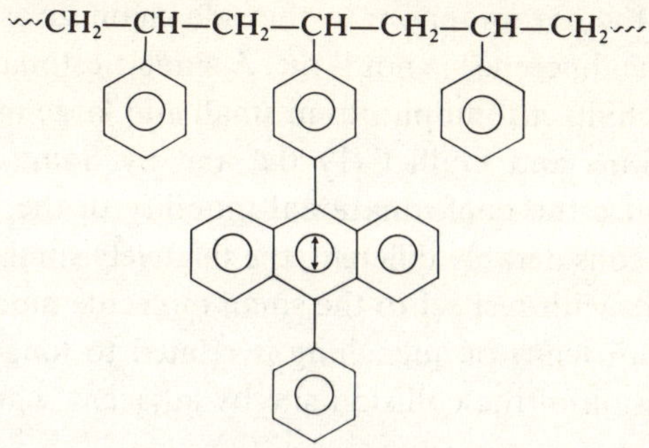

III PS-DPA

fluorescence $I(t)$ was purely single exponential. The mean relaxation time $\langle\sigma\rangle$ versus solvent viscosity for PS-DPA is shown in figure 6.7(*b*). As might be expected in this case, the authors did not find a good fit for the $r(t)$ curves with (6.37).

In a subsequent paper, Valeur and Monnerie (1976*b*) described experiments on the same polymer using carbon tetrachloride as the quencher in 1,2-dichloroethane and chloroform solution. This procedure, involving the use of static quenchers for the determination of rotational relaxations in macromolecules, had previously been investigated by Weill (1971) and Bentz *et al.* (1975). In this case, the autocorrelation function (6.32) gave an excellent fit in describing the mean anisotropy at various quenching levels. The mean anisotropy is defined

$$\bar{r}=(r_0/\tau)\int_0^\infty M_2(t)\exp[-t/\tau]\mathrm{d}t \qquad (6.41)$$

and for a fluorophore incorporated into a macromolecule chain the combination of this equation with (6.32) gives the expression

$$1/\bar{r}=(1/r_0)(1+(\tau/\beta)+\{(\tau/\beta)[1+(\tau/\beta)]\}^{1/2}) \qquad (6.42)$$

The mean intensity was measured at various levels of quenching, and it was shown that the Stern–Volmer law applied over the concentration ranges used in the experiments. Table 6.1 shows a summary of data on the lifetimes and quenching constants k_q for the model compound 9,10-dimethyl anthracene, PS-A (II), where the anthracene is in the backbone of the chain and PS-DPA (III) where the dimethyl anthracene is on a side chain.

It is interesting to note that the ratio of the quenching constant for the model compound versus the polymer chromophore in both solvents is nearly the same, about 3.8 for II and 3.3 for III. That is to say, the quencher is about four times as efficient in quenching an isolated model chromophore as it is with a chromophore attached to the polymer chain.

Furthermore, attaching the chromophore to the side chain gives slightly more accessibility, but the difference is not large. A similar estimate of the difference between quenching chromophores in small and large molecules has been made by Heskins and Guillet (1970*a*) and by Somersall and Guillet (1972) (§6.10). Since the conformational mobility of the polymer systems studied must be considerably different, the relatively similar values for the ratio of efficiencies with respect to the small molecule models may mean that the reduced efficiency of quenching is related to longer range effects such as the occlusion of the collision site by adjacent segments of the same polymer chain.

The mean relaxation time $\langle\sigma\rangle$ can be obtained from quenching experiments using the relationship:

$$\langle\sigma\rangle = \lim_{\tau\to\infty}(\bar{r}\tau/r_0) \tag{6.43}$$

Plots of $1/\bar{r}$ as a function of $1/\tau$ can be extrapolated to $1/\tau = 0$. The results obtained in this way were in excellent agreement with those obtained from the decay of anisotropy.

North (1974) has reviewed the method of obtaining relaxation times by fluorescence depolarization. He points out that there are several difficulties in this measurement, particularly in the steady-state procedures. In the first place, theories for the interpretation of depolarization often assume that rotational diffusion is isotropic, which may not be the case for localized movement of the group attached to the polymer chain. Secondly, the extent of movement of a fluorescent group may not represent the actual movement of the chain, depending on the relative size of the group and the chain segments and also whether the group is attached on a side group or near the middle or the end of the chain. Observations using continuous irradiation, i.e., the stationary state technique, which includes most polymers studied so far, may be in serious

Table 6.1. *Fluorescence lifetimes* τ_0 *and quenching constants* k_q *for polystyrene containing anthracene groups*[a]

	In chloroform			In 1,2-dichloroethane		
Compound	τ_0 (ns)	$k_q \times 10^{-9}$	ratio	τ_0 (ns)	$k_q \times 10^{-9}$	ratio
9,10-Dimethyl anthracene (model)	8.1	1.7	—	10.1	3.3	—
II, PS-A	6.7	0.44	3.9	7.3	0.9	3.7
III, PS-DPA	5.8	0.52	3.3	6.0	1.0	3.3

[a] Mean relaxation time $\langle\sigma\rangle$ is 4.3 ns in dichloroethane and 3.7 ns in chloroform.

error if the basic Perrin equation (6.22) is applied to systems where non-isotropic rotations are taking place. Furthermore, the model of cooperative three- and four-bond motion between nearby C–C bonds proposed by Valeur *et al.* (1975*a*) may not be correct. Liao and Morawetz (1980) have published data on excimer emission between main-chain phenyl groups in polyamide and polyethylene oxide copolymers which strongly suggest that these short-range cooperative motions of two bonds do not take place. Instead they propose that after one bond rotates, the strain exerted on the stationary part of the chain is compensated by very small distortions of a large number of bonds and bond angles in the region of the moving bond. Clearly, much more work is necessary to clarify this important aspect of polymer chain dynamics.

Relaxation times have been measured experimentally on a variety of polymers by different procedures. Table 6.2 lists a number of rotational relaxation times which have been determined by both stationary and transient measurements and where good agreement seems to have been obtained. North points out that the rotational relaxation times obtained with anthracene or naphthalene probes incorporated into poly(methyl methacrylate), poly(butyl methacrylate) and poly(p-chlorostyrene) are all in good agreement with the respective dielectric relaxation times (4, 5 and 5.5 ns, respectively). However, for very flexible chains such as poly(methyl acrylate) and poly(vinyl acetate) the dielectric relaxation times are much shorter, indicating that the rotation of the fluorescence probe is not in this case indicative of the actual relaxation of the polymer segments because the probe itself exerts a viscous drag on the chain to which it is attached.

6.8 Local viscosity in polymer environments

Nishijima and co-workers have also used depolarization of fluorescence to examine the microviscosity of polymer solutions and melts. According to (6.14), the rotational relaxation time ρ is directly related to the viscosity η of the medium. For ordinary solvents consisting of small molecules, the viscosity of the medium can usually be taken as identical to the viscosity of the solvent. In this case, the effect of molecular volume of the molecule can be calculated from the slope of a plot of $1/\rho$ versus T/η for the solvent.

However, in concentrated solutions or melts containing polymer molecules, the actual hydrodynamic frictional force exerted on rotating fluorescent molecules is not related to either the viscosity of the pure solvent or the viscosity of the solution. The local viscosity is a combination of effects due to the solvent and its interaction with segments of the polymer chain adjacent to the rotating chromophore. The viscosity determined by the application of (6.14) will then be the effective viscosity of the local environment surrounding the fluorescent molecule and may in fact vary with the nature of the fluorescent probe as well as the solvent and polymer. For example, figure 6.8 shows estimates of the local viscosity in poise estimated by (6.14) for solutions of polystyrene in benzene, containing naphthacene as the fluorescent probe (Nishijima and Mito 1967). It is important to note that there is very little change in the local viscosity with polymer concentration up to about 20% and this local viscosity is very similar to that experienced by the probe molecule in benzene alone. As pointed out in Chapter 4, this is a very important

Table 6.2. *Steady-state rotational relaxation times, 25 °C*

Polymer	Fluorescent species	Solvent	ρ (ns)
Poly(methyl acrylate)	9-Vinylanthracene copolymer	Toluene	1.2
	Dichlorofluorescein end-group	DMF	3.4
	Anthracene end-group	Toluene	0.3
Poly(vinyl acetate)	9-Vinylanthracene copolymer	Toluene	3.4
	Dichlorofluorescein end-group	DMF	1.4
	9-Methylanthracene end-group	Toluene	0.7
Poly(methyl methacrylate)	9-Vinylanthracene copolymer	Toluene	3.9
	Dichlorofluorescein end-group	DMF	5.4
Poly(*n*-butyl methacrylate)	9-Vinylanthracene copolymer	Toluene	4.0
	Bis(methylbenzoxazolyl)ethylene end-group	Toluene	2.2
	Dichlorofluorescein end-group	DMF	6.5
	9-Methylanthracene end-group	Toluene	0.8
Polystyrene	9-Vinylanthracene copolymer	Toluene	6.0
	Bis(methylbenzoxazolyl)ethylene end-group	Toluene	5.1
	Anthracene end-group	Toluene	1.7
	9-Methylanthracene end-group	Toluene	0.8
	2-Vinylnaphthalene copolymer	Toluene	5.5
Poly(*p*-chlorostyrene)	9-Vinylanthracene copolymer	Toluene	6.6
	9-Methylanthracene end-group	Toluene	1.4
Poly(*m*-chlorostyrene)	9-Vinylanthracene copolymer	Toluene	6.5
Poly(*p*-methoxystyrene)	9-Vinylanthracene copolymer	Toluene	5.0
Poly(*p*-bromostyrene)	9-Vinylanthracene copolymer	Toluene	5.5
Poly(*N*-vinylcarbazole)			
(M_w, 91 900)	9-Vinylanthracene copolymer	Toluene	26.2
(M_n, 3 700)	Dichlorofluorescein end-group	DMF	3.8
(M_w, 6 000)	Rhodamine-B end-group	DMF	31.4
(M_w, 53 500)	Carbazole residues of homopolymer	Toluene	52.6

Source: North (1975).

characteristic of polymer solutions because the macroscopic viscosity would have increased by several orders of magnitude under these conditions.

It is important in studies of polymer photochemistry that the appropriate viscosity is used to represent the viscous drag on a moving segment or particle. The microviscosity, that is, the viscosity measured by some suitable probe, rather than the macroscopic viscosity observed by conventional methods such as capillary or rotating cylinder viscometry, should be used. If data on the local viscosity are not available, then as a first approximation the best value to use is that of the pure solvent itself, as indicated in this figure. Figure 6.8 also shows a stepwise increase in viscosity at approximately 30 and 70% polymer, which was attributed by the authors to orientation effects in the polymer solution. However, it seems possible that other phenomena relating to excluded volume and the free volume of the polymer chains themselves may be responsible for this effect. In more recent studies Hirota and Nishijima (1980) repeated these experiments with 3-butyl perylene as the probe and found no evidence for a stepwise increase in internal viscosity.

Another important result involves the relation between the macroscopic melt viscosity of polyethylene and local viscosity (Nishijima and

Figure 6.8. Local viscosity and solution viscosity of polystyrene solution in benzene. (From Nishijima 1970.)

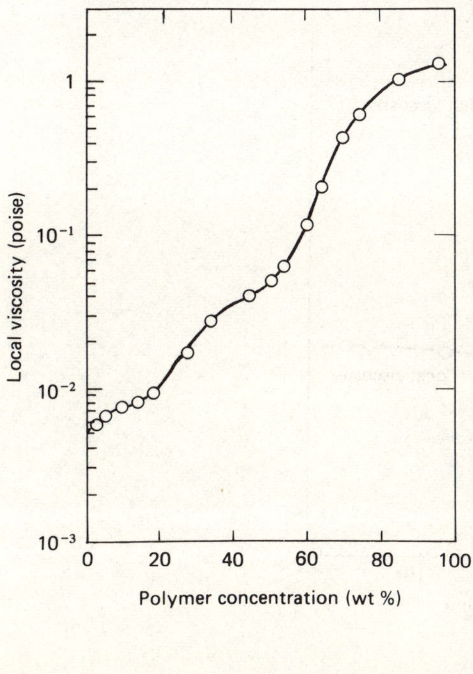

Saito 1967). Solutions of compound IV in linear hydrocarbons and polyethylene of varying molecular weights showed an increase in the local viscosity for molecular weights up to about 2000 and thereafter the local viscosity was independent of molecular weight (figure 6.9). However, not only is the local viscosity significantly lower than that of the melt viscosity for low molecular weight compounds, there is also a drastic increase in melt viscosity above 2000 molecular weight which is not reflected at all in the local viscosity. Based on this type of information, it is deduced that the local viscosity is related to the movement of segments of the polymer molecule of about molecular weight 2000, i.e., about 60 ethylene units. This is believed to be the approximate size of the moving segment which contributes to the glass transition temperature and also to the molecular movement involved in viscous flow.

It is interesting to compare this local viscosity with that estimated from the diffusion of naphthalene in polyethylene given by Heskins and Guillet (1970*b*). Applying (3.13), the local viscosity of polyethylene at 80 °C is about 1 poise, while Nishijima reports a value of 2×10^{-2} poise for the

Figure 6.9. Local viscosity and melt viscosity of *n*-paraffins and polyethylene at 150°C. (From Nishijima 1970.)

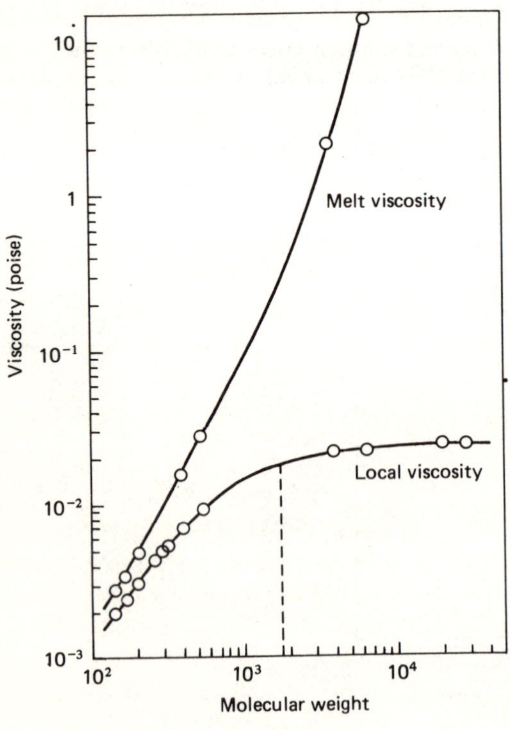

local viscosity of polyethylene at 150 °C. Accepting an activation energy of 16 kcal mol^{-1} as observed for diffusion of tetradecane in polyethylene and similarly for viscous flow in polyethylene melts, yields a calculated viscosity of ~1 poise at 80 °C, in quite remarkable agreement. The phenomenon observed here is believed to be general for all polymers and the concept of local viscosity should be used in all attempts to interpret transport or other kinetic phenomena in polymer systems.

Nishijima and co-workers have also reported (Nishijima 1964, Nishijima *et al.* 1965) the observation of multiple transitions in polymer solids employing fluorescent compounds such as Auramine-O which can undergo internal quenching by rotation. This method is claimed to be particularly sensitive to local motions of side groups in polymer chains in the solid.

Jarry and Monnerie (1979) have studied the inclusion of large fluorophores such as the diphenyl polyenes dissolved in polyisoprene. Studies of the depolarization of these compounds showed the onset of cooperative motion at T_g in polyisoprene. The frequency dependence of T_g was also shown to be important. That is, on the nanosecond time scale of the fluorescence depolarization experiment, the glass transition temperature is about 50° greater than that measured by calorimetric methods. These chromophores, as well as anthracene attached to the main chain of polyisoprene, were able to respond to changes in the glass transition, resulting from alternation in the degree of crosslinking or plasticization of the polymer. More recently, studies were made of the effect of orientation by drawing, however because of the high effective frequency of this procedure the results are difficult to correlate with the classical measurements of T_g (Monnerie 1981).

6.9 Measurement of diffusion by fluorescence quenching

The efficient quenching of luminescence by certain 'quenching' molecules often occurs at every collisional encounter. If a fluorescent or quenching group is attached to a polymer chain, estimates can be made of diffusion constants and local concentration effects by application of the Stern–Volmer equation:

$$\phi_0/\phi = 1 + k_q\tau[A] \tag{1.41}$$

where $[A]$ is the concentration of the quenching species. If the fluorescence lifetime in the absence of quencher, τ, is known, and the quencher reacts at every collision, then $k_q = k_{\text{diff}}$ and combining (1.41) with (3.13) gives

$$\phi_0/\phi = 1 + 2.2 \times 10^5 (T/\eta)\tau[A] \tag{6.44}$$

where η is the 'effective viscosity' of the medium. Alternatively, if one assumes that segment diffusion is very small compared to that of a small molecule reagent (Heskins and Guillet 1970*a*), the diffusion constant D of the small molecule can be obtained by combining (1.41) and (3.15) to give

$$\phi_0/\phi = 1 + (4\pi\rho' N/1000) D\tau[A] \tag{6.45}$$

The use of (6.44) and (6.45) assumes that there are no steric restrictions on the approach of the donor and acceptor species in polymer molecules, as compared with small molecule models. This assumption can usually be tested by selecting a range of small molecule analogs for comparison. When quenching does not occur at every collision, it is still possible to apply (6.44) and (6.45), provided that an efficiency factor is included, as defined in (1.42). If this factor does not depend on the attachment of the chromophore to a polymer chain, then these procedures will still be valid.

In dilute solutions of polymers there may be a difference between the concentration of A inside and outside the polymer coil, and hence these relations can be used to probe such effects (Moldovan and Weill 1971).

Among the earliest studies of diffusion in polymers by fluorescence quenching is the work of Heskins and Guillet (1968, 1970*a*, 1970*b*). The quenching of naphthalene fluorescence in both small molecule ketones and in ethylene-carbon monoxide polymers was used to determine both the diffusion constants of small molecules in solid polyethylene and the efficiency of energy transfer. It was concluded that the efficiency was about one-third as great when the carbonyl chromophore was contained in a polymer chain than for small molecule model compounds. This was independent of whether or not the polymer was in solution or in the solid phase. The diffusion rate constant k_{diff} for naphthalene in solid polyethylene at 80 °C was found to be $1.0 \times 10^8\,l\,mol^{-1}\,s^{-1}$ which was in excellent agreement with the value for octadecane in polyethylene determined by a radio tracer procedure. Later, Somersall and Guillet (1972) measured the fluorescence quenching of small ketones and poly(vinyl ketone)s with biacetyl and also the quenching of naphthalene fluorescence by the same ketone in the same solvent. They also showed that the quenching of naphthalene fluorescence was about one-third less efficient when the ketone was on a polymer chain as compared to a small molecule ketone while, on the other hand, the quenching of the ketone fluorescence by biacetyl was more efficient in the polymer than in small molecule models. They used this phenomenon to suggest that in the polymeric ketones there was some excitation transfer from carbonyl group to carbonyl group, which increased the effective radius over which quenching could occur.

Weill and co-workers (Moldovan and Weill 1971, Duportail *et al.* 1971) also studied the quenching of small molecule fluorescence by polymer quenchers in order to determine the fraction of the total solution volume occupied by the polymer coil. The inhomogeneous spatial distribution of the quencher caused a deviation from normal Stern–Volmer law. By using a gaussian distribution of the chain segment density in the macromolecular coil, it was possible to explain the quenching kinetics up to the concentration where interpenetration of the coils began. Further differences from linear Stern–Volmer relationships were explained by a difference in solvation of the fluorescing polymer by the solvent and the quencher molecule. The fluorescence data were interpreted in terms of the extent to which the quencher was concentrated or excluded from the polymer domain.

Local concentration effects of this type are especially pronounced in solutions of polyelectrolytes. For example, Taha and Morawetz (1971) studied the effect of sodium poly(vinyl sulfonate) (PVS) on the quenching of the quinine cation by Fe^{2+}, Ag^{+} and Br^{-} ions. The quenching efficiency was enhanced when small amounts of PVS were added because both the quinine 2^{+} and Fe^{2+} ions tended to cluster with the polyanion, thus giving a higher local quenching concentration. At high PVS concentrations the effect was reversed because the quinine ions were gradually excluded from the coil interior by the more compact inorganic ions. The anionic quencher Br^{-} showed exactly the opposite effect because it was confined to regions outside the polyanions. Similar phenomena were observed with the quenching of the ruthenium bipyridyl complex via Cu^{2+} ions in the presence of PVS (Meisel and Mathieson 1977). As expected, the presence of a polyion inhibits reactions between positively and negatively charged species. An important consequence of this effect is that when $Ru(bpy)_3^{2+}$ is excited in the presence of a neutral electron acceptor, the back reaction which normally takes place within the solvent cage is inhibited by the presence of PVS. The PVS attracts the $Ru(bpy)_3^{3+}$ ion produced, but repels the negatively charged product of the electron transfer (Meierstein *et al.* 1978).

Kirsh *et al.* (1975) also studied quenching between polymers containing fluorescing and quenching groups distributed at random along the chain. As the concentration of the two polymers was increased, the change in fluorescence intensity reflects the ability of polymer coils to interpenetrate. In these experiments, it was shown that the interpenetration of the flexible chain molecules was negligible until a critical polymer concentration had been achieved. Further studies of this type have involved quenching of phosphorescence and energy transfer experiments to determine the interpenetration of coils in solution and in the solid phase.

MacCallum and Rudkin (1978) have used a photoluminescence method to determine diffusion constants of gases in polystyrene and poly(methyl methacrylate). They studied the quenching of either excimer or additive fluorescence by oxygen in thin films after displacement of the nitrogen atmosphere over the sample by oxygen. The relationship between oxygen concentration $[O_2](t)$ and the diffusion coefficient D for a flat film of thickness l is given by

$$\frac{[O_2](t)}{[O_2](\infty)}=\frac{[I_N/I(t)-1]}{(I_N/I_0)-1}=(4/l)(Dt/\pi)^{1/2} \tag{6.46}$$

where $I(t)$ is the emission intensity at time t after replacement of the nitrogen by oxygen, and I_N and I_0 denote steady-state values in the absence and presence of oxygen, respectively. Values of the diffusion coefficients of 1.0 and 3.8×10^{-8} cm^2 s^{-1} were obtained at 20°C for PS and poly(methyl methacrylate).

6.10 Molecular orientation in solid polymers

Nishijima and co-workers have used the polarization characteristics of fluorescence to study orientation behavior in stretched and unstretched polymer samples. They used highly anisotropic molecules such as IV and V as probes for this orientation. In general, the oscillator system for fluorescing molecules is assigned along three perpendicular axes which are fixed to the molecule according to the electronic anisotropy in its particular electronic state. However, for molecules such as IV and V the fluorescence can be represented as resulting from a single linear

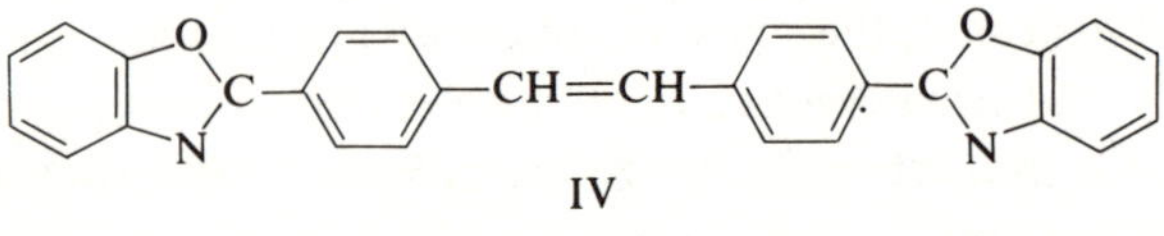

IV

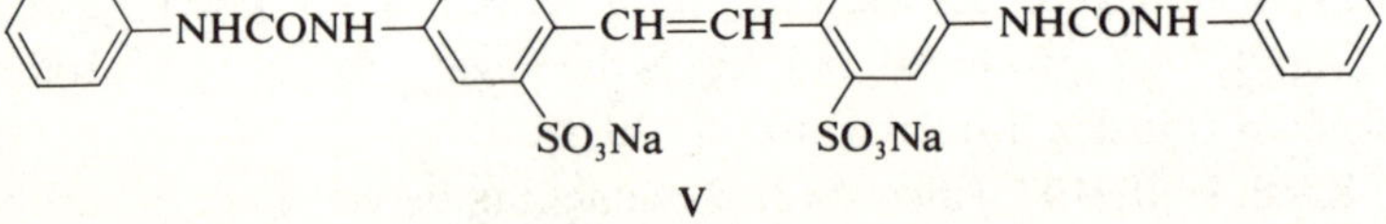

V

oscillator M which lies along the long molecular axis for electronic transitions between the ground state and the first excited state ($S_1 \rightarrow S_0$). In this case, the optical coordinates can be arranged, as in figure 6.2 with the linearly polarizing exciting light incident along the y axis with an electric vector P_1 in the direction of the z axis, and the polarized component of fluorescence measured through an analyzer in the direction of the y axis, with the electric vector of the component P_2 lying in the zx

plane at an angle ξ with P_1. The intensity of the polarized component of the fluorescence I_ξ emitted from the molecule is then given by

$$I_\xi = KM_1^2M_2^2 \tag{6.47}$$

where M_1 and M_2 are components of M in the directions of P_1 and P_2, respectively, and K is a proportionality constant. As pointed out by Nishijima *et al.* (1967), the angular distribution of the polarized component of fluorescence emitted from the system yields the fourth moment of the orientation distribution function, which provides direct information about the type of molecular orientation in the system.

Examples of the types of orientation which can be distinguished by determining the orientation pattern of polarized fluorescence have been calculated and are shown in figure 6.10. As can be seen, fluorescence methods can identify many types of orientation which are significantly different in detail but have the same orientation function f when determined from anisotropy measurements by birefringence, for example. Nishijima and co-workers have used this method to determine the orientation in crystalline and amorphous regions of a variety of polymers, including poly(vinyl alcohol), poly(vinyl chloride) and polypropylene. For example, an angular distribution of the polarized component of fluorescence in uniaxially stretched poly(vinyl alcohol) films containing 10^{-5} M of compound *V* at various draw ratios is shown in figure 6.11. Similar results were found when the chromophores were introduced by partial

Figure 6.10. Orientation patterns with corresponding orientation function (f) and angular distributions of polarized components of fluorescence. (From Nishijima 1970.)

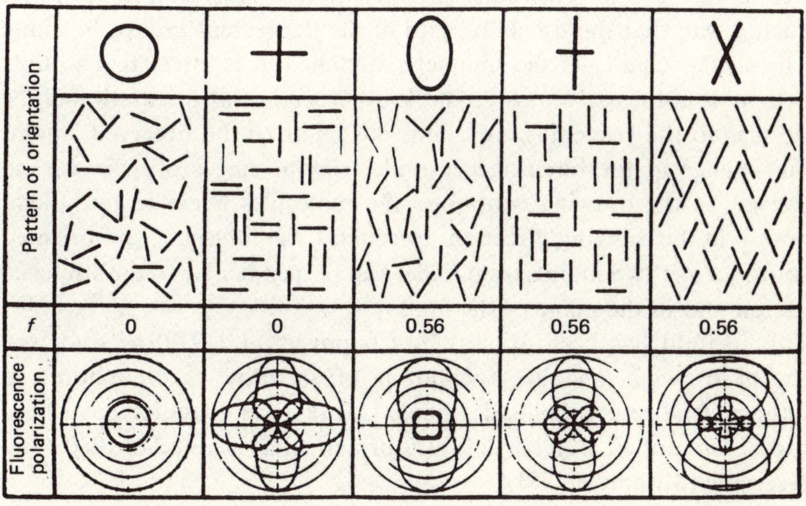

Figure 6.11. Angular distributions of the polarized component of fluorescence $I_{\parallel}$, emitted from the fluorescent compound dispersed in uniaxially stretched poly(vinyl alcohol) film at the draw ratio of (1) 1, (2) 1.08, (3) 1.3, (4) 1.6, (5) 2.0 and (6) 5.0. (From Nishijima 1970.)

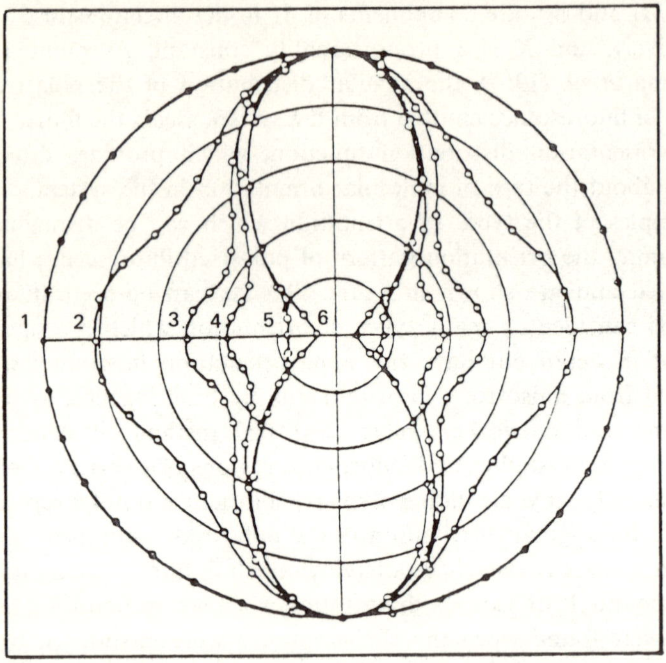

dehydration of either poly(vinyl alcohol) or poly(vinyl chloride). Thermal dehydrochlorination or dehydration normally introduces sequences of double bonds into the backbone of the chain and these emit fluorescence in such a way that the oscillator axes of the fluorescent groups lie along the molecular chain. As the sample is stretched, it is interesting to note that there is in the early stages a marked perpendicular orientation of the molecules to the stretching axis. This is similar to the observed 'anomalous' change in the birefringence under similar stages of deformation. However, as the drawing continues, the molecules begin to line themselves up in the drawing direction. In general, the molecules are oriented primarily along two main axes parallel and perpendicular to the stretched direction and in the plane of the film.

This method has been utilized by Chapoy *et al.* (1979) to study the effect of drawing on the orientation of intrinsic chromophores in poly(phenylene terephthalimide) fibers. More recent studies have been reported on drawn polyethylene (Fuhrman and Hennecke 1980) and polystyrene (Monnerie 1981).

7

Excimers and exciplexes

7.1 Introduction

7.1.1 *Donor–acceptor complexes*

In fluid solution, and indeed in the solid phase, molecules can form molecular complexes. If an electron acceptor A, which is a Lewis acid, is added to a solution of a molecule which is an electron donor D (i.e., a relative Lewis base), a weak complex is formed which can often be detected in the absorption spectrum. A new optical absorption band appears which is characteristic of neither A nor D, and this is known as a charge-transfer (CT) absorption transition. It is attributed to a donor–acceptor (DA) complex formed by the partial or complete transfer of an electron from D to A. An example of such a charge-transfer absorption spectrum is given in figure 7.1. The theory of donor–acceptor interaction

Figure 7.1. Charge-transfer absorption spectrum of donor–acceptor complex of 1,4-dimethylnaphthalene and *p*-chloranil. (After Aladekomo and Birks 1965.)

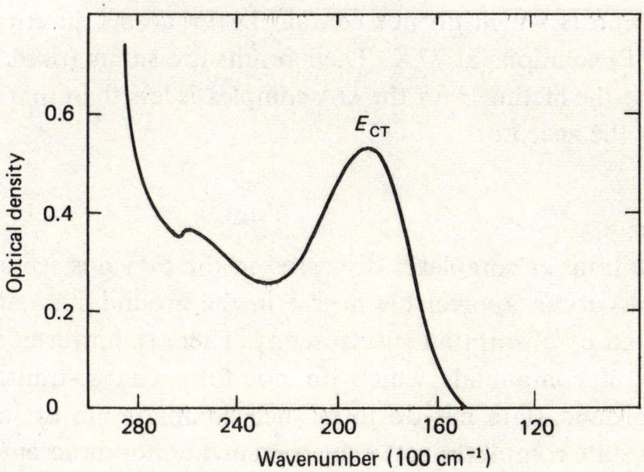

and charge-transfer spectra has been developed by Mulliken (1950*a*, *b*, 1951, 1952, 1955) in a series of papers, and an excellent summary appears in Birks (1970). The energy E_{CT} of the maximum of the absorption transition is related to the ionization potential of the donor, I_D, and the electron affinity of the acceptor. For example, for aromatic hydrocarbons a linear relationship has been observed by McConnell *et al.* (1953) for a series of DA complexes with a common acceptor following the equation

$$E_{CT} = A'I_D - B' \tag{7.1}$$

This linear relation has been verified for many sets of donor–acceptor complexes with a related series of donors and a common acceptor in a given solvent.

In addition to showing changes in the absorption spectrum, donor–acceptor complexes have also been shown to emit radiation as phosphorescence or fluorescence. Fluorescence is usually characteristic of the DA complex itself, while phosphorescence may be characteristic of either the DA complex or the individual donor or excited acceptor A. The fluorescence spectrum from a DA complex is a broad structureless band, while phosphorescence may exhibit structure if it comes from either the donor or the acceptor, but will be structureless if it comes from the DA complex. The fluorescence lifetimes are characteristic of the complex and are unrelated to those of either the donor or the acceptor, and are usually in the range of 2–10 ns. The spectra are structureless and are approximately mirror-symmetric to the corresponding CT absorption spectrum.

The phosphorescence lifetimes of DA complexes are usually less than those of the donor molecules, and are red-shifted from that of the donor due to the charge-transfer interaction. For example, Iwatawa *et al.* (1967) have observed phosphorescence lifetimes of DA complexes of durene and hexamethyl benzene with 1,2,4,5-tetracyanobenzene (TCNB) and phthalic anhydride, two acceptors which do not contain heavy atoms, in ethylether/isopentane (EP) solutions at 77 K. Their results are summarized in table 7.1. In this case the lifetime τ for the DA complex is less than that of either the donor or the acceptor.

7.1.2 *Exciplexes*

The charge-transfer complexes discussed in the previous section are those which exist to an appreciable degree in the ground state and hence can be detected by absorption spectroscopy. There is, however, an increasing number of compounds which do not form charge-transfer complexes in the ground state but do form such complexes in excited states. Such excited state complexes between an excited donor or acceptor and a ground state acceptor and donor, respectively, are described as *exciplexes.* Similarly, a complex between an excited molecule and a ground state molecule of the same species is called an *excimer*, the term being derived from *excited dimer.* An excimer is thus a one-to-one exciplex formed from two molecules or groups of the same species. It is important to distinguish between excited states of dimers which have appreciable binding energy in the ground state and exciplex (or excimers) which by definition have no binding in the ground state. An energy level diagram (after Weller 1970) which illustrates this point is shown in figure 7.2. The emission spectrum from two molecules A and B which are capable of forming an exciplex will depend on the distance r_{A-B} between the two groups.

Table 7.1. *Phosphorescence lifetimes* (s) *of donors* $(\tau_T)_D$, *acceptors* $(\tau_T)_A$, *and DA complexes* $(\tau_T)_{DA}$, *in EP solutions at 77 K*

Donor	$(\tau_T)_D$	Acceptor	$(\tau_T)_A$	$(\tau_T)_{DA}$
Durene	5.9	TCNB	3.2	1.9
Durene	5.9	PA	1.4	0.78
Hexamethylbenzene	5.6	TCNB	3.2	0.41
Hexamethylbenzene	5.6	PA	1.4	0.3

Figure 7.2. An energy diagram for an exciplex. E_{DM} is the exciplex binding energy. (After Weller 1970.)

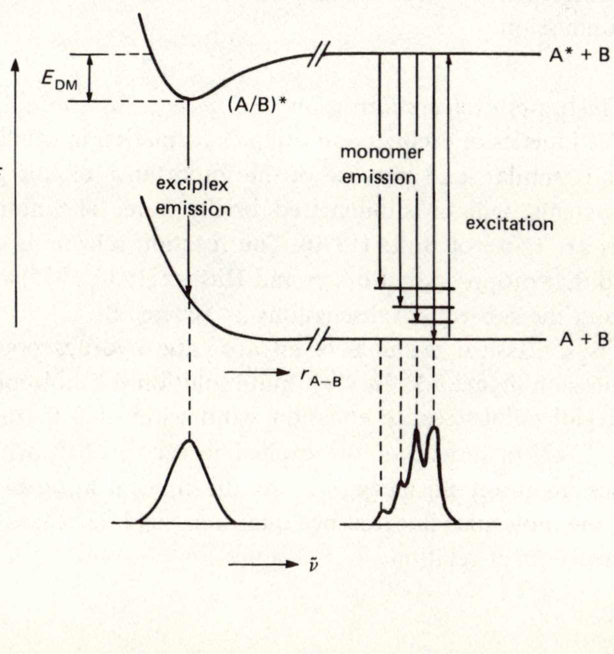

The excited state wavefunction of a one-to-one exciplex formed from a donor molecule D and an acceptor molecule A can be written in the general form:

$$\psi_E = \alpha\psi_1(D^+A^-) + \beta\psi_2(D^-A^+) + \gamma\psi_3(D^*A) + \delta\psi_4(DA^*) \qquad (7.2)$$

In this equation the first two terms on the right-hand side correspond to the charge resonance states and the last two terms to exciton resonance states. The relative contributions of these states to the total wavefunction will depend on the relative ionization potential and electron affinities of the molecules involved. A useful classification of the different types of excimers which can be formed with aromatic hydrocarbons has been given by Birks (1970).

The most common method of studying excimers and exciplexes is by means of their fluorescence spectra. For example, in relatively concentrated solutions of aromatic hydrocarbons such as pyrene, a new, short-lived broad band red-shifted from the normal fluorescence band is observed (figure 7.3). As can be seen in the figure, the intensity of the excimer fluorescence increases with the concentration of the aromatic hydrocarbon.

Similarly, exciplex formation can be followed by fluorescence spectroscopy. Figure 7.4 shows the fluorescence spectrum of solutions of diethylaniline with various concentrations of biphenyl. At concentrations of only 0.3 mol l^{-1} the fluorescence corresponding to biphenyl is almost completely quenched in favor of that of the exciplex. Note that all of the curves pass through a common point which is characteristic of this type of exciplex emission.

7.1.3 *Kinetics of exciplex formation*

The kinetics of excimer and exciplex formation in small molecule encounters is similar, and because of the importance of this process in polymer systems, will be summarized briefly here. The notation and derivations are those of Birks (1970). The reaction scheme is essentially identical to that proposed by Förster and Kasper (1954, 1955) and will be referred to in the subsequent discussions as Scheme I.

Pyrene is a classical example of an aromatic hydrocarbon showing excimer emission (figure 7.3). In very dilute solution the molecule shows a structured violet fluorescence emission band with an 0–0 transition at 27 000 cm^{-1}, characteristic of the excited molecule $^1M^*$, which has a fluorescence quantum efficiency, q_{FM}. As the concentration of pyrene is increased, the molecular fluorescence quantum yield decreases according to the Stern–Volmer relation:

$$\Phi_{FM} = \frac{q_{FM}}{1 + [^1M]/[^1M]_h} = \frac{k_{FD}[^1M^*]}{I_0} \tag{7.3}$$

where $[^1M]_h$ is the half-value concentration at which $\Phi_{FM} = 0.5q_{FM}$. The parameter k, which is equal to $1/[M]_h$ is the Stern–Volmer coefficient of concentration quenching. As the monomer fluorescence is quenched, a broad structureless blue fluorescence appears with a peak maximum corresponding to an energy of about 72 kJ below that of the monomer M_0. On the other hand, the excimer fluorescence quantum yield Φ_{FD} increases with the concentration of pyrene monomer according to the relationship

$$\Phi_{FD} = \frac{q_{FD}}{1 + [^1M]_h/[M]} = \frac{k_{FD}[^1D^*]}{I_0} \tag{7.4}$$

where q_{FD} is the excimer fluorescence quantum efficiency defined as the

Figure 7.3. Fluorescence spectra of pyrene solutions in cyclohexane. Intensities normalized to a common value of Φ_{FM}. A, 10^{-2} M; B, 7.75×10^{-3} M; C, 5.5×10^{-3} M; 3.25×10^{-3} M; E, 10^{-3} M; G, 10^{-4} M. (After Birks and Christophorou 1963.)

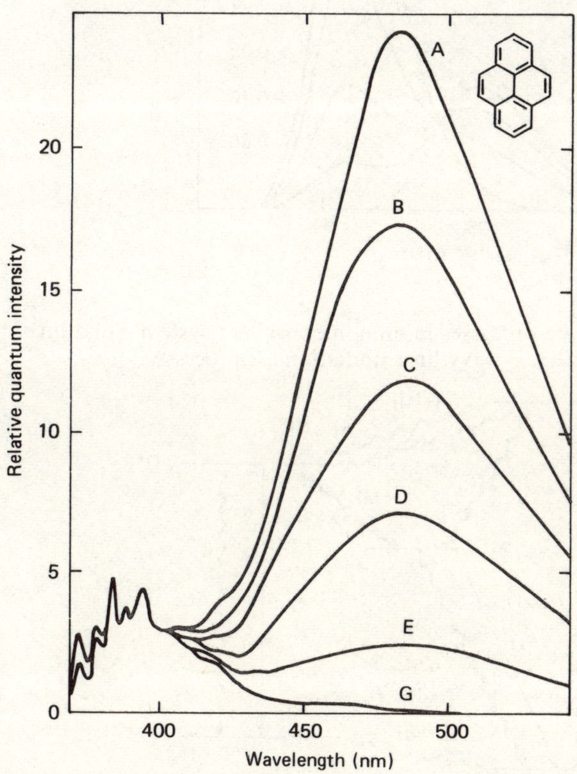

limiting value of Φ_{FD} as $1/[^1M]\rightarrow 0$. From this we can write

$$\Phi_{FD}/\Phi_{FM} = (q_{FD}/q_{FM})([^1M]/[^1M]_h) \tag{7.5}$$

A kinetic scheme can be derived using the various processes and rate constants as defined in the Jablonski diagram (figure 7.5). To simplify the analysis, the following supplementary quantities are defined:

$$k_M = k_{FM} + k_{TM} + k_{GM} = k_{FM}/q_{FM} = 1/\tau_M \tag{7.6}$$

$$k_D = k_{FD} + k_{TD} + k_{GD} = k_{FD}/q_{FD} = 1/\tau_D \tag{7.7}$$

Figure 7.4. Exciplex fluorescence. Fluorescence spectra of toluene solutions of diethylaniline for various added concentrations (C_A) of biphenyl. (After Knibbe and Weller 1976.)

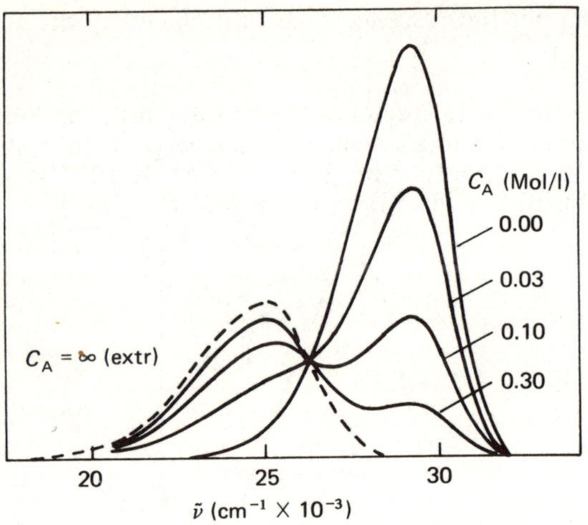

Figure 7.5. Rate processes in monomer/excimer system. Solid lines, radiative processes; wavy lines, radiationless processes.

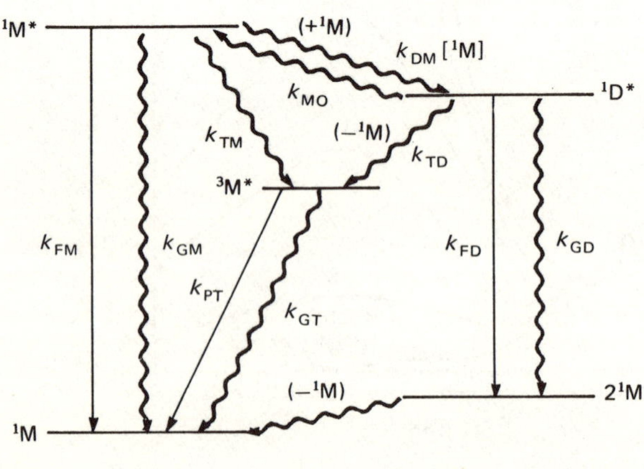

where τ_D is the excimer fluorescence lifetime and q_{FD} is the excimer fluorescence quantum efficiency. For steady-state excitation with light of intensity I_0, the rate equations are:

$$d[^1M^*]/dt = I_0 - (k_M + k_{DM}[^1M]) + k_{MD}[^1D^*] \tag{7.8}$$

$$d[^1D^*]/dt = k_{DM}[^1M][^1M^*] - (k_D + k_{MD})[^1D^*] \tag{7.9}$$

Under photostationary conditions where $d[^1M^*]/dt = d[^1D^*]/dt = 0$ the following relationships can be obtained:

$$\frac{[^1D^*]}{[^1M^*][^1M]} = \frac{k_{DM}}{k_{MD}+k_D} = K_{eq} \tag{7.10}$$

$$K = 1/[^1M]_h = (k_D/k_M)K_{eq} \tag{7.11}$$

$$\frac{\Phi_{FD}}{\Phi_{FM}} = \left(\frac{q_{FD}}{q_{FM}}\right)\left(\frac{[^1M]}{[^1M]_h}\right) = \left(\frac{k_{FD}}{k_{FM}}\right)K_{eq}[^1M] = K_1[^1M] \tag{7.12}$$

From steady-state observations of the fluorescence quantum yield, one can obtain the parameters q_{FM}, q_{FD}, Φ_{FM}, Φ_{FD}, K_1 and K_{eq}. However, to evaluate the individual rate constants, it is necessary to make fluorescence lifetime measurements.

7.1.4 *Lifetime measurements*

If we consider excitation by a δ function light flash which produces an initial concentration $[^1M^*]_0$ of excited molecules at $t = 0$, the rate equations are:

$$\frac{d[^1M^*]}{dt} = k_{MD}[^1D^*] - X[^1M] \tag{7.13}$$

$$\frac{d[^1D^*]}{dt} = k_{DM}[^1M][^1M^*] - Y[^1D^*] \tag{7.14}$$

where

$$X = k_M + k_{DM}[^1M] \tag{7.15}$$

$$Y = k_D + k_{MD} \tag{7.16}$$

Solving these and applying the initial conditions $[^1M^*] = [^1M^*]_0$, $[^1D^*] = 0$, at $t = 0$, we obtain:

$$[^1M^*] = \frac{[^1M^*]_0}{\lambda_2 - \lambda_1}[(\lambda_2 - X)e^{-\lambda_1 t} + (X - \lambda_1)e^{-\lambda_1 t}] \tag{7.17}$$

$$[^1D^*] = \frac{k_{DM}[^1M][^1M^*]_0}{\lambda_2 - \lambda_1}(e^{-\lambda_1 r} - e^{-\lambda_2 t}) \tag{7.18}$$

where

$$\lambda_{1,2} = \tfrac{1}{2}[X + Y \mp \{(Y - X)^2 + 4k_{MD}k_{DM}[^1M]\}^{1/2}] \tag{7.19}$$

The quantum intensity of molecular fluorescence at time t, which defines the molecular fluorescence response function is

$$i_M(t)=\frac{k_{FM}[^1M^*]}{[^1M^*]_0}=\frac{k_{FM}(\lambda_2-X)}{\lambda_2-\lambda_1}(e^{-\lambda_1 t}+Ae^{-\lambda_2 t}) \tag{7.20}$$

where

$$A=(X-\lambda_1)/(\lambda_2-X) \tag{7.21}$$

and the excimer fluorescence response function is

$$i_D(t)=\frac{k_{FD}[^1D^*]}{[^1M^*]_0}=\frac{k_{FD}k_{DM}[^1M]}{\lambda_2-\lambda_1}(e^{-\lambda_1 t}-e^{-\lambda_2 t}) \tag{7.22}$$

7.2 Experimental studies of excimer and exciplex kinetics of small molecules

7.2.1 *Experimental methods*

A number of experimental methods have been developed to determine the values of the rate constants from experimental procedures. These include pulse fluorometers, phase and modulation fluorometers and photon sampling fluorometers, a number of which are now commercially available. The most powerful methods developed in the recent past have involved the use of pulsed lasers as light sources. Detailed discussions of the treatment of data by these various methods are beyond the scope of this book. Extensive studies have been made of the conditions under which excimers are formed and dissociated in solution. An excellent review has been published by Klöpffer (1973).

The absence of excimer emission at room temperature does not imply the absence of excimer formation because if the excimer has only a small enthalpy of formation, rapid dissociation may occur by thermal motion before emission can take place. Furthermore, if the lifetime of the excited monomer is short, fairly high concentrations of the excimer will be needed in order to observe its formation. Concentration quenching of fluorescence is often caused by excimer formation, followed in some cases by rapid photochemical reaction, for example in the case of anthracene which forms a photodimer.

Since the formation of small-molecule excimers in solution is frequently diffusion controlled, it will be affected by the viscosity of the solvent. In general, increasing the temperature will favor association due to the reduction in solvent viscosity but at high temperature the situation will be reversed and thermal dissociation of the excited dimer to the excited and non-excited monomer will predominate.

Case 1. If the temperature is high enough so that dissociation and association are rapid compared to normal deactivation processes, an equilibrium will be set up between the excimer and the monomer, which leads to a decay which is single exponential and includes the sum of both monomer and excimer emission.

Case 2. At low temperatures, on the other hand, when excimer dissociation is negligible, two exponential decays are observed and the decay law for the monomer will follow the sum of two exponentials according to (7.20), while that for the excimer will include the difference according to (7.22).

A rise in pressure will increase the rate of excimer formation and if the experiments are carried out in the dissociation equilibrium range, one can calculate the distance between the molecular planes in the excimer from the pressure-induced decrease in the molar volume. From measurements such as these the distance of the molecular planes in the naphthalene excimer have been estimated at approximately 3 Å (Braun and Förster 1966). Experimental studies have shown that nearly all planar aromatic molecules form excimers and theoretical studies have identified the excimer structure as being a coplanar sandwich structure of the type shown in figure 7.6. As indicated from (7.2), the interaction energy E_{DM} must be described in terms of a combination of both charge resonance and excitation resonance interactions. Its significance is further illustrated by the energy level diagram (figure 7.2).

7.2.2 *Intramolecular excimers*

Because of the importance of excimers to both photophysical and photochemical processes, and the ease with which they can be studied by luminescence measurements, they have been widely investigated in recent years. Of particular interest to the study of polymeric molecules has been the extended series of investigations into excimer formation in bichromophoric molecules in which the excimer is formed between two similar chromophores within the same molecule. These

Figure 7.6. Geometry of the naphthalene excimer.

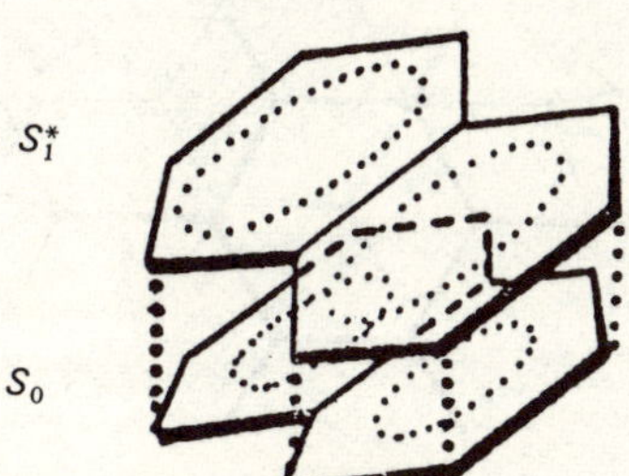

intramolecular excimers offer the special advantage that they can be studied at very low concentrations in solution where normal bimolecular excimers do not form. The first studies published on intramolecular excimers were those of Hirayama (1965) on diphenyl and triphenyl alkanes, and a paper on paracyclophanes and vinyl polymers by Vala, Haebig and Rice (1965).

A further suggestion that the fluorescence of polystyrene (PS) might be caused by inter- and intramolecular excimers was published in 1963 by Yaneri, Bovey and Lumry (1963).

One can, in general, distinguish two types of molecular structures in which the formation of excimers would be possible. One type includes molecules in which the two chromophores are connected by a flexible, atomic chain. In such a molecule only certain conformations of the flexible chain will be favorable to excimer formation and, in order to exhibit excimer emission, these conformations must be reached in appreciable quantities within the lifetime of the excited state of one of the chromophores. In molecules of the second type, a sandwich structure is fixed by a more-or-less rigid link in both the excited and the ground state, such as in the paracyclophanes. The probability of excimer formation in this case will obviously be relatively independent of solvent and other experimental conditions. The two types of excimers are shown schematically in figure 7.7. Polymer systems can show one or both of these types of excimers.

In Hirayama's early studies of the intramolecular excimer formation in

Figure 7.7. (*a*) Probable excimer structure of 1,3-diphenyl propane; (*b*) (4,4)-*para*cyclophane. (From Klöpffer 1973.)

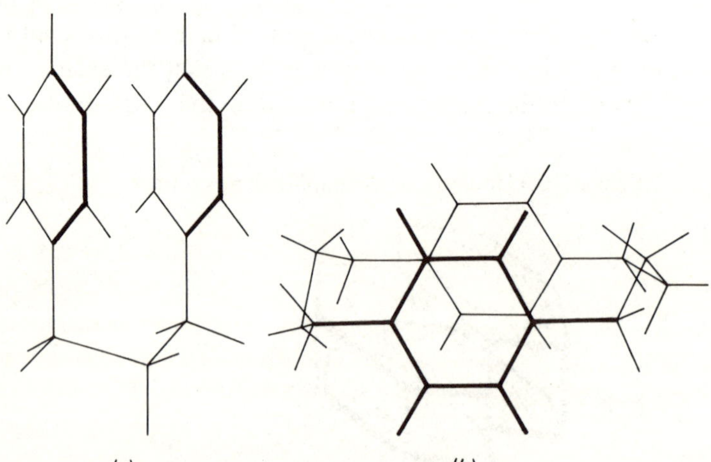

(*a*) (*b*)

di- and triphenyl alkanes, it was observed that excimers were formed only if the interacting chromophores are separated by three carbon atoms. This so-called $n=3$ rule was later confirmed for a number of chromophores other than the benzene ring. Recently, however, a number of exceptions to this general rule have been experimentally discovered which confirm the expectation that this rule cannot apply generally in many bimolecular systems, and certainly not in polymer systems. For 1,3-diphenylpropane, the excimer should have a gauche–gauche conformation (figure 7.7) and in the ground state this will be a repulsive conformation since the aromatic rings have thicknesses of about 3.2 Å. Studies of molecular models of the gauche–gauche arrangement suggest that the spacing of the rings in the excimer should be 2.54 Å, however it seems unlikely that such close spacings would actually occur and as indicated previously, the spacing of 3 Å has been estimated for the naphthalene excimer from pressure experiments.

Crystallographic studies on ground-state charge-transfer complexes have also shown typical interplanar distances of 3.2 Å which are also consistent with the measured values for excimers. The reason for the $n=3$ rule, thus, seems to be that for values of $n<3$, there are no conformations available to the molecule which will allow a completely coplanar arrangement of the aromatic rings. For values of $n>3$, although such conformations do exist, their probability is so much lower with respect to alternative conformations that the concentration of excimer sites is too low to be observed.

Longworth and Bovey (1966) studied meso and *d,l*-2,4-diphenyl pentane, model dimers for PS, whose ground state conformations are known in solution from NMR analysis. The possible conformations are shown in figure 7.8. The most stable conformations in solution are structures *a* and *b*, which are clearly not suitable for excimer interaction. Only structures *d*, *e* and *f* have conformations suitable for excimer formation and these are shown to have energies considerably higher than structures *a* and *b*. The fact that the *d,l*-2,4-diphenyl pentanes do not show excimer emission is presumably due to the fact that the energy required to form these higher energy conformations provides too great a barrier to the rotational movements necessary to form the excimer within the lifetime of the excited state of the phenyl chromophore. More recent work (Bokobza *et al.* 1977) indicates that the meso compound does form an excimer.

A similar effect was observed by Klöpffer (1969*a*, *b*) comparing 1,4-bis-carbazolyl butane with 1,3-bis-carbazolyl propane. Again, for $n=4$, no excimer emission is shown in the carbazolyl derivative. Furthermore, in solid solution at 77 K, all excimer-forming derivatives show only monomer fluorescence. Since the rotational motion required to move from one conformation to another is believed to be completely inhibited at 77 K, this evidence strongly suggests that molecular motion is required for the formation of the excimer. Furthermore, the excimer conformations are not those of lowest energy.

Further confirmation of this postulate was demonstrated by Hoyle *et al.* (1978) using time-resolved fluorescence spectroscopy. Figure 7.9 shows the emission spectra from 1,3-bis-carbazolyl propane taken at various times after the initiating light pulse. It can be seen that initially only emission from the monomer is observed and then after several nano-seconds the emission from the excimer begins to appear. Since the rotational relaxation times for the backbone C–C bonds are of the order of 2–4 ns, this supports this theory of excimer formation.

Since the original work of Hirayama, a number of investigators have studied bichromophoric systems to further elucidate the structural fea-

Figure 7.8. Conformations of 2,4-diphenylmethanes: (*a*) preferred ground-state conformation of *meso* isomer (V); (*b*), (*c*) preferred ground-state conformations of *d*, *b* isomer (IV); high-energy conformations favourable for excimer interaction of phenyl groups are (*d*) and (*e*) for *meso* isomer and (*f*) for *dl* isomer. (From Longworth and Bovey 1966.)

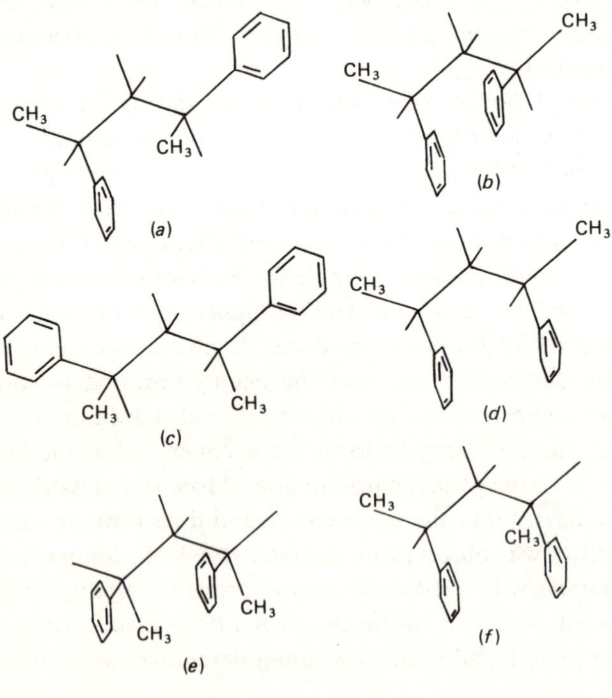

tures of molecular exciplexes. For example, Chandross and Dempster (1970) carried out a study of symmetrical and unsymmetrical dinaphthyl alkanes and showed that only the symmetrical isomers exhibit strong excimer formation. Molecular models indicate that in the unsymmetrical dinaphthyl compounds only about 25% overlap of the naphthalene rings is possible and hence the stabilization energy is not sufficient to lead to excimer formation and emission.

When excimers or exciplexes are formed from large conjugated ring systems such as pyrene, there is evidence that the $n=3$ rule no longer applies. For example, Kühnle and Zachariasse (1976*a*, *b*) found excimer formation for $n=2$ to 6 and 8 to 16 in the 1,*n*-dipyrenyl alkanes. Also, there is a blue shift in the excimer maximum with decreasing temperature in poly(1-vinyl pyrene) and it has been postulated that excimer formation in pyrene is possible from a non-equilibrium or displaced position of the aromatic rings. It was concluded that the pyrene chromophore is large enough so that even a partial overlap of the ring structures leads to excimer formation.

The second type of excimer occurs in molecules exemplified by the paracyclophanes which have the general formula:

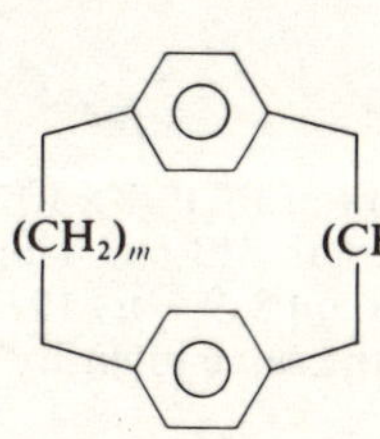

I

They were synthesized by Cram and Steinberg (1951) who investigated the changes in the uv absorption spectra when *m* and *n* were altered. If $m=n=2$, closest contact is achieved between the benzene rings, and the electron clouds penetrate one another so that even the rings are distorted. The smallest distance between the benzene rings is 2.8 Å and the largest is 3.1. If $m=n>4$, no interaction between the chromophores is observed in the ground state from the uv absorption spectra. However, if *m* or $n<3$, there is a red shift in the spectrum, indicative of a ground state interaction between the two benzene rings.

The emission spectra from the paracyclophanes which show strong ground state interaction cannot be attributed to excimers since dissociation into the non-interacting ground state in these molecules is not possible. In 2,2′-paracyclophane the emission spectrum is approximately a mirror image of the long wavelength absorption band which is characteristic of the interactive ground state. Furthermore, both absorption and fluorescence at low temperatures show vibrational fine structure, while pyrene excimer emission, for example, is structureless, even at 4 K. On the other hand, 4,4′-paracyclophane shows true excimer fluorescence and no monomer component. Since the ground-state interplanar distance is too large to allow excimer formation without closer approach of the rings, the high efficiency of excimer formation is astonishing. On the other hand, paracyclophanes with *m* or $n > 4$ do not show excimer fluorescence, which may be due to the greater instability of the particular chain conformation favorable for excimers.

7.2.3 *Intramolecular exciplexes*

Recently there has been considerable interest in intramolecular exciplexes formed in molecules containing two chromophores of different types. For example, Chandross and Thomas (1971) have studied molecules of the general structure II and have demonstrated fluorescence from

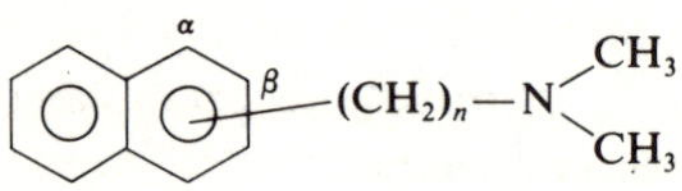

II

Figure 7.9. Time resolved fluorescence spectra of BCP (2×10^{-4} M) in degassed benzene at 23°C: excitation wavelength 313 nm. Time settings from lamp maximum (ns): (*a*) 0–0.23, (*b*) 8–9.4, (*c*) 19–24, (*d*) 139–290. The spectra are adjusted to fit on the same intensity scale. (From Hoyle *et al.* 1978.)

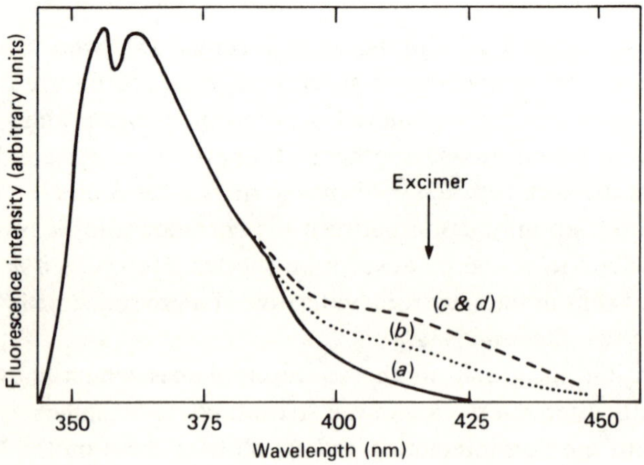

exciplexes formed from naphthalene and the dimethyl group. They showed that exciplex formation was more favorable when the naphthalene ring was substituted in the α position rather than in the β, and that the highest efficiency was obtained with $n=3$. However, the steric requirements were not so specific as in the case of excimers since exciplex emission was observed for both $n=2$ and $n=4$ in certain cases. In general, exciplexes are characterized by greater charge-transfer components and greater solvent dependence due to the more polar nature of the complex.

Itoh *et al.* (1977) observed emission from both an intermolecular exciplex and a charge-transfer complex in the system 9,10-dicyanoanthracene-$(CH_2)_3$-naphthalene and in the corresponding intermolecular system. At room temperature there was no difference between the absorption spectrum of this compound and the sum of the spectra of the respective components, so emission was attributed to the exciplex state. However, below about 170 K a new emission arose from an absorption seen in both the excitation and absorption spectrum which was attributed to an intermolecular ground-state charge-transfer complex. The existence of two different states was ascribed to the fact that the internal conversion from a Franck–Condon charge-transfer state to an identical fluorescence state is sterically forbidden. Further calculation showed that the most stable geometries for the two states involved a transition that is possible for two free molecules, but forbidden by the methylene bridge. Studies by Brimage and Davidson (1972) of naphthalene and diethylamine chromophores attached through methylene bridges showed that the compound where $n=1$, in which exciplex formation was sterically forbidden, had the same reactivity towards quenching by tertiary alkyl amines in solution as 1-methyl naphthalene. However, the reactivity of compounds $n=2$ and $n=3$ was negligible, apparently because of highly efficient quenching of the naphthalene S_1 state by intermolecular exciplex formation.

Of course, not all exciplexes decay by the emission of radiation. For example, it is now quite well established that a non-emitting excimer is involved in the dimerization of anthracene and an extensive discussion of this subject by Lewis and Ware suggests that a non-radiative excimer or exciplex state may be the precursor for many photochemical reactions.

Further confirmation of the importance of molecular mobility in the formation of exciplexes comes from work by Farid *et al.* (1979) who studied bichromophoric molecules of the general structure:

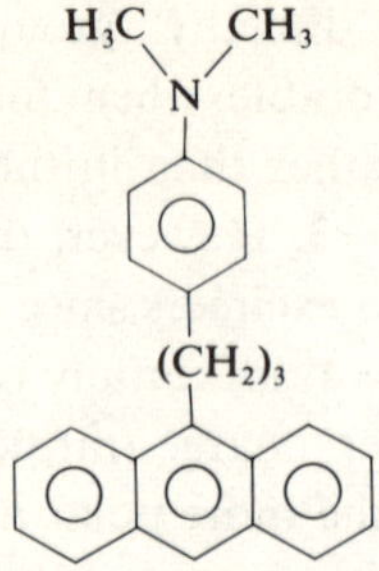

III

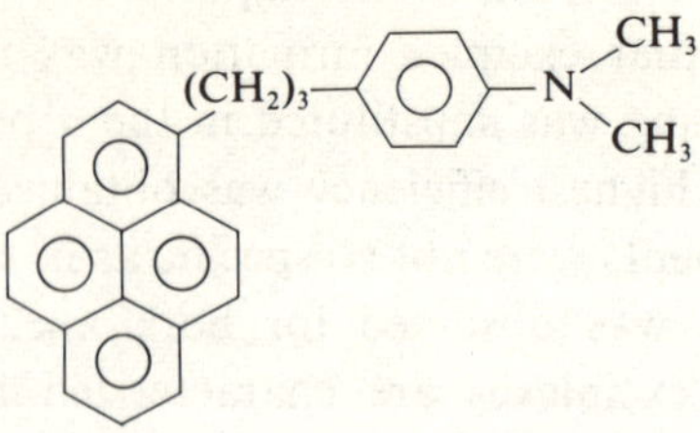

IV

These compounds, which were first synthesized and studied by Okada *et al.* (1972) and Masaki *et al.* (1976) show strong exciplex emission in fluid solvents such as toluene. Exciplex emission intensity is relatively higher in IV than in III. However, neither of these compounds, when dissolved in PS, showed any evidence of exciplex emission. This suggests that the aniline group is normally in a conformation well removed from the aromatic chromophore and that the PS matrix effectively inhibits the motions which are necessary to form the exciplex.

Based on the lifetimes and intensities of the exciplex emission, Farid *et al.* (1979) estimated the frequency of folding of such molecules to be of the order of $2 \times 10^8\ s^{-1}$ when these molecules are dissolved in fluid solvents such as toluene. From the lack of exciplex emission in these compounds when dissolved in solid PS below its glass transition, they concluded that the rotational processes required to reach the exciplex configuration are at least two orders of magnitude slower in the polymer matrix than in a mobile fluid.

7.2.4 *Triplet excimers and exciplexes*

Triplet excimers appear to occur less commonly than singlet excimers and, in fact, the first experimental evidence for triplet excimers was made in 1966 by Castro and Hochstrasser (1966). At low temperatures a new structured emission was seen from pyrene crystals where the required sandwich configuration was present. At temperatures greater than 200 K an observed broad band was assigned to a different type of triplet excimer. Later, Avouris and El-Bayoumi (1974) reported evidence for the existence of an intermolecular triplet excimer in 1,3-diphenyl propane during the softening of a glassy matrix.

In order to study triplet excimers, phosphorescence measurements must be made using techniques where the prompt emission is eliminated. There are two major difficulties about the observation of intramolecular triplet excimers. The first is that the binding energy of triplet excimers is expected to be very small because both excitation resonance and charge transfer should be weaker in the triplet state than in the singlet state. This consideration would be expected to be more important in inter- rather than intramolecular excimers and hence the most promising experiments will likely be those involving intramolecular excimer formation. Secondly, it is very difficult to observe phosphorescence in the liquid solutions which are necessary to allow the formation of the triplet excimer.

Since the contribution of resonance to the binding energy of triplet excimers is small, an effective excimer interaction is to be expected only if the charge-transfer interaction is greater than in aromatic compounds like benzene and naphthalene, that is, if the difference between ionization energy and electron affinity is small. For this reason, triplet excimers have been observed in halogenated benzene derivatives which have a greater electron affinity than benzene. Studies by Klöpffer (1973) suggest that the low temperature phosphorescence of poly(*N*-vinylcarbazole) (PVCz) can be attributed to a triplet excimer. In this case, the basic unit of the polymer is a good electron donor with an ionization potential of 7.4 eV and it has a relatively high triplet state energy, 292 kJ. The phosphorescence is a broad, structureless band centered at about 20 000 cm^{-1}. The phosphorescence decay is non-exponential, as shown in figure 7.10. After the initial rapid decay, all of the samples shown in figure 7.10 show a single exponential ratc corresponding to a lifetime of 8 s, which is attributed by Klöpffer to the decay of the triplet excimer.

7.3 Excimers in polymer systems

7.3.1 *General considerations*

Synthetic polymers prepared by the free radical addition of vinyl monomers have a general structure of the form given below:

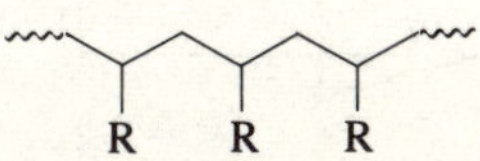

V

in which the substituents R are separated by three carbon atoms. Links with two or four carbon atoms which will occur due to head-to-head or tail-to-tail additions occur only as defects and are rarely observed in polymer systems. If the groups, R, are capable of forming excimers or exciplexes it is obvious that this structure is very similar to those which obey Hirayama's $n=3$ rule, and, provided that the chain is sufficiently flexible so that free rotation is possible about the carbon–carbon bond, one should expect to find excimer emission from such polymers.

The most common polymer of this type is, of course, PS (VI) which is the polymer in which intramolecular excimers were first observed. It is therefore not surprising that the spectra, both in emission and absorption of solutions of PS, have many of the features consistent with those of the 1,3-diaryl propanes (VII) which closely resemble the structure of the two

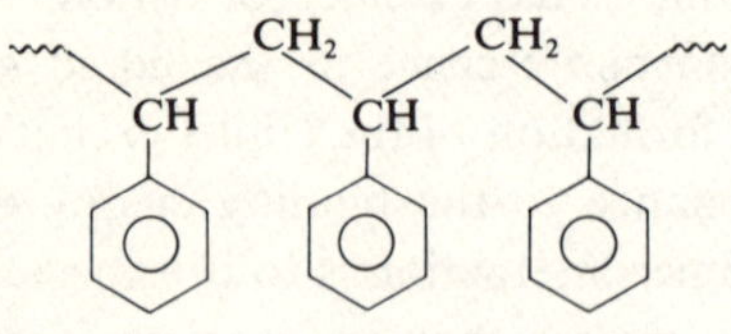

VI

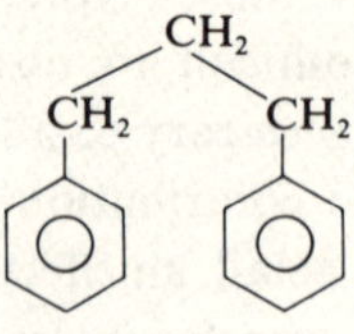

VII

neighboring side groups of the polymer. In solid solution and in low temperature glasses, emission from PS is primarily due to the monomer (phenyl side group) emission, while in fluid solution, both monomer and

Figure 7.10. Phosphorescence decay of solid poly(vinylcarbazole). Samples at 77 K. (From Klöpffer 1973.)

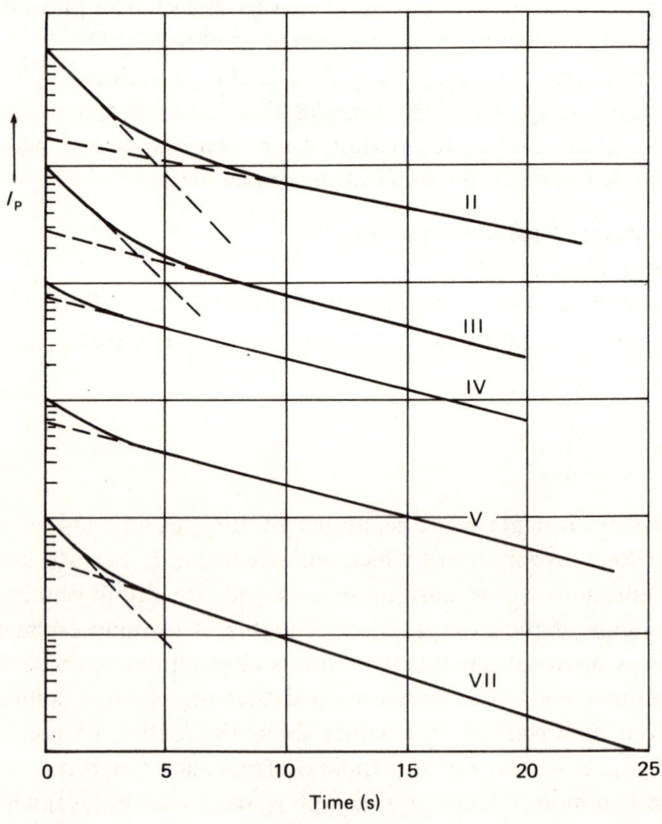

excimer fluorescence are observed. The quantum yield is lower than in corresponding monomeric model compounds.

Excimer formation in polymers has been extensively studied because of its connection with energy migration and molecular mobility in polymer systems. Pre-eminent among early investigations are those of Klöpffer, Fox and co-workers, David and Gueskens, and Frank and Harrah. The vinyl polymers and copolymers which have been studied include PS, poly(1-vinylnaphthalene) (P1VN), poly(2-vinylnaphthalene) (P2VN), poly(1- and (2-naphthylmethacrylate) poly(vinyl toluene), PVCz, poly(acenaphthalene) and poly(1-vinylpyrene).

Excimers are observed in all of these aromatic polymers under a variety of conditions and, particularly in bulk polymers, excimer emission will predominate over the normal emission from monomeric chromophores. There is, however, one major difference between the polymeric excimers and those of the dimeric analogs of the type described first by Hirayama, in that at least in some polymers, an important part of the excitation of the excimer site appears to be the result of singlet energy migration.

It is now generally accepted that in certain solid aromatic vinyl polymers excimer emission occurs after an initial excitation of an aromatic chromophore, followed by intramolecular singlet energy migration, either along the polymer chain, or intermolecular transfer between chromophores in different chains in the bulk phase until the excitation is competitively trapped at a chain conformation which is geometrically suitable for excimer formation. Such a chain conformation is termed an *excimer-forming site.* In fluid solutions of low viscosity, interconversion of chain conformations proceeds relatively rapidly and the lifetime of a particular conformation is limited by collisional processes and the magnitude of rotational barriers with respect to the thermal energy. If the residence time of the random hopping singlet excitation on a particular chromophore at a non-excimer-forming site is long enough, solvent collisions with polymer will cause a rotational sampling of chain conformations leading to excimer formation. However, if the residence time is too short and/or the rotational sampling rate too slow (such as in high viscosity solutions or at very low temperatures), excimer sampling must occur by competitive trapping at a suitable excimer site formed prior to the arrival of the excitation.

Further evidence for this mechanism of excimer formation is given by time-resolved fluorescence spectroscopy. Figure 7.11 (Hoyle *et al.* 1978, Ghiggino *et al.* 1978*c*) shows that the emission from P1VN in solution observed after an interval of 0.2 ns after the initiating pulse is primarily that of the monomeric naphthalene unit and that the excimer emission only begins to predominate after times of 4–8 ns. Studies of the dimer 1,3-dicarbazolyl propane showed the same effect. At longer time intervals, the ratio of excimer to monomer fluorescence did not change, even though they have different lifetimes, because the dissociation equilibrium of excimer to the monomer maintains a constant ratio of the two species.

In the solid state, however, the rotational freedom of the molecule is much reduced and large scale configurational changes are unlikely within the time scale of the experiment. However, there is a possibility that adjacent chromophores will be in a marginal excimer-forming site which is not yet geometrically suitable but with small rotations can reach the desired conformation. Thus one can distinguish two separate types of excimer formation processes in solid polymers, one in which excitation moves to a pre-formed excimer site in the solid, and the second, a temperature-independent excitation migration to a marginal excimer site containing a small geometric barrier. Detailed discussion of energy migration processes will be postponed until chapter 9, and we will consider here primarily those factors which lead to the possibility of excimer formation regardless of the mode of excitation.

7.3.2 *Theoretical treatments of excimer formation in polymer solutions*

Although a great deal of information can be gleaned from studies of the steady-state emission from excimer-forming polymers, an under-

Figure 7.11. Time resolved fluorescence spectra of PVN (4×10^{-3} M) in degassed benzene at 23°C: excitation wavelength 313 nm. Time settings from lamp maximum (ns): (*a*) 0–0.23, (*b*) 8–9.4, (*c*) 19–35, (*d*) 110–290. The spectra are adjusted to fit on the same intensity scale. (From Hoyle *et al.* 1978.)

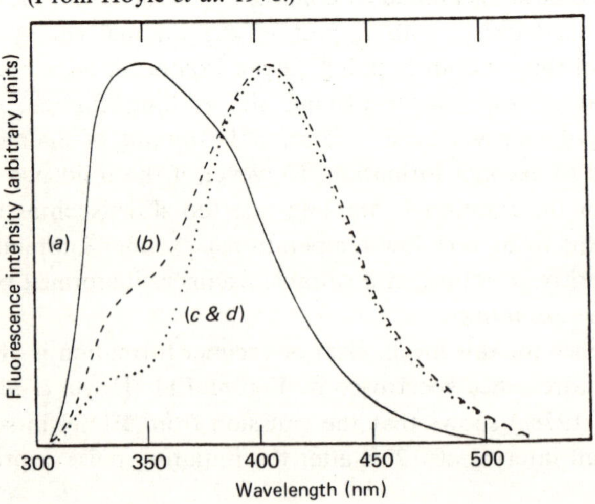

standing of the detailed mechanism requires measurements of the rates of emission of various species. In early studies of excimer emission from polymers, it was generally assumed that the Birks mechanism outlined in §7.1.3 could be applied to polymer systems. However, strictly speaking, this scheme is suitable only for the description of small molecules in dilute solution where the formation of excimers occurs by diffusion-limited processes. In bichromophoric and polymeric compounds, intramolecular excimer sites are created by a sequence of bond rotations and mass diffusion may play a relatively small role except at high concentration. A general scheme (scheme II) for the kinetics of emission from intramolecular excimers in bichromophoric compounds has been developed by

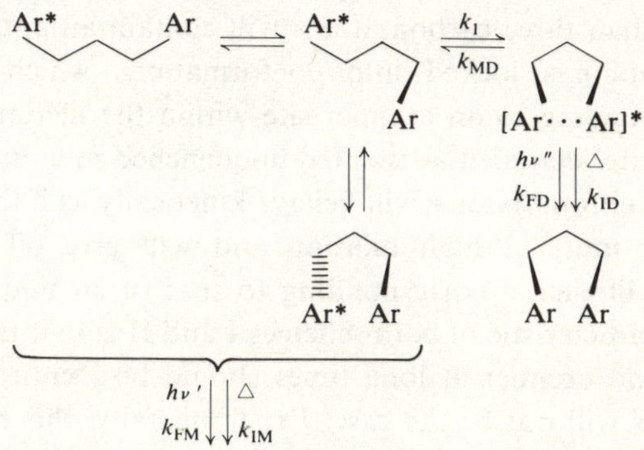

Scheme II

Klöpffer (1973) and improved by Morawetz and his collaborators (Goldenberg *et al.* 1978). Morawetz postulates that excimer formation can only occur in 1,3-bichromophoric compounds via the trans-gauche conformer. If the interconversion of three unassociated conformers is rapid, the Morawetz scheme reduces to Birks kinetics with the exception that

$$k_{DM}[M] = \alpha k_a \tag{7.23}$$

where α is the fraction of conformers present in the trans-gauche conformation before excitation, and k_a is the rate constant for conformational change.

Because scheme II assumes rapid bond rotation it may only apply to polymeric systems at relatively high temperatures where this rapid motion is a reasonable assumption. Furthermore, in polymers like the naphthyl acrylates and methacrylates whose chromophores are separated by more than three atoms, a full description of the kinetics in terms of individual bond rotation is prohibitively complex. Holden and Guillet (1980) have suggested that, in certain polymers, bond rotations are relatively slow and the rate of interconversion of the starting conformations is often lower than the rate of emission from the excited states. In this case a third mechanistic scheme is required to explain the observed effects (scheme III).

$$\begin{array}{ccccc} \text{Isolated unquenched monomer units} & \underset{k_2}{\overset{k_1}{\rightleftharpoons}} & \text{Quenched monomer units} & \underset{k_{MD}}{\overset{k_{DM}}{\rightleftharpoons}} & \text{Excimer} \\ h\nu' \; k_{FM}\downarrow \; \downarrow k_{IM} \; \Delta & & h\nu' \; k_{FM}\downarrow \; \downarrow k_{IM} \; \Delta & & h\nu'' \; k_{FD}\downarrow \; \downarrow k_{ID} \; \Delta \\ M & & M & & M+M \end{array}$$

Scheme III

A feature of scheme III is that any polymer containing chromophores separated by more than three carbon atoms will contain an appreciable fraction of chromophores locked into conformations which cannot achieve that corresponding to an excimer site within the lifetime of the excited state. These are identified as isolated unquenched monomer units in scheme III. These chromophores will behave kinetically as if they were isolated monomeric units at high dilution and will give off normal fluorescence with a lifetime τ corresponding to that of an unquenched monomer unit. A characteristic of both schemes I and II is that the decay time of monomer and excimer at long times should be identical while, with scheme III, this will not be the case. Experimentally, this has been observed, but often ignored in a number of studies of polymer excimers.

Holden and Guillet (1980) have shown that this kinetic analysis is consistent with the observed decay rates for monomer and excimer emission in poly(2-(1-naphthyl)ethyl methacrylate) where adjacent chromophores are 11 atoms apart

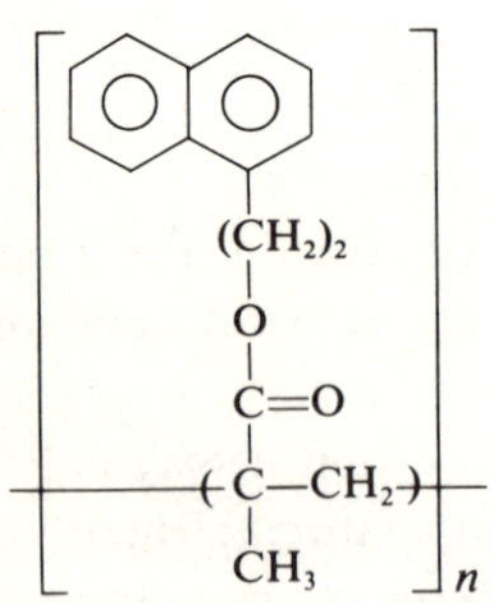

VIII

Additional examples of polymers obeying scheme III are poly(1-naphthyl methacrylate) and poly(2-naphthyl methacrylate) (Holden, Wang and Guillet 1980). In these polymers the monomer and excimer lifetimes differ considerably but the monomer values were the same within experimental error as those of model compounds containing single naphthalene chromophores.

Alternatively, a chromophore in a polymer may be sufficiently far from its neighbors that it cannot transfer its energy to an excimer site. In copolymers of a fluorescent monomer with a diluting monomer such as methyl methacrylate (MMA), fluorophores isolated by MMA units will not form nearest-neighbor excimers, so that they will have longer fluorescence decay times than fluorophores in sequence. Such considerations can be used to explain the results of Ito *et al.* (1977).

As a consequence of the kinetic processes described in these three schemes, the decay of monomer fluorescence in nearly all excimer-forming polymers is kinetically non-single exponential. However, in polystyrene and poly(α-methyl styrene) exponential decays were obtained in fluid solution at room temperature (Ghiggino *et al.* 1978*c*) and it is possible that in this case the steric effect of the chain substantially reduces the tendency of the excimers to dissociate.

The total fluorescence quantum yields of polymers are often small in solution at room temperature (~0.05). It has been suggested that this may be because the chromophores in polymers can undergo temperature-dependent self-quenching, as observed for concentrated solutions of large aromatic molecules. The effect would appear as an extra non-radiative decay mechanism for a photoexcited monomer, and a shortening of the monomer decay time relative to that of the excimer. Conclusive evidence for concentration quenching in a polymer has not been found, although fluroescence decay curves published by Ghiggino *et al.* (1978*b*, *c*) suggest that P1VN may exhibit the effect. Polystyrene may be another example.

It should be pointed out that when interpreting transient fluorescence measurements, it is not necessary to define or assume a particular kinetic scheme. Provided χ_R^2 and other criteria show that the decays are adequately described by dual exponentials, the data should be reported in a form such as (7.24) and (7.25) from which rate constants can be derived once one has established which scheme is valid.

$$I_{FM}(t) = A_1 \exp(-\lambda_1 t) + A_2 \exp(-\lambda_2 t) \quad (7.24)$$

$$I_{FD}(t) = A_3[-\exp(-\lambda_1' t) + \exp(-\lambda_2' t)] \quad (7.25)$$

Some authors have analyzed polymer steady-state emission by observing the effect of quenchers, such as carbon tetrachloride, on emission intensity. In this case, it is frequently observed that the Stern–Volmer plots (1.41), as a function of quencher concentration are often strongly non-linear. It should be pointed out that when both monomer and excimer decays are non-exponential a Stern–Volmer kinetic analysis (which requires an exponential decay of the emitting species) is no longer valid. However, the idea of a fluorescence lifetime τ which is defined as the time required for the population of the excited state to decay to 1/e of its original value is useful and valid, even when the decays are not single exponential. Experimentally, if one can observe significant amounts of monomer emission at long times following the excitation of the chromophore it is usually valid to attribute this to a feedback step such as excimer dissociation. If the long decay times, $\tau_2 = \lambda_2^{-1}$ of the monomer and the excimer coincide, then the Birks mechanism can usually be applied. For example, table 7.2 shows representative values of the rate constants according to scheme I for several polymers and model compounds. It is important to note that the rate constants for formation and dissociation are much lower for intramolecular excimers than for those in small molecule complexes.

In a system of the complexity of a homopolymer in fluid solution the assumption that the rate of excimer formation can be described by a single time-independent rate constant k_{DM} is probably only a crude approximation. In most polymer systems, excimer formation will require electronic energy transfer to the sites as well as bond rotation to create the site, implying that there should be at least two components to the value of k_{DM}. Recently, Ng and Guillet (1981) have reported that the decay parameters for many naphthyl-containing polymers can be better fitted with a sum of three exponentials of the form:

$$I_F(t) = A_1 \exp(-t/\tau_1) + A_2 \exp(-t/\tau_2) + A_3 \exp(-t/\tau_3) \qquad (7.26)$$

rather than (7.24) which is the sum of two exponentials. The use of the three-exponential fit improves both consistency between the monomer

Table 7.2. *Rate constants for excimer formation, excimer dissociation and non-radiative decay for polymers and dimers in fluid solution at 25°C*[a]

		$10^{-7} \times$ rate constant (s^{-1})			
Polymer	Solvent	k_M	$k_{DM}[M]$	k_{MD}	k_D
Poly(1-vinylnaphthalene)	CH_2Cl_2	1.8	9.7	1.8	2.4
Poly(1-vinylnaphthalene)	THF	1.8	8.3	2.7	3.1
Poly(1-naphthyl methacrylate)	CH_2Cl_2[b]	15	59	2.0	14
1,3-Biscarbazole propane	α-MeTHF	6.8	14.8	5.7	2.3

[a] Rate constants defined in Scheme 1: $k_M = k_{DM} + k_{IM}$, $k_D = k_{FD} + k_{ID}$.
[b] At 23 °C.
Source: Holden and Guillet (1980).

and the excimer curves and decreases the error function χ_R^2, the reduced sum of squares of residuals. Similar conclusions have been reported by Holden and Guillet (1980), Roberts *et al.* (1980), Phillips *et al.* (1980) and DeSchryver *et al.* (1981), and it seems likely that future work will bring physical significance to these more complicated fitting functions for the decay of transient emission from polymeric systems.

7.3.3 *Effect of temperature on excimer emission*

For dilute solutions of excimer-forming polymers, excimer emission is usually very weak at low temperatures, below 123 K. As the temperature is increased, the intensity of excimer emission generally increases to a maximum and then falls off. This is illustrated in figure 7.12 which shows representative behavior of three different excimer-forming systems, P2VN, a typical excimer-forming polymer, 2-ethynaphthalene (EN), a model small molecule, and 1,3-di-2-naphthylpropane, a dimeric model compound. The intermolecular excimer from EN shows a maximum in the I_D/I_M ratio at 183 K but those of the intramolecular excimers, if they exist, are at much higher temperatures, well above 25°C. In spite of this, authors have interpreted the temperature dependence using the Birks scheme.

Solutions of the differential equations (7.8) and (7.9) for steady-state illumination give the following relations for the intensity of monomer, and excimer emission.

$$I_M = \frac{k_{FM} I_a (k_D + k_{MD})}{(k_M + k_{DM}[M])k_D + k_{MD}k_M} \tag{7.27}$$

$$I_D = \frac{k_{FD} I_a k_{DM}[M]}{(k_M + k_{DM}[M])k_D + k_{MD}k_M} \tag{7.28}$$

Figure 7.12. Temperature dependence of monomer (I_M) and excimer (I_D) emission intensities, and of I_D/I_M for (*a*) poly(2-vinylnaphthalene) (molecular weight 4.3×10^{-3}, 10^{-4} M in repeating units), (*b*) 1,3-di-2-naphthylpropane (10^{-4} M), and (*c*) 2-ethylnaphthalene (0.26 M) in α-methyl THF. (From Holden and Guillet 1980.)

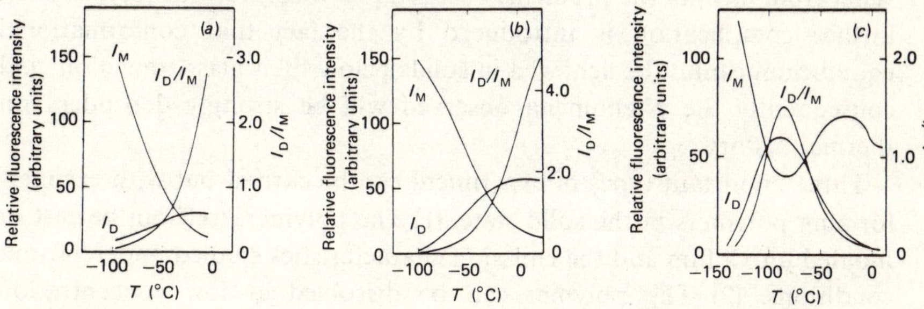

$$\frac{I_D}{I_M} = \frac{k_{FD}k_{DM}[M]}{k_{FM}(k_D + k_{MD})} \tag{7.29}$$

where

$$k_M = k_{FM} + k_{IM}$$

$$k_D = k_{FD} + k_{ID}$$

It is usually assumed that at low temperatures k_{MD}, the rate constant associated with the dissociation of the dimer to monomer is negligible so that the temperature dependence can be ascribed solely to the rate of formation of the dimer corresponding to k_{DM}. At high temperatures, the assumption is made that the reverse is true, so that k_{MD} is very much greater than k_D. In this case, the ratio I_D/I_M becomes a measure of a pseudo-equilibrium between the monomer and the excimer, and thermodynamic parameters can be derived, such as the binding enthalpy in the excimer. Estimates have been made using this assumption for a variety of polymers, and values ranging from 2–5 kcal mol^{-1} have been calculated for the binding energy of the excimer. However, recent information suggests that it is unlikely that there is a high temperature limit above which one can assume $k_{MD} \gg k_D$ and in fact the two quantities are of comparable magnitude in many polymers at room temperature. For example, inspection of the rate constants of table 7.2 shows that in no case of an intramolecular excimer was this condition satisfied. Furthermore, increasing the temperature does not usually bring the polymer into an equilibrium region, because both k_{MD} and k_D increase with temperature. Cases where a maximum in I_D/I_M are observed experimentally are probably not due to an equilibrium association-dissociation of monomer to dimer, but to a slightly higher activation energy for k_D than for k_M and k_{DM}. For this reason great care should be taken in interpreting data on steady-state emission intensities for polymers.

7.3.4 *Excimers in solid polymers*

Theoretical treatments of excimer formation in solid polymers suffer from most of the problems outlined previously for fluid solutions. A further complication is introduced by the fact that conformational equilibrium cannot be achieved in solids below their glass transition, and consequently the phenomena observed will be strongly dependent on thermal history.

Three important types of experiment can be carried out with excimer-forming polymers in the solid state. (1) The polymer itself can be cast or molded into a film and the emission characteristics studied under various conditions. (2) The polymer can be dissolved at low concentration (usually less than 0.5 wt-%) in another glass-forming polymer (usually PS or PMMA) which does not absorb or emit in the regions of interest. (3) The polymer can be dissolved in typical glass-forming liquids such as EPA or methyl tetrahydrofuran (MTHF) (§5.1), which are fluid to quite low temperatures (~130 K) and then form clear, rigid matrices. Studies of excimer-forming polymers in polymer matrices whose glass transitions are above 25°C (case no. 2) usually show excimer emission, even at 77 K (Fox 1972), while the same polymers in EPA or MTHF glasses at 77 K (case no. 3) will show no excimer emission (Fox 1972, Frank and Harrah 1974, Holden and Guillet 1980). This is strong evidence that excimer-forming sites do not occur in the thermodynamically most stable conformations of most polymer chains. Hence they represent higher energy conformations which are thermally populated. The number of these sites present in the solid polymer will then depend on the temperature at which molecular motion is reduced to such a degree that interconversion of these to the more stable conformations cannot occur. Although there is not much experimental evidence to support this, it is usually assumed that this restriction in motion in polymeric glasses occurs at the glass transition temperature T_g or the casting temperature of the film, whichever is the lower. According to this argument, the reason that no excimer emission is shown by excimer-forming polymers in EPA type glasses is that the glass transition is so low (~130 K) that substantially all of the higher energy excimer-forming conformations have been converted to lower energy conformers (but not necessarily those of lowest energy) under conditions where no further conformational change is possible due to the restricted mobility imposed by the glassy matrix.

The problem of relating the intensity of emission from excimer sites to their concentration would be simple except for the unknown effect of energy migration on the population of both excimer and monomer sites. Since the extent of energy migration is expected to be strongly affected by both the nature of the repeating chromophores in the polymer and on their average separation (see chapter 9), a general solution to this problem remains to be developed. The earliest and still the most elegant theory was developed by Frank and Harrah (1974) for studies of P2VN in PS matrices. They used a statistical model based on the rotational isomeric state approximation to formulate an expression for the fraction of excimer-forming sites in solid polymers. Films of P2VN and PS were prepared by evaporation of fluid solutions at a constant temperature. In fluid solution, rapid interconversion of polymer chain conformations takes place through rotations about the methylene linkages if sufficient thermal energy is available to surmount the rotational barrier. Higher energy polymer chain conformations will be favored by increasing the solution temperature.

If a polymer solution is allowed to evaporate at a temperature below the glass transition, the resulting cast film will have a thermal distribution of chain conformations which is characteristic of the temperature of casting. Under these conditions it was proposed that ΔE, the energy of the excimer site conformation with respect to the ground state conformation can be obtained from an Arrhenius plot of I_D/I_M against $1/T_{cast}$ where T_{cast} is the temperature at which the film was cast. Figure 7.13 shows data obtained by Frank and Harrah plotted in this form, from which they calculated that the excimer site conformation energy for P2VN is 10.5×1.2 kJ mol^{-1}. More recent work by Frank and Gashgari (1979) and by Semerak and Frank (1981) indicates that P2VN is not soluble in PS above concentrations of about 0.05 wt-%. Since these measurements were made at 0.2% this value may include processes other than conformational changes in the polymer and thus may be in error.

Frank and Harrah's kinetic scheme for activated excitation migration

Figure 7.13. Excimer to monomer intensity ratio versus the reciprocal casting temperature for solid films of 0.20% P2VN in a PS host. Monomer intensity was measured at 3275 Å. The points are average values of ratios obtained from spectra run at 295 K. The error bars indicate the range of results obtained for different samples. (From Frank and Harrah 1974.)

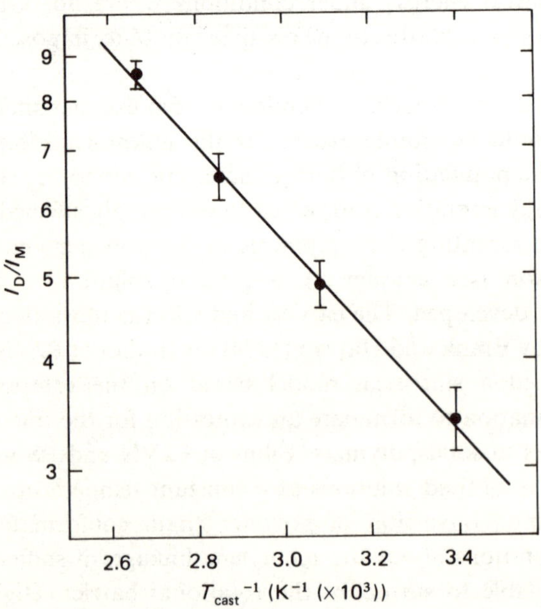

utilized the Birks scheme (scheme I), §7.1.5, from which they derive expressions for the steady-state intensities I_D and I_M for the excimer and monomer emission, respectively, as given below:

$$I_D = \frac{k_{FD}(k_M/k_D)I_0 f_R[1+(k_E/k_M)]}{k_M + k_E f_R + k_M(k_{MD}/k_D)} \tag{7.30}$$

$$I_M = \frac{k_{FM} I_0[1 - f_R + (k_{MD}/k_D)]}{k_M + k_E f_R + k_M(k_{MD}/k_D)} \tag{7.31}$$

and the ratio

$$R = \frac{I_D}{I_M} = \left(\frac{k_{FD}k_M}{k_{FM}k_D}\right)\left[\frac{f_R + 1 + (k_E/k_M)}{1 - f_R + (k_{MD}/k_D)}\right] \tag{7.32}$$

where f_R is the fraction of aromatic groups in excimer-forming sites and k_E is the rate constant associated with energy migration. These expressions can be simplified using the assumption that the deactivation of the excimer does not proceed through the excited monomer so that $k_{MD} \ll k_D$ over the temperature range studied. This conclusion has been reached from consideration of the intensity data for P2VN, both in solid films and in solution (David *et al.* 1972*a*, *b*). Furthermore, if one assumes that the energy migration rate is much faster than the overall rate of monomer decay, so that $k_E \gg k_M$, (7.32) reduces to:

$$R = \frac{I_D}{I_M} = \left(\frac{k_{FD}k_E}{k_{FM}k_D}\right)\left(\frac{f_R}{1 - f_R}\right) \tag{7.33}$$

Although the validity of this assumption for PVN remains in question, it is probably valid in a number of other polymers such as PVCz. It has been shown that the ratio of emission rate constants for the excimer and monomer k_{FD}/k_{FM} is independent of temperature for pyrene and for 9-methyl-9,10-dimethyl anthracene. David *et al.* (1972*a*, *b*) have shown that k_{FD} and k_{FM} are independent of temperature for solutions of both P1VN and poly(acenaphthalene). Since in this theory the fraction of excimer-forming sites, f_R, is assumed to be fixed by the casting temperature and independent of the film temperature, the entire temperature behavior below 180 K could be attributed to changes in the energy migration rate constant k_E.

The concentration of excimer-forming sites is actually composed of two factors: $f_R = P_E f_E$, where P_E is the probable number of aromatic rings in an excimer site, and f_E, the fraction of excimer-forming sites. If the concentration of excimer-forming sites is low enough, there will be only two aromatic rings per site and $P_E = 2$. However, as the concentration increases, the possibility of excimer sites containing more than two rings in a suitable conformation must be considered. For example, a series of three rings would give rise to two excimer-forming sites, four rings to three sites, etc. Thus P_E is given by:.

$$P_E = 2N_1 + \tfrac{3}{2}N_2 + \tfrac{4}{3}N_3 + \ldots \tag{7.34}$$

where N_n is the probability of finding n adjacent excimer-forming sites. N_n can be given by the expression:

$$N_n = nf_E^{n-1}(1 - f_E)^2 \tag{7.35}$$

f_E, the fraction of excimer-forming sites, can be determined by assuming a Boltzman distribution over the states of the system if the state energies of each particular site conformation are known. Frank and Harrah calculate f_E on the basis of the rotational isomeric state approximation in which each skeletal bond in the backbone of the polymer chain occurs in one of several possible discrete rotational states. The analysis of the conformational state degeneracy assumes only intermolecular interactions between adjacent aromatic chromophores. Selecting the minimum conformational unit as a three-skeletal carbon rotational dyad gives nine discrete chain segment conformations with nine possible conformational energy states. For an isotactic polymer these states are shown in figure 7.14. The further assumption that the only conformational states which are appreciably populated are the ground and excimer states, leads to an expression for f_E of the form:

$$f_E = \frac{W_E \exp(-\Delta E/RT)}{W_G + W_G \exp(-\Delta E/RT)} \tag{7.36}$$

where W_E and W_G are the degeneracies of the excimer and ground states, respectively, and ΔE is the energy of the excimer site conformation with respect to the ground state. The conformational analysis indicates that a syndiotactic polymer will have a doubly degenerate ground state and a doubly degenerate excimer state, while an isotactic polymer has a double degenerate ground state and a singly degenerate excimer site. Using the experimentally determined conformational energy of 10.5 kJ, the calculated values of excimer-forming sites for P2VN are shown in table 7.3. Similar arguments were used (Frank and Harrah 1974) to estimate that the number of excimer-forming sites in solid PVCz is about 10^{-3} moles per base mole of polymer, whereas in poly(4-vinyl biphenyl) the fraction is of the order of 0.1. There is still some question as to the validity of these calculations because of the problems of compatibility of both polymers and quenchers in solid polymer matrices alluded to earlier.

The conformation energy, ΔE, should be distinguished from the rotational barrier obtained from solution measurements. David *et al.* (1972*a*, *b*) and Harrah (1972) have determined a rotational barrier for

Figure 7.14. Conformations for a three skeletal carbon rotational dyad of isotactic polymer chain sequences based on a threefold rotational potential. The skeletal carbons and hydrogens have been omitted; C denotes the remainder of the chain on either side of the dyadic segment; and R represents a substituent aromatic ring. Steric interactions are denoted by N for no appreciable inhibition, C–R for a chain–ring interaction and C–C for a chain–chain interference. The standard *trans–gauche* nomenclature is used to identify the nine conformations. (After Frank and Harrah 1974.)

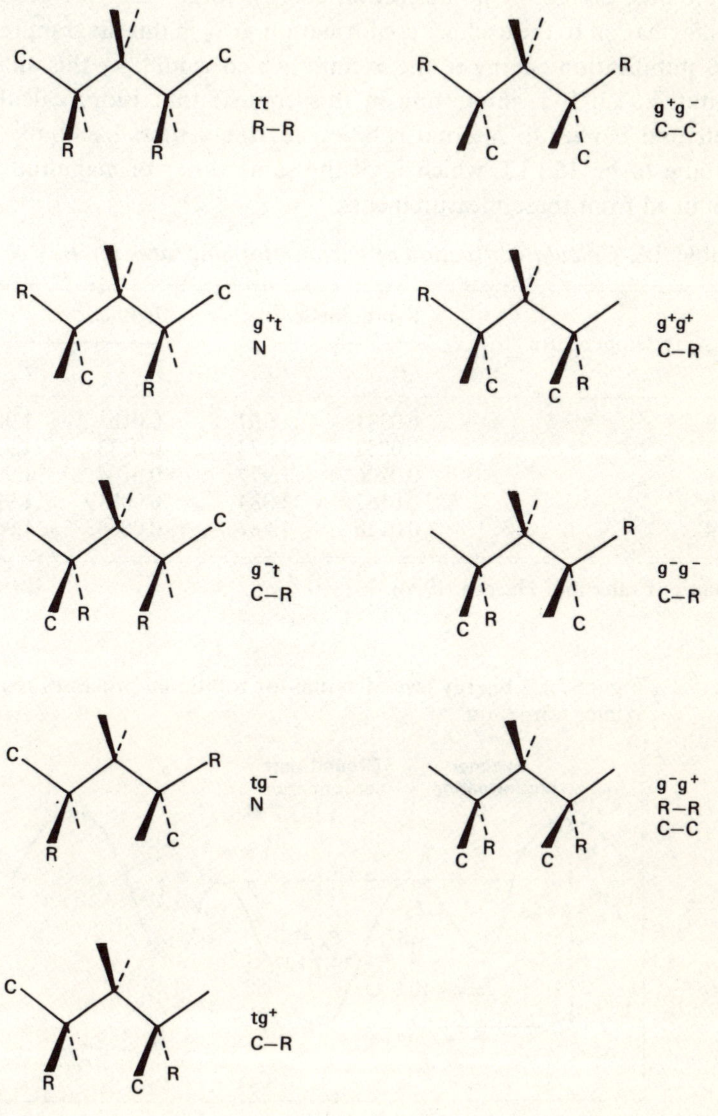

P2VN in solution in MTHF and *n*-propyl benzene to be 14.7 and 15.1 kJ, respectively. Solution measurements give the equivalent of an activation energy for overcoming the rotational barriers in solution. The differences between these two energies are shown schematically in figure 7.15. ΔE_1 corresponds to the conformational site energy of 10.5 kJ which is the energy difference between the rotational potential minima corresponding to an excimer conformation and the preferred ground state conformation minimum, whereas ΔE_A represents the energy barrier which must be overcome, essentially an activation energy, to go from the ground-state conformation to the excimer conformation. ΔE_2 in this diagram represents the stabilization energy of the excimer which would, by this analysis, be about 4.6 kJ. It is interesting in this context that Flory calculates the potential barrier to internal rotation for the central backbone bond in butane to be 15.1 kJ, which is of the same order of magnitude as that obtained from these measurements.

Table 7.3. *Calculated fraction of excimer forming sites for P2VN*

Casting temperature (K)	Syndiotactic		Isotactic	
	f_E	P_E	f_E	P_E
379	0.0331	1.967	0.0168	1.983
353	0.0262	1.974	0.0133	1.937
331	0.0206	1.979	0.0104	1.990
312	0.0162	1.984	0.00819	1.992
294	0.0128	1.987	0.00643	1.994

Source: Frank and Harrah (1974).

Figure 7.15. Energy level diagram for rotational processes required for eximer formation.

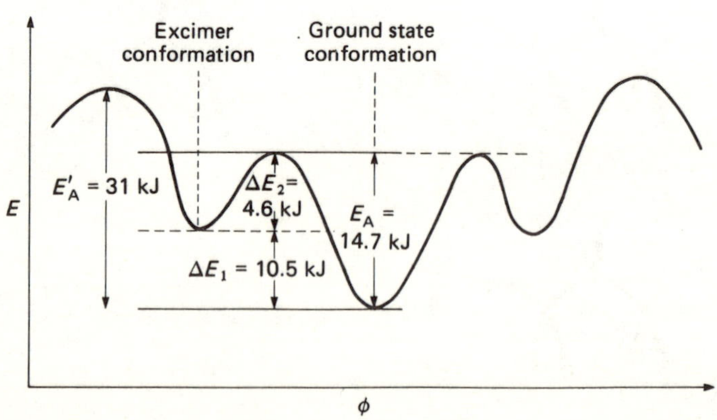

As shown in figure 7.16, excimer emission from solid polymers shows a maximum intensity at a particular temperature, T_{max}. The analysis given above applies only to the low temperature region, that is, below I_{max}. Above I_{max} a different kinetic treatment was developed and the result of this gives an expression for $I_{D_{max}}/ID$ of the following form:

$$(I_{D_{max}}/I_D) - 1 = (k^0_{ID}/k_{FD}) + (k'_{ID}/k_{FD})\exp(-E_{ID}/RT) \qquad (7.37)$$

where the non-radiative decay constant k_{ID} is given by

$$k_{ID} = k^0_{ID} + k'_{ID}\exp(-E_{ID}/RT) \qquad (7.38)$$

The first term in (7.37) is small and usually neglected. Plots in this form usually give a linear relationship from which a value of E_{ID} can be

Figure 7.16. Excimer to monomer intensity ratio versus the reciprocal temperature for neat PS films cast at 295 K and with molecular weights of 2100 (triangle), 10 000 (square), 33 000 (diamond), and approximately 50 000 (circle). Monomer intensity was measured at 2800 Å and excimer intensity at 3350 Å. The data are all normalized to unity. (From Frank and Harrah 1974.)

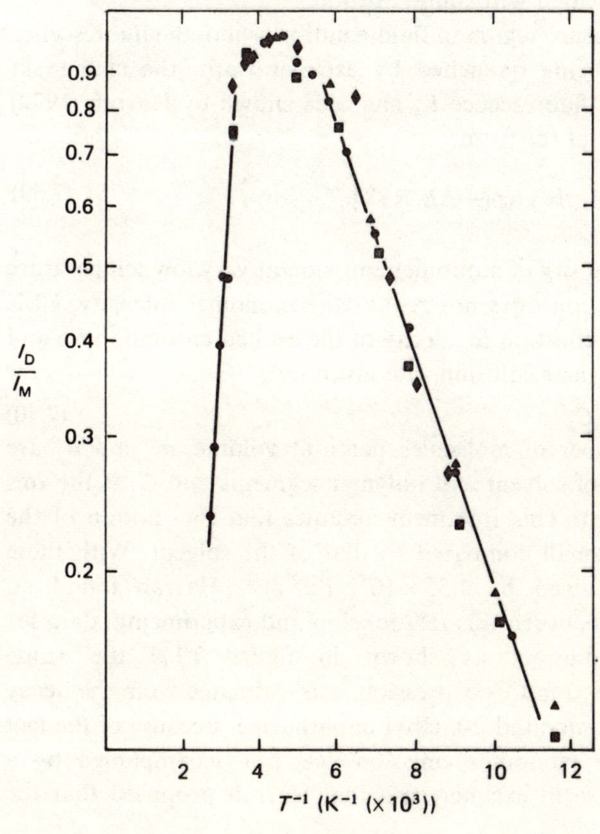

calculated to be 13.9 and 22.3 kJ mol^{-1} for solid P2VN and PS systems, respectively. On the other hand, similar data on quenching of fluorescence in fluid solutions by Fox *et al.* (1972) and by Harrah (1972) give values of about 31 kJ for the excimer binding energy. These values are about 17 kJ larger than the solid solution data. Frank and Harrah (1974) suggest that this difference can be explained on the basis that the excimer dissociates to ground state monomers followed by a suitable chain rotation to minimize the repulsive forces between the rings.

One could, however, explain the results on the basis of figure 7.15 if one assumed that in the solid state one is only sampling conformational barriers between the ground state and the excimer conformation and hence the activation energy would be E_A, corresponding to the rotational barrier between those two states. In solution one would be sampling all rotational states with an energy barrier corresponding to E'_A, the difference in energy between the ground state conformation and the largest rotational barrier in the backbone chain. There is, however, some question as to the validity of the Birks scheme of excimer dissociation in solid polymers, and these values should therefore be treated with some caution.

In the high temperature region in fluid solution where the fluorescence of the monomer is being quenched by excimer traps, the change in intensity of monomer fluorescence I_M has been shown by Harrah (1972) to follow an equation of the form:

$$\frac{I_M}{I_M(0)} = [1 + (Z_{sp}/k_D)\exp(-\Delta E/RT)]^{-1} \tag{7.39}$$

where $I_M(0)$ is the intensity of monomer emission at very low temperature where excimer formation does not reduce the monomer intensity, k_D is the unperturbed rate constant for decay of the excited chromophore and Z_{sp} is the solvent-polymer collision rate given by:

$$Z_{sp} = \pi N \sigma_1 \sigma_2 \bar{C}_1 \tag{7.40}$$

where N is the number of molecules per unit volume, σ_1 and σ_2 are collisional diameters of solvent and polymer segments and C_1 is the rms velocity of the solvent. This treatment assumes that the motion of the polymer segment is small compared to that of the solvent. With these assumptions Z_{sp} is given by $1.55 \times 10^{10}\, T^{1/2}\, s^{-1}$. Harrah found an excellent correlation between this relationship and experimental data for P2VN in dilute solution, as shown in figure 7.17; the value $k_D = 5.9 \times 10^{-5}\, s^{-1}$ used in this expression, was estimated from the decay time for the model compound 2-methyl naphthalene. Because of the fact that the decrease in monomer emission was not accompanied by a corresponding increase in excimer emission, Harrah proposed that the dissociation of the excimer gives rise to two ground state molecules rather than a ground state and an excited monomer, as is the case in small molecule models. He also showed that the excimer emission could be fitted very well by an equation of the form

$$\frac{I_D}{I_D(\max)} = 1 + \left(\frac{Z_{sp}}{k_{DE}}\right)\exp(-\Delta E/RT) \tag{7.41}$$

where k_{DE} is the excimer emission rate constant measured at room temperature in rigid PS solution ($1.56 \times 10^7\ s^{-1}$) with a binding energy for the excimer of 31 kJ which is, within experimental error, the value found by Aladekomo and Birks (1965) for ethyl naphthalene (28 kJ).

When a glassy polymer sample coated at a particular temperature is rapidly heated to a higher temperature near T_g, the ratio I_D/I_M increases, because the increase in free volume of the matrix allows changes in polymer conformations which may permit more aromatic rings to achieve excimer-forming conformations. However, the process takes some time to occur. Frank (1975) has reported the observation of relaxation processes in PS containing small amounts (0.2%) of P2VN in the region of the glass transition, by such measurements. The ratio of I_D/I_M initially fixed by the temperature at which the film is cast, will change as a function of time due to molecular motion in the relaxation region, that is the glass transition region of the host polymer. For example, figure 7.18

Figure 7.17. Experimental and theoretical temperature dependence of monomer chromophore fluorescence intensity: calculated with $\Delta E = 15.1$ kJ/mol (solid line); experimental points (cross) for P2VN. (From Frank and Harrah 1974.)

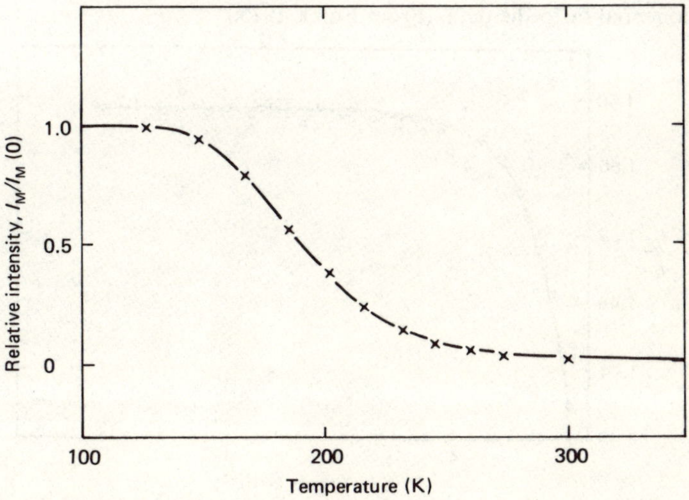

shows the time dependence of emission from P2VN in PS cast from benzene at 22°C and held at 73°C. This shows the approach to relaxation of the polymer as a function of time and the solid line is a single exponential fit to the data. Unfortunately, the kinetic analysis is complicated due to the incompatibility of the two polymers at this concentration.

This type of experiment may provide a useful way of probing molecular motion in polymers below their glass transition, but theoretical treatments are hindered because of the difficulty of establishing the actual number of excimer-forming sites in solid polymers.

7.4 Experimental studies of polymer excimers

7.4.1 *Polystyrene*

Polystyrene (PS) was one of the first excimer-forming polymers to be studied experimentally. Figure 7.19 shows the room temperature absorption and fluorescence spectrum for atactic PS. The broad structureless emission band shown at 335 nm was first attributed by Basile (1962) and Yaneri *et al.* (1963) to excimer emission. It is also obvious from the figure that the excimer emission is quenched more efficiently by oxygen than is monomer fluorescence which occurs at about 280 nm. Vala *et al.* (1965) provided further supporting evidence for Basile's assignment and provided a theoretical analysis of the emission, which suggested that it came from a state which was configurationally mixed with charge-transfer and neutral exciton states. The excimer band is

Figure 7.18. Time dependence for 0.20% P2VN-PS film cast from benzene at 22°C and held at 73°C. The solid line is a single exponential fit to the data. (From Frank 1975.)

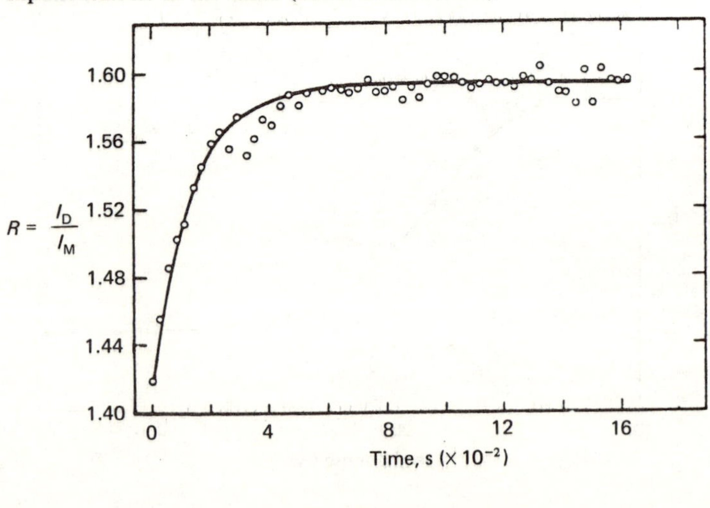

centered at 340 nm and the ratio of excimer to monomer is independent of polymer concentration at concentrations less than 0.1 %, thus indicating that the excimers must be due to intra- rather than intermolecular interactions. In PS it is now established that the excimers are formed between neighboring side groups and, characteristically, oxygen and other quenching agents reduce both monomer and excimer fluorescence intensities to a different degree. Studies of the quenching of emission by oxygen show that monomer fluorescence can arise from excited side groups which have been created either by direct absorption or by dissociation of excimers, or by both processes. If the latter case obtains, a kinetic analysis by scheme I gives the following values for the rate constants in isotactic PS: $k_{DM} = 4 \times 10^9$, $k_{FD} = 2 \times 10^6$, $k_{GD} = 7 \times 10^7$, and $k_{MD} = 10^7\,s^{-1}$. Comparison of these data with those for diphenyl propane shows similar values, except that k_{DM} and k_{FD} of the isotactic polymer are substantially higher than for model compounds. However, the validity of this type of analysis is probably questionable because more recent evidence suggests that the excimer does not dissociate to monomer at these temperatures.

Several authors have observed an increase in the ratio of excimer emission with increased molecular weight. This effect, illustrated for PS in figure 7.20, is pronounced at low molecular weights. This has usually been taken as confirmation of the requirement for an energy migration step in the population of excimer sites. However, it is also possible that the probability of formation of an excimer site across a chain loop may also increase with molecular weight. Furthermore, MacCallum (1981) has

Figure 7.19. Room-temperature absorption and fluorescence spectra of atactic polystyrene in 1,2-dichloroethane (1×10^{-3} M). The three fluorescence curves denote: nitrogen solution (dotted line), aerated solution (broken line), oxygenated solution (full line). From Vala *et al.* 1965.)

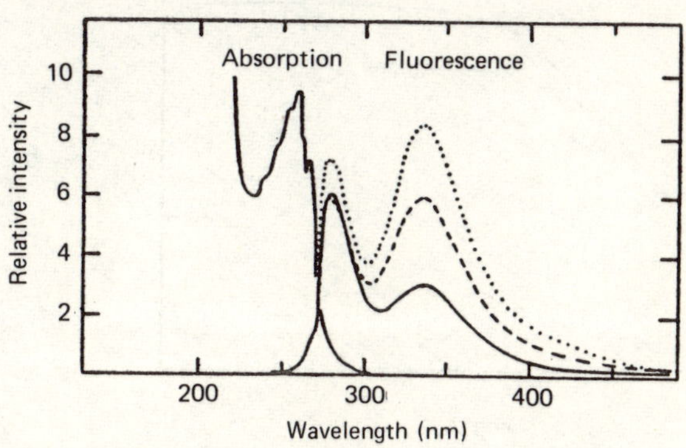

recently shown that the results can also be satisfactorily explained by assuming that the end groups of the chain have different quantum yields for emission than chromophores in the middle. In view of these uncertainties, it is clear that much additional work needs to be done to clarify this issue.

In the solid state the ratio of excimer to monomer fluorescence as a function of temperature for PS films cast at 295 K was independent of molecular weight. The emission goes through a maximum at 210 K for PS and 160 K for P2VN. The absolute intensity of the monomer emission increases as temperatures decrease, which was attributed to a decrease in the rate of energy transfer to the excimer trap.

Effect of tacticity on excimer formation in polystyrene. David *et al.* (1974) studied the ratio of excimer to monomer emission in PS at 77 K and showed (table 7.4) that while in atactic PS the amount of excimer emission increased slightly with orientation, in isotactic PS there was from 10 to 100 times as much excimer emission as monomer emission, depending on the degree of tacticity. Since isotactic PS is known to exist in a 3–1 helix in the crystalline state and the spacing of these rings is of the order of 6 Å it is suggested that this isotactic arrangement provides many more excimer sites than in the random atactic PS. Since at 77 K there would be no appreciable molecular motion, it is tempting to identify this isotactic conformation as one providing the maximum number of excimer sites.

Figure 7.20. Variation of the ratio of excimer to monomer fluorescence intensity with number-average molecular weight $\bar{M}_n$ for anionic polystyrene in 1,2-dichloroethane. (From Ishii *et al.* 1978.)

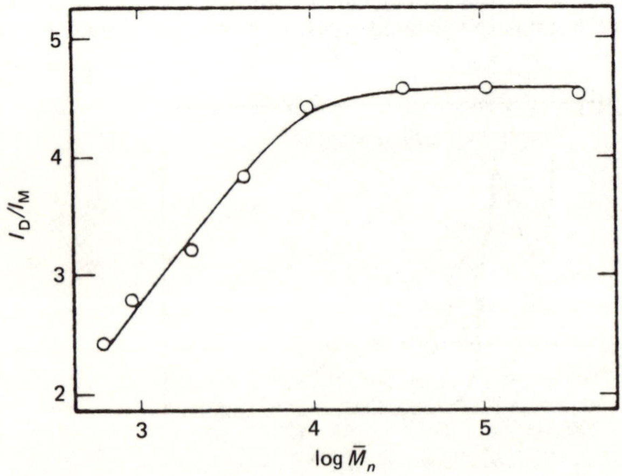

7.4.2 *Poly(1-vinylnaphthalene) and poly(2-vinylnaphthalene)*

These two polymers have been studied more carefully than perhaps any others, and early work is discussed in §7.3.3. In solution at low temperatures the decrease in excimer emission follows an Arrhenius expression (figure 7.21) when plotted in the form of $\log(I_D/I_M)$ versus T^{-1} from which activation energies of 14.4 kJ mol^{-1} are obtained for P2VN. This can be compared with the value of 5.7 kJ mol^{-1} for PS from similar measurements. Since for these polymers excimer intensities reach

Table 7.4. *Ratio of normal to excimer fluorescence intensity* (I_D/I_M) *in polystyrene at 77 °C*

Polystyrene type	I_D/I_M
Atactic, unoriented	1.43
Atactic, oriented (F=0.07)	1.67
Atactic, oriented (F=0.1)	2.08
Isotactic (25% crystalline)	10
Isotactic (35% crystalline)	100

Figure 7.21. Arrhenius plots for the determination of excimer activation energies. (From Fox *et al.* 1972.)

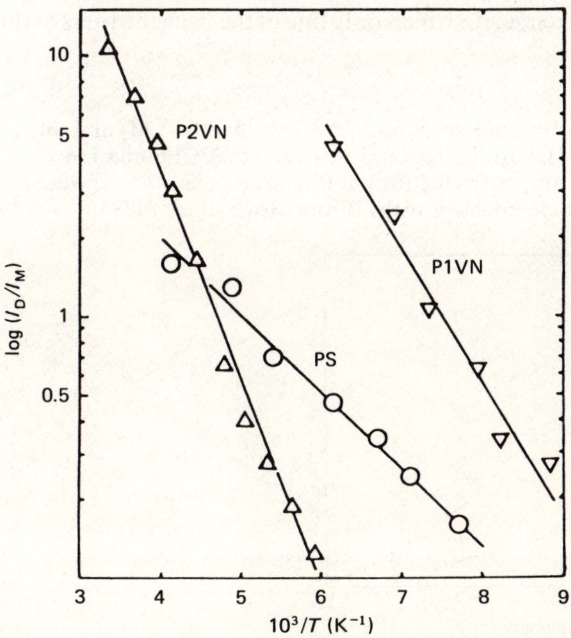

a maximum not far below room temperature, these activation energies thus correspond to the region in which the intensity of emission relates primarily to the rate of excimer formation and is not altered appreciably by dissociation to the monomer. Although these activation energies cannot be directly related to the rotational motion of a specific chromophore, they are of the right order of magnitude for cooperative motions of side groups with backbone carbon atoms and the larger value for the naphthalene chromophore is to be expected from steric considerations.

7.4.3 *Poly(vinylcarbazole) and its analogs*

Among the more important excimer-forming chromophores is the carbazole molecule. Extensive studies on polymers containing this group have been reported, the most important of which are those involving poly(vinylcarbazole) (PVCz). This polymer, which has important industrial applications as the base for a xerographic process, has been studied extensively (Klöpffer 1969*a*, *b*, 1973). The emission spectrum of the excimer is much broader than is characteristic of excimers in other systems, as shown in figure 7.22 which compares the emission from PVCz with that of a bichromophoric model, 1,3-bis-carbazolylpropane. The emission spectrum contains contributions from three species with maxima at 350, 370 and 430 nm. Johnson (1975) has proposed mechanisms to explain these broad emissions on the basis of two types of excimers, one from the normally eclipsed form and the other a partially eclipsed, so-called 'high energy' excimer, in which only one of the benzene rings of the carbazole overlaps.

Figure 7.22. Fluorescence spectra of (*a*) PVCz (5×10^{-4} M) and (*b*) BCP (2×10^{-4} M), both in degassed benzene at 23°C: excitation wavelength, 330 nm; corrected for phototube response. The spectra are not on the same intensity scale. (From Hoyle *et al.* 1978.)

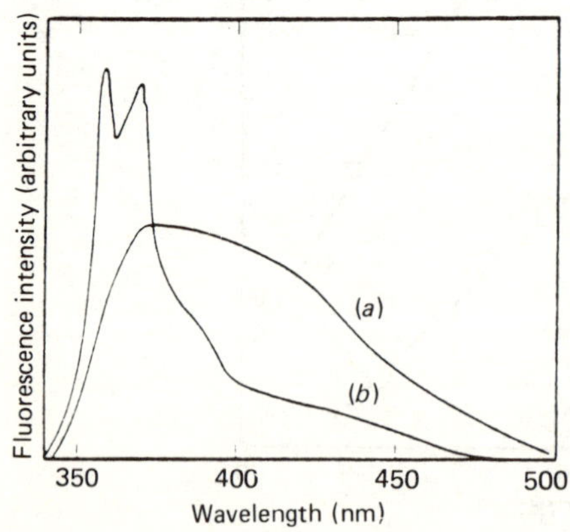

The two-site excimer was first proposed by David *et al.* (1972*a*, *b*). Further information on these structures can be obtained by time-resolved fluorescence spectroscopy (Hoyle *et al.* 1978) which showed that the high energy excimer, the partially eclipsed form, was present at the earliest times observable, i.e., less than 0.2 ns after absorption of a photon, while the emission from the low-energy sandwich excimer was virtually absent at this early time scale. Figure 7.23 shows the emission spectrum observed at approximately 0.2, 10, 25 and 300 ns from the lamp maximum. The observed shift of the emission from the high energy to the low energy sandwich excimer suggests that a significant amount of time is required for the formation of the sandwich excimer, and that this is presumably rate-limited by rotations about the backbone chain of the PVCz. Recent studies by De Schryver *et al.* (1982) on model compounds suggest that these two excimers arise from isotactic and syndiotactic dyads in the polymer chain.

When the emission from PVCz in solution is quenched by the addition of dimethyl terephthalate (DMTP) there is evidence of a new emission caused by an exciplex formed between PVCz and DMTP. Use of time-resolved fluorescence spectroscopy has shown (Hoyle and Guillet 1979) that there are two distinct components to this exciplex emission, one of which is attributed to an excited trimer (sometimes called an 'exterplex'):

$$\overline{\overset{+}{(D-D}\cdots A^{-})^{*}}$$

Figure 7.23. Time resolved fluorescence spectra of PVCz (5×10^{-4} M) in degassed benzene at 23°C: excitation wavelength 313 nm. Time settings (ns): (*a*) 0.23, (*b*) 8–9.4, (*c*) 19–35, (*d*) 182–323. The spectra are adjusted to fit the same intensity by scale. (From Hoyle *et al.* 1978.)

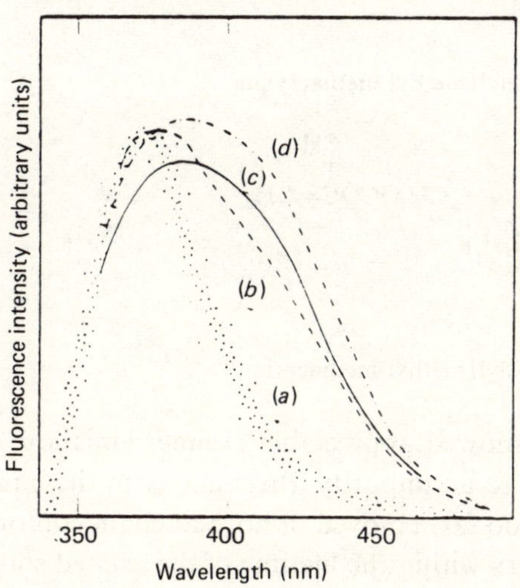

consisting of a charge-transfer complex between two PVCz donor units and one DMTP acceptor. Time-resolved emission spectra supporting this interesting hypothesis are shown in figure 7.24.

Keyanpour-Rad *et al.* (1978) reported the synthesis of a series of carbazole-containing polymers from monomers of the structures shown below.

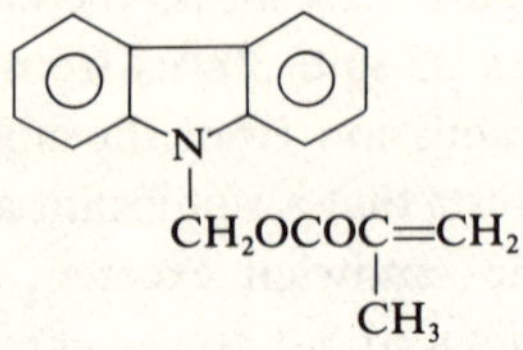

IX

(9-carbazolyl)methyl methacrylate

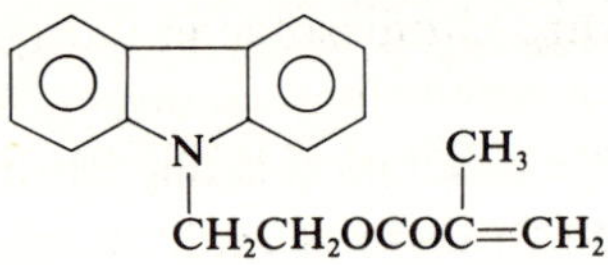

X

2-(9-carbazolyl)ethyl methacrylate

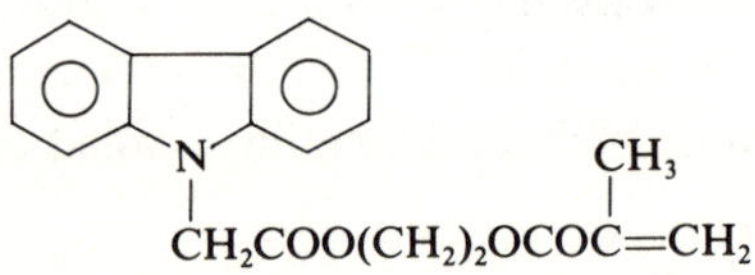

XI

2-(methacroyloxy)ethyl (9-carbazolyl)acetate (IIIa)

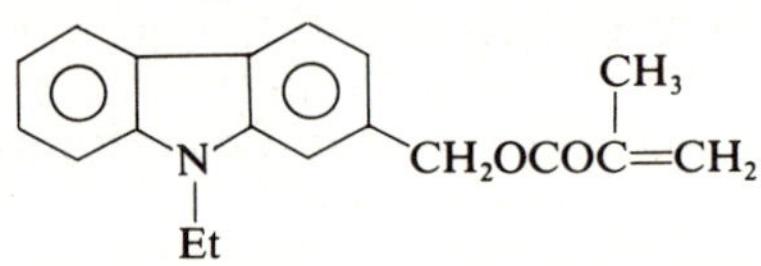

XII

[2-(9-ethyl)carbazolyl] methyl methacrylate

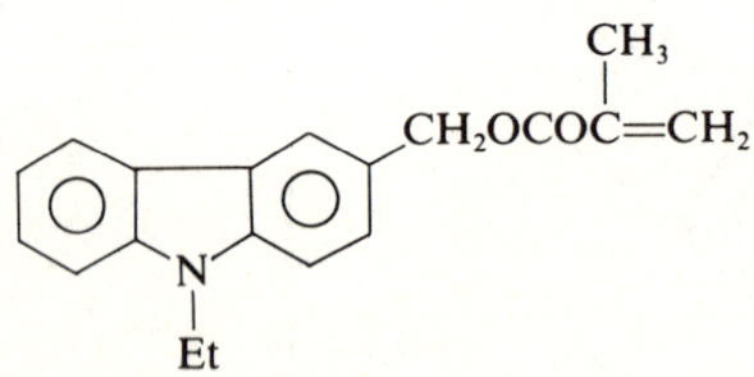

XIII

[3-(9-ethyl)carbazolyl] methyl methacrylate

Only polymers from XII showed appreciable excimer emission, and this was later shown to be due to an impurity. Presumably in these molecules the carbazole groups are too far removed in normal chain conformations to achieve excimer geometry within the lifetime of the excited state. These results should be contrasted with those of similar polymers containing the naphthalene in place of carbazole (§7.4.4) where extensive excimer emission is observed.

7.4.4 *Polymers from aromatic acrylates and methacrylates*

Excimer emission is not restricted to the vinyl aromatic polymers. For example, early studies by Bairamov *et al.* (1971) and by Somersall and Guillet (1973) showed that excimer emission predominated in aromatic esters such as poly(1-naphthyl methacrylate) in the solid state. Somersall and Guillet (1973) established that in solution, the ratio of excimer to monomer emission varied considerably with the solvent power, being high in poor solvents and low in good solvents (figure 7.25). This correlated with the change in effective volume of the random coil in solution estimated from the intrinsic viscosity, as shown in table 7.5. It was demonstrated that this was not a polar effect since similar increases in excimer emission could be observed by adding a non-solvent to the solvent, thus compressing the molecular coils: this did not depend on whether the non-solvent was polar, such as methanol, or non-polar, such as hexane (Aspler and Guillet 1979).

The strong solvent effect in naphthyl ester polymers, which appears to be absent in PS and P1VN was explained as being due to the fact that in the naphthyl esters a significant portion of the excimers is being formed

Figure 7.24. Time resolved fluorescence spectra of PVCz (5×10^{-4} M)–DMTP (0.07 M) in degassed benzene at 23°C; excitation wavelength 330 nm. Time settings (ns): (*a*) 0–0.23, (*b*) 3.7–8.5, (*c*) 7.8–13.1, (*d*) 13.1–17.7, (*e*) 17.7–25.1. The spectra are not on the same absolute scale. (From Hoyle and Guillet 1979.)

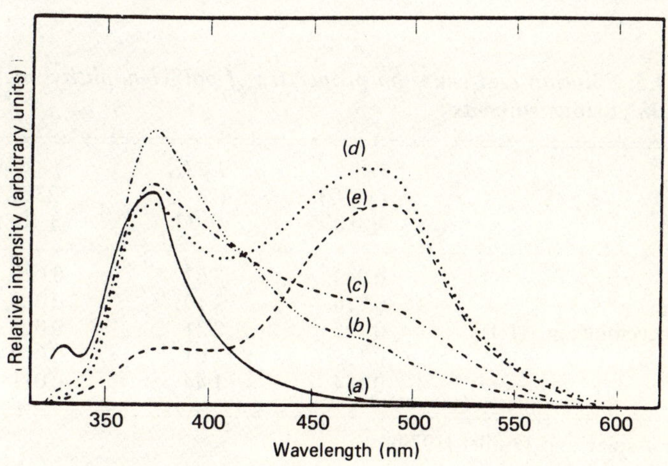

Figure 7.25. Fluorescence of PNMA in different solvents (25°C). In benzene (full line); in EtOAc (dashed line); in chloroform (dotted line); in chloroform–cyclohexane (1:1) (chain line). (From Somersall and Guillet 1973.)

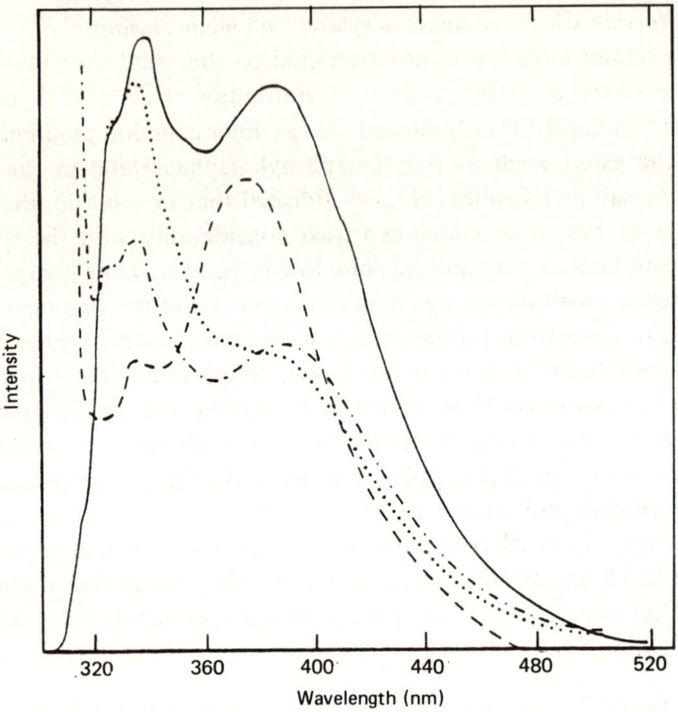

Table 7.5. *Solution and emission properties of poly(1-naphthyl methacrylate) in various solvents*

Solvent	[η] (30 °C) $\pm$0.005	I_D/I_M (23 °C) $\pm$0.1	ϕ_F, total (23 °C) $\pm$10%
EtOAc	0.040	3.67	0.020
Toluene	0.116	3.19	0.010
$CHCl_3$-cyclohexane (1:1)	0.125	2.81	0.015
CH_2Cl_2	0.158	2.61	0.010
$CHCl_3$	0.184	1.44	0.011

Source: Aspler and Guillet (1979).

across loops in the polymer chain, rather than from adjacent monomer units. As the random coil of the polymer is expanded in a good solvent, the possibility of the intrachain contacts will be reduced with a consequent reduction in the total number of excimer-forming sites. However, later studies (chapter 9) suggest that another important factor is the greater efficiency of energy migration in the compact coil than in the expanded coil. It is important to note that, whereas PS and P1VN follow the Hirayama rule, i.e., the chromophores are separated by only three carbon atoms, there will be at least seven atoms separating the two chromophores in the naphthyl esters and, hence, the number of conformations of the chain leading to excimer-forming sites is expected to be significantly lower.

One should therefore consider PS and P1VN as examples of polymers where excimers can form at nearly every point along the chain so that the concentration of excimer sites in fluid solution, where mobility allows the necessary rotation, could be quite high. Naphthyl esters, on the other hand, will be examples of polymers where the probability of adjacent excimer formation is quite low and the high intensity of excimer emission is primarily due to efficient energy migration to the few excimer-forming sites which exist within the total polymer chain. This explanation is also consistent with the results of Aspler and Guillet (1979) who showed that the ratio of excimer to monomer emission increased with the molecular weight of the chain, suggesting again that the excitation hunts through a large part of the total chain in order to find a suitable excimer-forming site.

Because of their ease of synthesis, the naphthyl acrylates and methacrylates have been studied extensively in recent years. Both the 1- and 2-naphthyl acrylates and methacrylates have been synthesized and their solution properties determined (Merle-Aubry *et al.* 1980). The decay times reflect the fact that acrylates are more flexible than methacrylates (table 7.6).

The naphthalene groups can be further removed from the polymer chain by insertion of methylene groups such as in poly(1-naphthylmethyl methacrylate) and poly(1-naphthylethyl methacrylate) and, as expected, the intensity of excimer emission decreases. However, it is still evident, even in polymers where more than 13 atoms separate the two chromophores (figure 7.26).

Further information about excimer formation in polymers has come from studies of copolymers of styrene with dinaphthyl fumarate (Holden and Guillet 1980). In these polymers, energy migration should be absent, at least at low concentrations of the naphthalene-containing chromophore, and the model should approximate the situation in which two naphthyl acrylate units are adjacent. Further, at low concentration of the fumarate monomer, it is likely that the only formation of excimers would be from adjacent chromophoric units. It was shown in this case that the intensity of excimer emission was higher than that observed in small model dimers such as meso-bis(1-naphthylmethyl)-2,3-dimethylbutanedioate, which suggests that the conformational effect of the long polymer chain actually forces the two naphthalene groups into con-

Table 7.6. *Decay parameters of naphthalene-containing polymers in degassed THF at 25 °C*[a]

Polymer	τ_1 (ns)	τ_2 (ns)	A_1/A_2	τ model (ns)
P1NA	1.4	9.0	6.8	11.8
P1NMA	1.9	7.9	7.7	10.2
P2NA	2.7	18.6	2.0	16.9
P2NMA	6.0	20.5	2.4	16.4

[a] $\lambda_{ex} = 280$ nm, $\lambda_{em} = 325$ nm, fitting function $I_M(t) = A_1 \exp(t/\tau_1) + A_2 \exp(t/\tau_2)$.
Source: Merle-Aubry *et al.* (1980).

Figure 7.26. Effect of side-chain length on extent of excimer formation for the homologous series (VI) in THF at 10°C. (From Nakahira *et al.* 1979.)

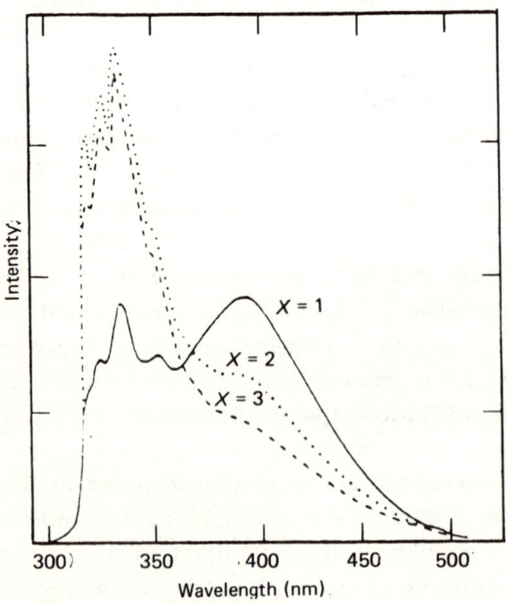

figurations more favorable to excimer formation than in the small molecule dimer. Similar conclusions were drawn by Bokobza *et al.* (1977) who synthesized dimer, trimer and tetramer models of PS and showed that only in the tetramer was the intensity of excimer emission comparable to that in PS itself.

Study of the decay curves from the isolated naphthalene pairs in copolymers of dinaphthyl fumarate and styrene (Holden and Guillet 1980) showed that monomer decays were non-single-exponential, with a minor component $\tau_1 = 10 \pm 2$ ns corresponding to naphthalene quenched by excimer formation, and a major component $\tau_2 = 44$ ns corresponding to naphthalene units in dead-end configurations, unable to form excimer during the lifetime of the excited state.

The rise time of excimer fluorescence agreed closely with the decay of monomer ($\tau \sim 10$ ns) and was considerably slower than in poly(1-naphthylmethyl methacrylate). The slower rise time and lower overall intensity of excimer emission is a dramatic demonstration of the importance of energy transfer to excimer excitation in most naphthalene-containing polymers, a factor which is presumed to be absent in this copolymer.

7.4.5 *Miscellaneous polymers and copolymers*

As outlined in the previous discussion it is not possible to assume that in most polymer systems the intensity of excimer emission is directly proportional to the concentration of excimer sites. The reason for this is that in addition to the concentration, one must also take into account the probability that the site can be populated by energy transfer, either by a hopping mechanism from chromophore to chromophore or by a collisional interaction during the lifetime of the excited state. However, it is possible to study excimer formation in polymers in the absence of energy transfer by synthesizing molecules containing isolated dichromophoric groups. For example, Morawetz (1979) has examined the temperature dependence of excimer formation in a polyamide, XIV, and its small-molecule model compound, XV.

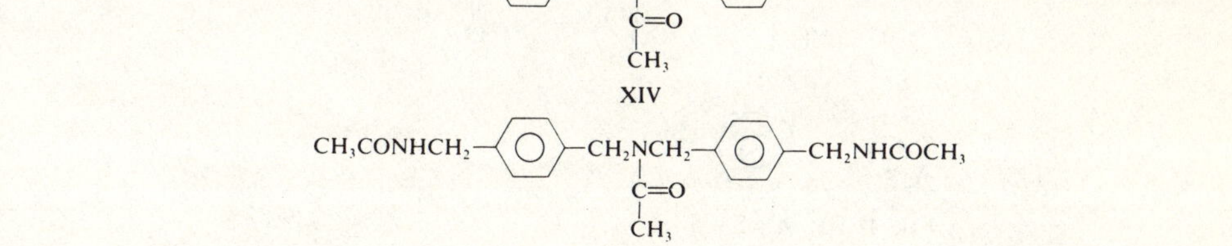

XIV

XV

It was shown that Arrhenius plots of the ratio I_D/I_M versus the reciprocal of temperature were linear and identical slopes were obtained for the two compounds. Morawetz suggested that the similar activation energies meant that the excimers were formed by similar processes in the two compounds and rejected a mechanism of excimer formation for the polymer which involved cooperative rotation of the two main-chain bonds. It is expected that this crankshaft motion would have a higher activation energy than rotation about a single C–C bond.

Another type of study involves the synthesis of polymers containing random or alternating units of a fluorescent monomer M_1 with a non-fluorescent monomer M_2. For example, David *et al.* (1973) and Alexandru and Somersall (1977) assumed that excimers were formed only between nearest neighbors and correlated the ratio I_D/I_M with $f_{M_1M_1}l_{M_1}$ where l_{M_1} is the average length of M_1 sequences. However, some care must be taken in the interpretation of these results to assure that energy migration is properly accounted for. It seems likely that energy migration may well not depend on concentration in the same way as the number of excimer-forming sites. For example, Wang and Morawetz (1975) observed excimer emission in stilbene-maleic anhydride and acenaphthalene-maleic anhydride alternating copolymers showing that in these polymers at least, excimers form from non-nearest neighbors.

On the other hand, alternating copolymers of vinylcarbazole with fumaronitrile or diethyl fumarate do not show any excimer emission at all (Yokoyama *et al.* 1975).

7.4.6 *Applications to structural studies*

In polymers where energy migration does not contribute to the population of excimer or exciplex sites their formation can be used to study conformational and other changes in polymer molecules. For example, Tazuke and Matsuyama (1977) have studied the interaction of a side chain with the backbone of a polyester by observing the extent of exciplex formation in the compounds such as XVI.

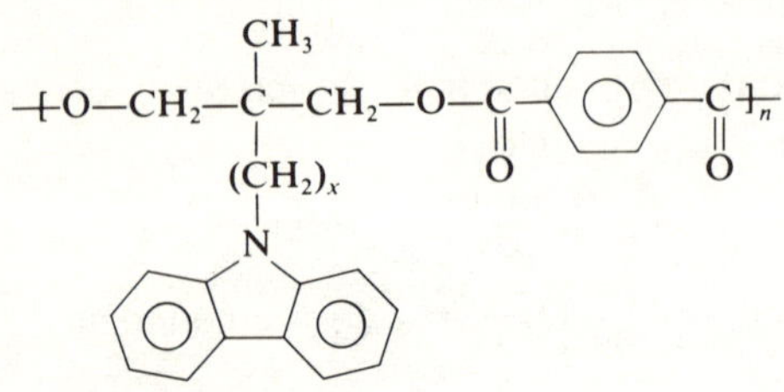

$x = 0, 4, 10 \quad n = 2-5$

XVI

Similarly, the fraction of chain conformations which can cyclize by forming large loops end-to-end, decreases as the length of the polymer chain increases. For example, Cuniberti and Perico (1977) synthesized a number of monodisperse polyethylene oxides terminated with pyrene groups. Figure 7.27 shows that the relative intensity of the excimer is inversely proportional to the molecular weight of the poly(ethylene oxide) separating the two chromophores. Data obtained from such chain cyclization studies can be compared with Monte Carlo and other calculations to refine mathematical models of polymer chain conformation.

Studies of this type have been extended by Winnik and co-workers (Redpath and Winnik 1982, Winnik *et al.* 1984) to determine cyclization rate constants (k_1) for both polystyrene and poly(dimethyl siloxane) chains terminated by pyrene groups. This was done by kinetic studies of

Figure 7.27. Effect of molecular weight M of polyethylene oxide on the extent of intramolecular excimer formation between terminal pyrene groups in THF at 20°C. M=(1) 19 500, (2) 8900, (3) 6500, (4) 3800 and (5) 1650. The dotted line corresponds to a monomeric model compound. (From Cuniberti and Perico 1977.)

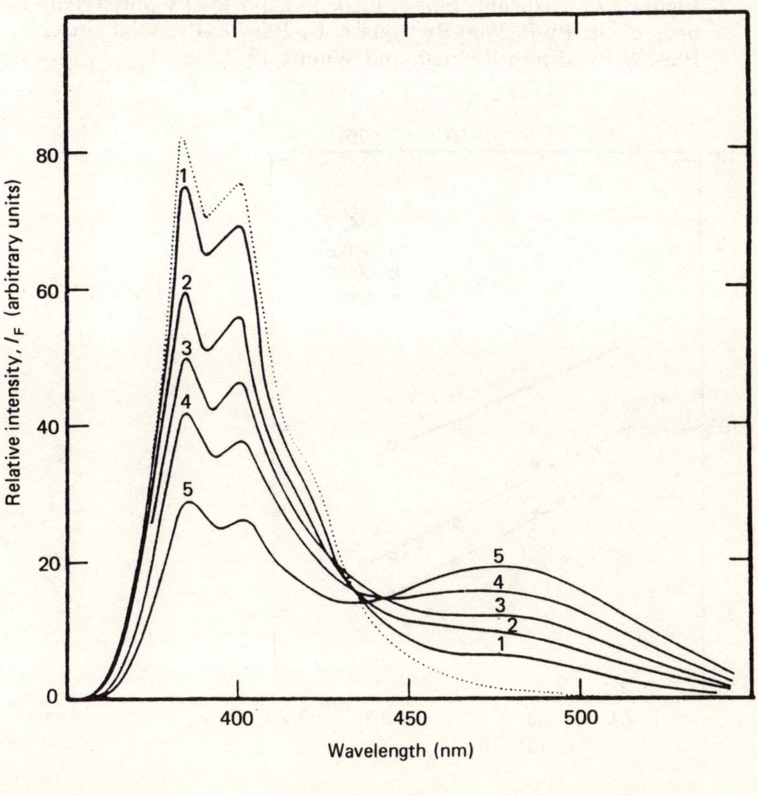

the rates of formation and decay of excimer emission. The rate constants are defined below:

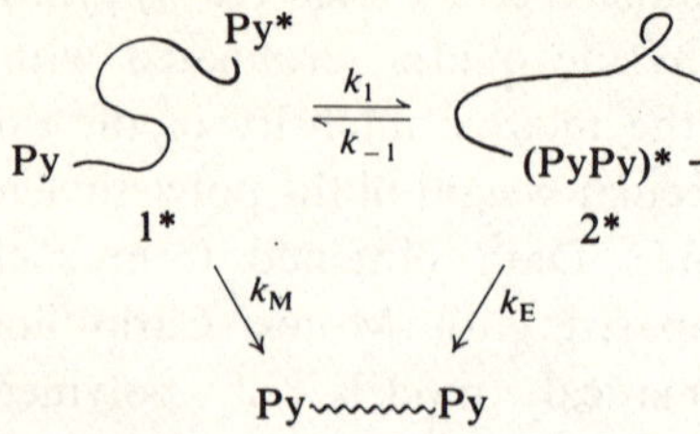

In carefully prepared, monodisperse polystyrene samples ($\bar{M}_n = 3900$, 6600 and 9200) the cyclization rate constant k_1 depended on chain length (figure 7.28), while the excimer dissociation constant (k_{-1}) was independent of M. The activation energy for k_1 was 14 kJ mol^{-1} while that for k_{-1} was significantly higher (44 kJ mol^{-1}) and rather similar to that observed for the dissociation of small model compounds containing the pyrene group (figure 7.29).

Figure 7.28. Arrhenius plot of log(k_1) vs. $1/T$ for Py-polystyrene-Py: open circle, Py-PS3900-Py: square, Py-PS6600-Py; solid circle, Py-PS9200-Py. (From Redpath and Winnik 1982.)

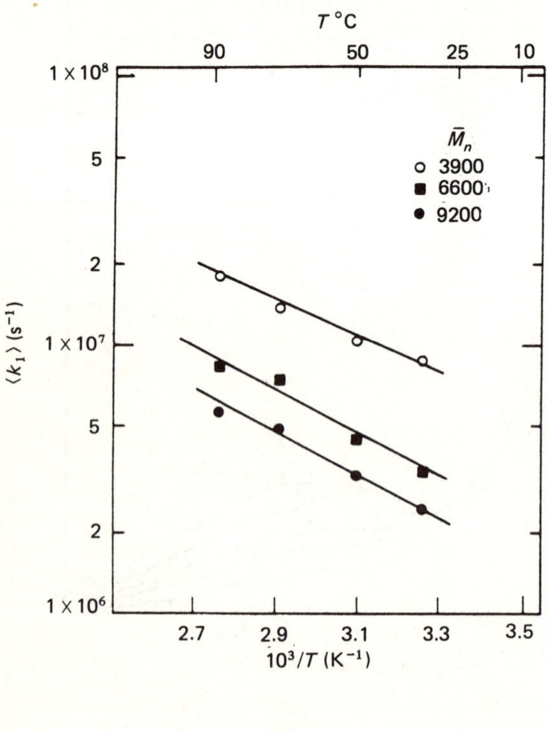

In studies of pyrene-terminated dimethyl siloxane polymers it was observed that k_1 decreased substantially when polystyrene was added to the solution of labelled polymer in toluene and cyclohexane. It was suggested that the presence of added polymer reduced the conformational mobility of the pyrene-containing polymer, possibly by complex formation or hydrodynamic screening.

A number of studies have been made of excimer-forming polymers dissolved in solid polymer matrices, such as polystyrene and poly(methyl methacrylate) (see §7.3.4). Since most polymer mixtures are incompatible, the polymer matrix usually acts as a poor solvent and increases the ratio of excimer to monomer emission of the minor component in a manner similar to the effect observed in poor liquid solvents for the same polymers. In homopolymer films, of course, the local chromophore concentration is very high and usually excimer emission predominates.

Frank and Gashgari (1979) have used the variation in excimer intensity (I_D/I_M) with matrix composition as the criterion of polymer compatibility. In their experiments, P2VN was dispersed in a variety of polymers. The ratio I_D/I_M was found to decrease sharply as the solubility parameters of

Figure 7.29. Arrhenius plots of log k_{-1} vs. $1/T$: open circle, Py-PS3900-Py; square, Py-PS6600-Py; solid circle, Py-PS9200-Py. (From Redpath and Winnik 1982.)

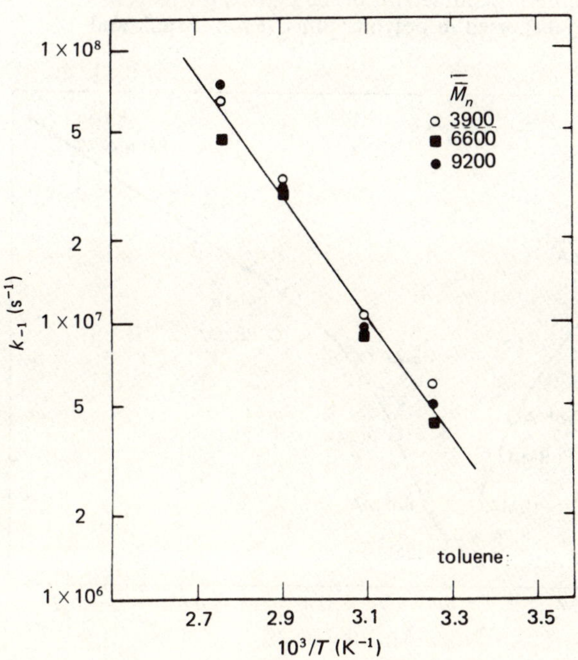

the matrix and P2VN approached each other, as shown in figure 7.30. This technique could possibly be used to determine the solubility parameter of an unknown excimer-forming polymer.

An interesting extension of this idea was recently reported by Soutar (1981) who showed a sharp minimum in the ratio I_D/I_M for P1VN and its copolymer with MMA in mixed solvent systems. It was shown that this corresponded to compositions where the solubility parameter of the solvent mixture exactly matched that of the polymer (figure 7.31).

7.5 Conclusions

Studies of excimer emission from polymers containing excimer-forming sites is a useful tool for the study of a variety of interactions in macromolecules. However, in the absence of a general theory which quantifies the relative roles of energy transfer and excimer site concentration, conclusions from these data should be drawn with caution. In solid polymers, where the chromophores exist at high concentration it seems clear that energy migration may well occur over extensive regions of the polymer by both inter- and intramolecular processes. However, the migration length or extent of energy migration may well vary from one polymer to the next and so the extent of excimer emission between two

Figure 7.30. Variation of ratio of excimer to monomer emission intensity with solubility parameter δ of the matrix, for poly(2-vinylnaphthalene) dispersed in polymer films. (From Frank and Gashgari 1979.)

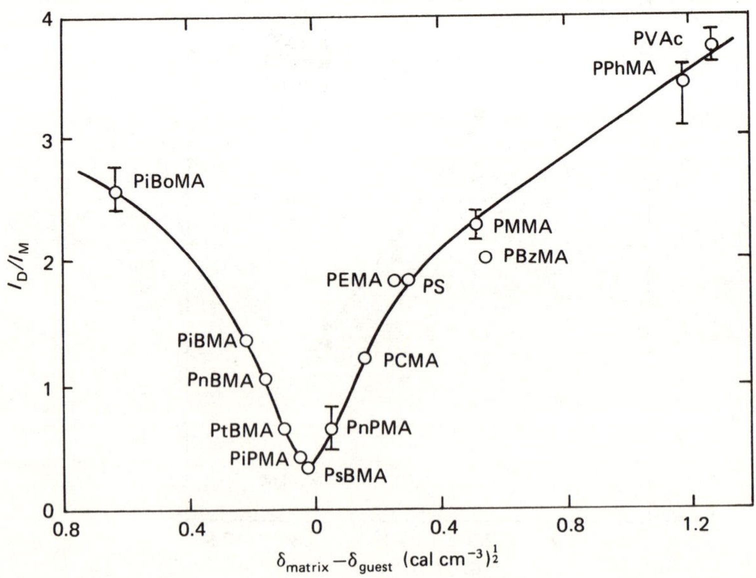

polymers which differ in structure or the mode of attachment of the chromophore cannot be directly related to the concentration of excimer sites. Similarly, in solution at high concentration, intermolecular energy transfer will also be highly effective in populating relatively low concentrations of excimer sites. This can be expected to change drastically as a function of concentration. Furthermore, strong solvent effects can be expected from changes in the effective hydrodynamic volume of the individual polymer chains.

At high dilution energy migration will be primarily intramolecular but will be strongly affected by the extension of the coil. Solvation also affects the number of loops across the chain and possibly also the number of conformations suitable for the formation of an excimer. In copolymers, and indeed in homopolymers, where the chromophores are removed by more than three carbon atoms, it seems likely that there will be chromophores isolated in conformations which cannot achieve excimer geometry within the lifetime of the excited state. In this case, complicated decay characteristics will be observed which can be best explained by use of scheme III.

In spite of these reservations, the formation of excimers in polymer systems provides a useful qualitative device for observing group inter-

Figure 7.31. I_D/I_M as a function of solubility parameter δ for the mixed solvent systems toluene/methanol and toluene/cyclohexane. (From Soutar 1981.)

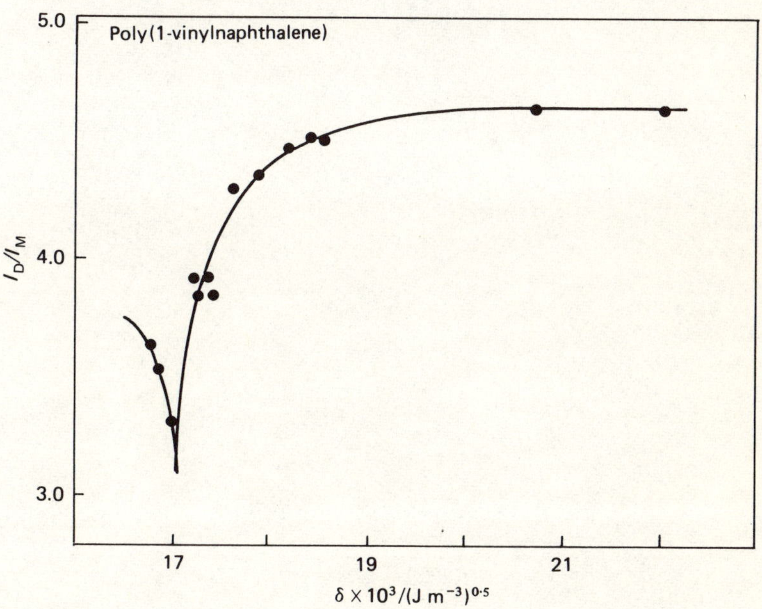

actions between individual chromophores on polymer chains. The data thereby allow the test of particular models of chain conformation, and can supplement the type of information obtained from more classical methods such as light scattering and NMR spectroscopy. Once a more definitive theoretical model is available to explain quantitatively the effects of energy migration and its role in the population of excimer sites, it seems clear that this phenomenon will provide a great deal of interesting and important information about the intimate details of the structure of polymers, both in solution and in the bulk phase.

8

Phosphorescence

8.1 Introduction

Phosphorescence is defined as emission from the triplet state. As outlined in chapter 1, direct population of the triplet state by absorption is spin forbidden and hence most phosphorescent emission will arise as a result of intersystem crossing from the excited singlet state. Since internal conversion from the excited triplet to the ground state singlet is equally spin forbidden, the radiative lifetime of the triplet is usually significantly longer than that of the singlet, being of the order of seconds rather than microseconds. Because of their longer lifetimes, molecules in the triplet state can readily lose their energy by the usual radiationless processes, by energy transfer to other molecules such as oxygen, and by chemical reaction. In mobile solution in conventional solvents at room temperature, collisions with the solvent or impurities can efficiently deactivate the triplet and hence phosphorescent emission can be observed at high intensity only when the rates of these radiationless processes are reduced, e.g., by dispersing the substance in a rigid medium or by cooling to low temperature. In much phosphorescence work, clear organic glasses are used as solvents at 77 K which reduces the effect of minor amounts of impurities. However, with careful purification, phosphorescence can also be observed from room temperature solutions of small molecules such as biacetyl and benzophenone.

8.2 Phosphorescence emission spectra

Much useful information can be obtained from phosphorescence emission measurements. By utilizing a rotating slit around the sample, as in a conventional phosphorimeter, one can eliminate prompt fluorescence emission and observe only the phosphorescence radiation.

A plot of emission intensity versus wavelength can be obtained by selecting a particular wavelength for excitation and measuring the distribution of intensities as a function of wavelength. Alternatively, one can obtain an excitation spectrum by measuring the intensity of the emitted light at a particular wavelength in the phosphorescence band and varying the wavelength of the excitation light. Excitation and emission spectra for a number of monomeric and polymeric ketones in THF-ether glass at

Figure 8.1. Phosphorescence spectra of aliphatic and polyvinyl ketones in THF/ether glass at 77 K. (*A*) Phosphorescence excitation, λ_{em} 430 nm; (*B*) phosphorescence emission, λ_{ex} 290 nm. (1) PMVK; (2) PMIPK; (3) di-isopropyl ketone; (4) copolymer MA/MVK (6%); (5) acetone. (From Somersall 1973.)

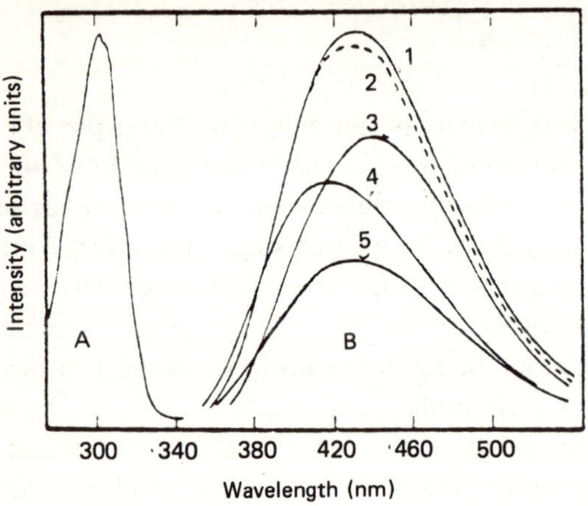

Figure 8.2. Phosphorescence excitation and emission spectra of ketone copolymers (films). (From Andrews 1978.)

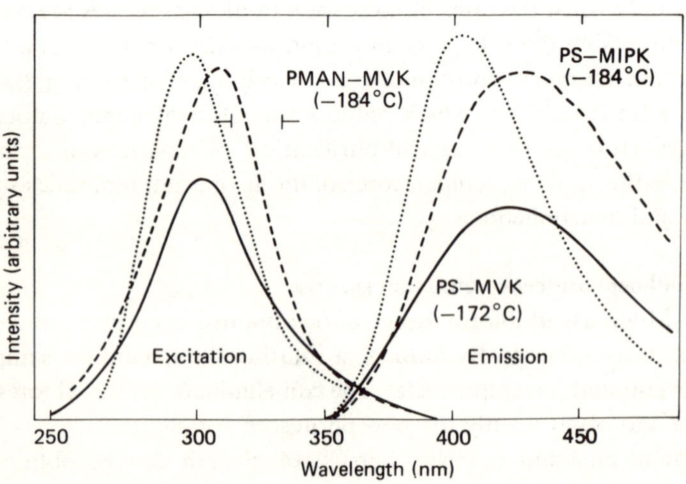

77 K are shown in figure 8.1, and for solid films in figure 8.2. From the short wavelength cutoff of fluorescence and phosphorescence emission spectra, one can estimate the singlet and triplet energies, respectively. Usually, it is possible to associate this cutoff with the 0–0 transition in both absorption and emission and hence assignments of the triplet and singlet energies can be made directly. From measurements of this type it has been estimated that the energy of the n-π^* triplet for aliphatic ketones is usually only about 10 kJ below the excited singlet, whereas for the π-π^* triplet, the difference is often as much as 120 kJ. This may be a consequence of the difference in spatial overlap of the paired and unpaired electrons in either case.

8.3 Transient measurements

Measurements of the rate of decay of phosphorescence can be easily made by a variety of procedures. The phosphorescence lifetime τ is longer for π-π^* states than for n-π^* states of ketones because the $T \rightarrow S_0$ transition borrows intensity from either of the allowed transitions for the highly spaced π-π singlet level above, or from a neighboring n-π^* singlet.

Because of impurity quenching, it is usually difficult to determine the absolute quantum yield of phosphorescence directly by spectrofluorimetry. One method which has been developed to find the fraction of triplet molecules which phosphoresce is by quenching with a highly fluorescent acceptor. The intersystem crossing yield (ϕ_{ISC}) can be measured by flash photolysis, fluorescence quenching, triplet sensitized isomerization, electron spin resonance and other methods. The phosphorescence quantum yield is then given by the product of both terms:

$$\phi_P = \phi_{ISC} q_P \tag{8.1}$$

where q_P = the quantum yield of emission from the acceptor.

Using such procedures, it has been estimated that the phosphorescence quantum yields for n-π^* states are higher, of the order of 0.1 to 0.5, than for π-π^* states (ϕ_P of the order of 0.1 or less) presumably because the intersystem crossing efficiency is much higher in the former case. The yields of phosphorescence may be increased by addition of heavy atoms or paramagnetic species, but the effect is more pronounced for π-π^* states than for n-π^* states.

8.4 Kinetics of phosphorescence

In the usual case where phosphorescence is derived from excitation of the singlet and intersystem crossing to the triplet, the rates can be described by a rather simple kinetic scheme. In this derivation we will use the notation of Birks (1970). The six relevant rate processes are shown below along with appropriate rate constants:

$S_0 \xrightarrow{h\nu} S_1$	Excitation		(8.2)
$S_1 \rightarrow S_0 + h\nu$	Fluorescence	k_{FM}	(8.3)
$S_1 \rightsquigarrow S_0$	Internal conversion	k_{GM}	(8.4)
$S_1 \rightarrow T_1$	Intersystem crossing	k_{TM}	(8.5)
$T_1 \rightarrow S_0 + h\nu''$	Phosphorescence	k_{PT}	(8.6)
$T_1 \rightsquigarrow S_0$	Internal conversion	k_{GT}	(8.7)

In the simplest case of dilute solution and monomeric chromophores, three processes will compete for the energy of the singlet excitation, namely fluorescence, intersystem crossing to the triplet, and internal conversion to the ground state singlet (8.3, 8.5 and 8.4). The total rate of these processes is given by k_M:

$$k_M = k_{FM} + k_{TM} + k_{GM} \tag{8.8}$$

In the absence of chemical reaction or external quenching, only two processes are involved in the depopulation of triplet state excitation energy, namely phosphorescence and internal conversion, and the total rate of these two processes is then given by

$$k_T = k_{PT} + k_{GT} \tag{8.9}$$

For steady-state illumination with low light intensity of I_0 einstein l^{-1} s^{-1} of a dilute solution of molar concentration $[M]$ the rate equations are given by:

$$\frac{d[^1M^*]}{dt} = I_0 - k_M[^1M^*] \tag{8.10}$$

$$\frac{d[^3M^*]}{dt} = k_{TM}[^1M^*] - k_T[^3M^*] \tag{8.11}$$

The triplet quantum yield ϕ_{TM} is defined by

$$\phi_{TM} = \frac{k_{TM}}{k_M} = \frac{k_T[^3M^*]_s}{I_0} \tag{8.12}$$

and the phosphorescence quantum efficiency q_{PT} is the ratio of the number of phosphorescence photons emitted to the number of molecules excited into the triplet and is given by:

$$q_{PT} = k_{PT}/k_T \tag{8.13}$$

The phosphorescence quantum yield ϕ_{PT} is the ratio of the number of phosphorescent photons emitted to the number of molecules excited into S_1, i.e., the number of absorbed photons and is given by:

$$\phi_{PT} = \frac{k_{PT}[^3M^*]_s}{I_0} = q_{PT}\phi_{TM} \tag{8.14}$$

Excitation by a delta function light flash or simply after cutting off the source of illumination will cause the triplet population to decay exponentially according to:

$$[^3M^*]=[^3M^*]_0e^{-k_Tt}=[^3M^*]_0e^{-t/\tau_T} \qquad (8.15)$$

where τ_T, the triplet lifetime or phosphorescence lifetime is given by:

$$\tau_T \equiv 1/k_T \qquad (8.16)$$

Evaluation of the rate parameters requires determination of the phosphorescent lifetime, τ_T, and the phosphorescence quantum efficiency q_{PT}. Although, in principle, the first excited triplet could be populated directly from the ground state singlet by absorption and q_{PT} thus determined directly, the very low $S_0 \rightarrow T_1$ absorption coefficient and the resultant feeble phosphorescence makes such measurements difficult in practice. Consequently, phosphorescence is normally excited by intersystem crossing from S_1 and the phosphorescence quantum yield ϕ_{PT} is measured by a method similar to those for the determination of ϕ_{FM}. Evaluation of q_{PT} from ϕ_{PT} requires knowledge of the triplet quantum yield ϕ_{TM}.

For a large number of aromatic hydrocarbons in dilute solution, the sum of ϕ_{TM} and ϕ_{FM} is very nearly equal to 1. However, this will not, in general, be true for polymers since in most of the systems studied, photochemical reaction also competes for the excitation energy of the excited triplet and singlet states.

8.5 Phosphorescence as a probe of molecular mobility in solid polymers

Because of the wide range of phosphorescent lifetimes which may vary from fractions of a millisecond to seconds, phosphorescent emission can give useful information about kinetic effects which are unobservable by fluorescence. However, up to the present time only a few studies have taken advantage of this principle. The earliest work involved dissolving various phosphorescent probes in polymer matrices and studying the temperature dependence of both the intensity and polarization of the emitted photons. These matrix effects were discussed in chapter 5. Among the earliest attempts to study phosphorescence emission from chromophores attached to polymer chains is that of Bousted (1970), who related the observed changes in intensity of phosphorescence with temperature in oxidized polyethylene to structural changes in the bulk polymer.

A more comprehensive study was reported by Somersall *et al.* (1974) on a variety of polymers containing small amounts of ketone and naphthalene chromophores. The polymers were coated as thin films which were inserted between two quartz plates in a brass cell holder, which

Figure 8.3. Arrhenius curves for polymer fluorescence. (From Somersall *et al.* 1974.)

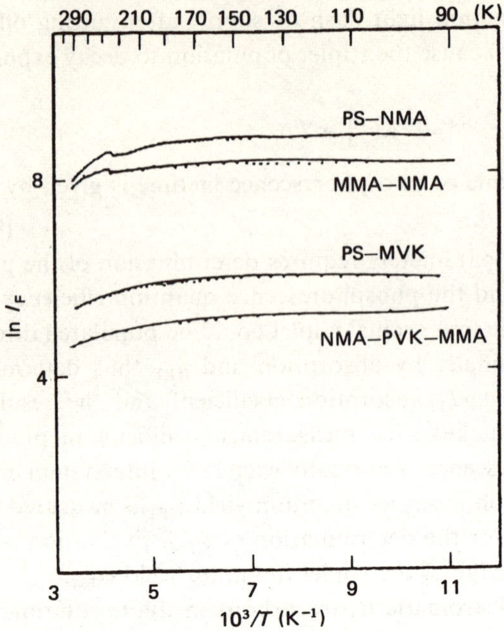

Figure 8.4. Phosphorescence intensity as a function of temperature for styrene copolymers. (From Somersall *et al.* 1974.)

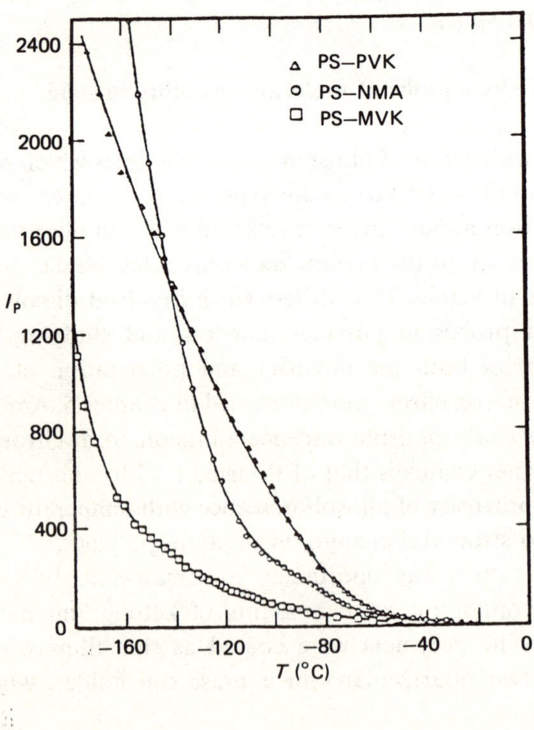

could be placed in the phosphorimeter. The cell was cooled with a stream of cold nitrogen to temperatures as low as 78 K and emission intensity was monitored as a function of temperature as the cell warmed to 25°C. Whereas fluorescence emission showed very little change with sample temperature (figure 8.3), the phosphorescence intensity changed drastically as the sample warmed to room temperature (figure 8.4). It was observed that not only were there large differences with temperature but also with the nature of the polymer. When these data are plotted in Arrhenius form (figure 8.5) typical plots show two or more straight line portions which intersect at temperatures corresponding to transitions relating to internal motions of the polymer chain.

8.6 Theoretical treatment of phosphorescence decay in solid matrices

The experimentally determined rate constant for the decay of a triplet chromophore attached to a polymer chain is given by (8.9) which is the sum of two terms. The rate constant for phosphorescence emission k_{PT} is generally regarded as being independent of temperature over the temperature range of the experiment, while the rate constant k_{GT} for nonradiative deactivation can be separated in two terms, a temperature-

Figure 8.5. Arrhenius curves for phosphorescence of the copolymer MMA-NMA-PVK (80:10:10). (From Somersall *et al.* 1974.)

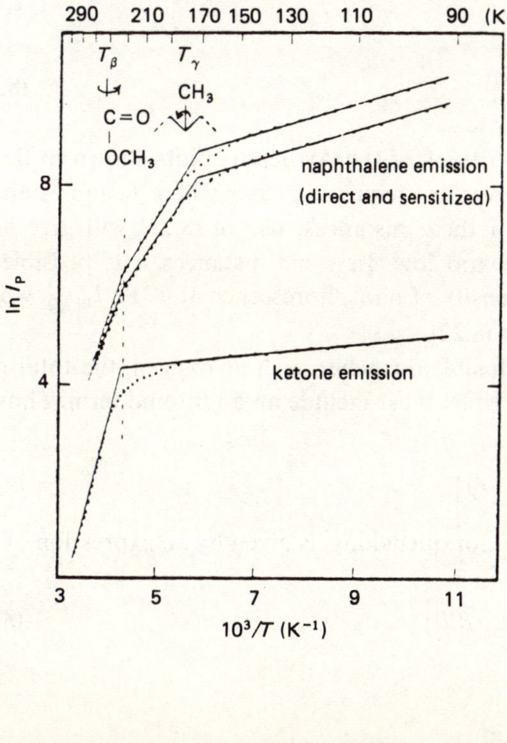

dependent and a temperature-independent component. Thus we can write:

$$k_T = k^0_{PT} + k^0_{GT} + A'_{GT} \exp[-\Delta E_{GT}/RT] \tag{8.17}$$

Here the superscripted rate constants are now independent of temperature. Rearrangement of this equation gives:

$$\ln(k_T - k^0_T) = \ln A'_{GT} - \Delta E_{GT}/RT \tag{8.18}$$

where k^0_T is the value of k_T as $T \rightarrow 0\,K$. For steady-state illumination at low phosphorescence intensity, I_P, we have:

$$I_P = BI_A\phi_{PT} \tag{8.19}$$

$$= BI_A q_{PT}\phi_{TM} \tag{8.20}$$

$$= BI_A(k_{PT}/k_T)(k_{TM}/k_M) \tag{8.21}$$

where B is an instrumental factor assumed to be constant over the conditions of the experiment. If we assume that k_T is the only temperature-dependent rate constant, it follows that:

$$\ln\left(\frac{1}{I_P} - \frac{1}{I^0_P}\right) = \ln\left(\frac{A'_{GT}k_M}{Bk^0_{PT}k_{TM}I_A}\right) - \frac{\Delta E_{GT}}{RT} \tag{8.22}$$

here ΔE_{GT} is the activation energy for the temperature-dependent part of the internal conversion rate constant k_{GM}. In the situation where I^0_P is very much greater than I_P it is possible to approximate (8.22) by the expression:

$$\ln I_P = \text{const} + \frac{\Delta E_{GT}}{RT} \tag{8.23}$$

This explains the linear portions of the Arrhenius plots shown in figure 8.5. This equation is, however, inadequate in cases where I_P and I^0_P are of comparable magnitude. In these instances, use of (8.23) will give activation energies which are too low. In some instances it is possible to approximate I^0_P by the intensity of phosphorescence at 77 K, $I_{P(77)}$, which then can be substituted in (8.22).

In the presence of a diffusible quencher such as oxygen, the total rate constant for decay of the triplet must include an additional term. Thus k_T will be given by:

$$k_T = k_{PT} + k_{GT} + k_Q[Q] \tag{8.24}$$

where k_Q, the rate constant for quenching, is given by an expression of the form

$$k_Q = A_Q \exp[-\Delta E_Q/RT] \tag{8.25}$$

The expression for the intensity of phosphorescence now becomes

$$\frac{1}{I_{\mathrm{P}}}-\frac{1}{I_{\mathrm{P}}^{0}}=\frac{k_{\mathrm{M}}}{BI_{\mathrm{A}}k_{\mathrm{PT}}k_{\mathrm{TM}}}(A'_{\mathrm{GT}}\exp[-\Delta E_{\mathrm{GT}}/RT]+[\mathrm{Q}]A_{\mathrm{Q}}\exp[-\Delta E_{\mathrm{Q}}/RT]) \qquad (8.26)$$

When quenching becomes the dominant mechanism for triplet decay this equation can be approximated by

$$\frac{1}{I_{\mathrm{P}}}=\frac{k_{\mathrm{M}}[\mathrm{Q}]A_{\mathrm{Q}}}{BI_{\mathrm{A}}k_{\mathrm{PT}}k_{\mathrm{TM}}}\exp[-\Delta E_{\mathrm{Q}}/RT] \qquad (8.27)$$

and hence a plot of the logarithm of the phosphorescence intensity versus $1/T$ according to

$$\ln I_{\mathrm{P}}=\mathrm{const}+\frac{\Delta E_{\mathrm{Q}}}{RT} \qquad (8.28)$$

will give a straight line with a positive slope equal to $\Delta E_{\mathrm{Q}}/RT$. If the concentration of the diffusible quenching molecule is constant throughout the experiment, then the numerical value of ΔE_{Q} will be equal to the activation energy for diffusion ΔE_{D}. With gaseous quenchers such as oxygen, maintaining a constant partial pressure over a very thin film of the polymer does not assure that the concentration of the quencher is constant throughout the experiment and ΔE_{Q} will also include the temperature coefficient of the solubility of O_2 in the polymer.

8.7 Experimental studies of phosphorescence decay in solid polymers

Using these kinetic expressions it is possible to explain the data of Somersall *et al.* (1974). Figure 8.3 shows the Arrhenius curves for phosphorescence from a polymer containing both ketone and naphthalene probes. It is notable that there is no change in the emission intensity from 90 K up to room temperature, thus confirming the independence of k_{FM}, k_{GM} and k_{TM} of temperature, at least under these experimental conditions. The much higher luminescence efficiency of the naphthalene chromophore is also evident from these results although substantial quenching seems to occur in the presence of added carbonyl groups such as those from poly(phenyl vinyl ketone). Figure 8.6 shows that the straight line Arrhenius plots for the phosphorescence emission from a wide variety of polystyrene samples intersect in the region of the γ-transition. In the case of polystyrene, the γ-transition can be related to rotational motions of the phenyl rings in the solid polymer. Correlation of this motion with small motions of the backbone methylene groups allows a mechanism for hole formation through which oxygen can diffuse to the excited chromophore.

Above the γ-transition, the activation energies calculated from the slopes of the Arrhenius plots agree reasonably well with the values for oxygen diffusion in polymers measured at higher temperatures (20–30 kJ mol^{-1}). In polyethylene two transitions were observed, one associated with the γ-transition at 165 K identified with the crankshaft motion of methylene groups in the polymer chain, and another at 100 K which has yet to be identified. Similar studies of copolymers with poly(methyl methacrylate) (PMMA, figure 8.7) showed evidence of the β-transition associated with the motion of the ester group at 220 K and the γ-transition associated with the α-methyl group at 170 K.

Data from Somersall *et al.* (1974) on a variety of polymers are summarized in table 8.1.

The identification of solid phase transitions from the intersection of straight line portions of the Arrhenius curves, while useful, must be applied with care. More detailed studies by Andrews (1978) using a nitrogen cryostat showed that the temperature at which the two Arrhenius plots intersect depends in many cases both on the partial pressure of oxygen and the sample dimensions as well as other experimental conditions. It was demonstrated that the high temperature region of the Arrhenius plot was

Figure 8.6. Arrhenius curves for polystyrene phosphorescence intensity in polystyrene and its copolymers containing keto or naphthalene groups. (From Somersall *et al.* 1974.)

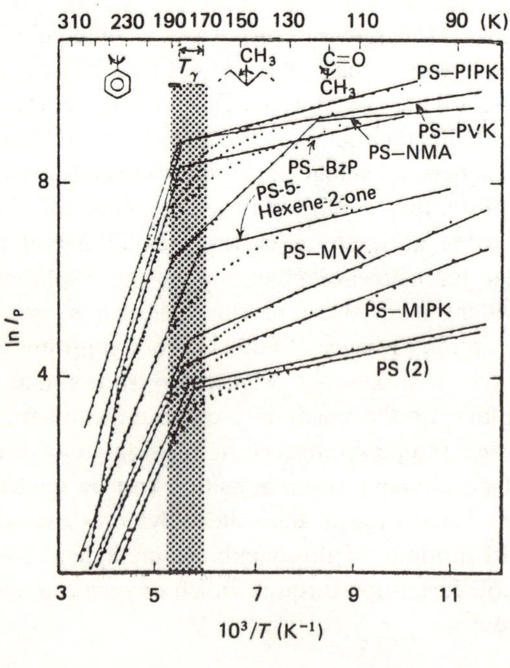

indeed caused by oxygen quenching and that the phosphorescence intensity obeyed a Stern–Volmer law typical of diffusion controlled quenching. Figure 8.8 (page 208) illustrates the effect of oxygen pressure on the emission from PMMA containing small amounts of copolymerized naphthalene groups. It is obvious that although deviations from linear Arrhenius behavior are shown even with the traces of oxygen remaining after extensive pumping under vacuum, the transition temperature as determined by the intersection of the two Arrhenius straight lines varies from 230 K in totally evacuated samples to 140 K in those exposed to 1 atmosphere of oxygen. These results can be explained by reference to (8.26). In the absence of oxygen or other quencher, the temperature variation of the phosphorescence will depend primarily on the first term inside the parentheses, and for the naphthalene chromophore in the absence of oxygen, there appears to be no change in the phosphorescence intensity up to 230 K, at which point a sharp discontinuity occurs. This can be associated with the β-transition in PMMA relating to motion of ester groups. These motions could perturb the excited state of the naphthalene chromophore sufficiently to provide an additional mode of deactivation of

Figure 8.7. Arrhenius curves for poly(methyl methacrylate) phosphorescence intensity in PMMA and copolymers containing keto or naphthalene groups. (From Somersall *et al.* 1974.)

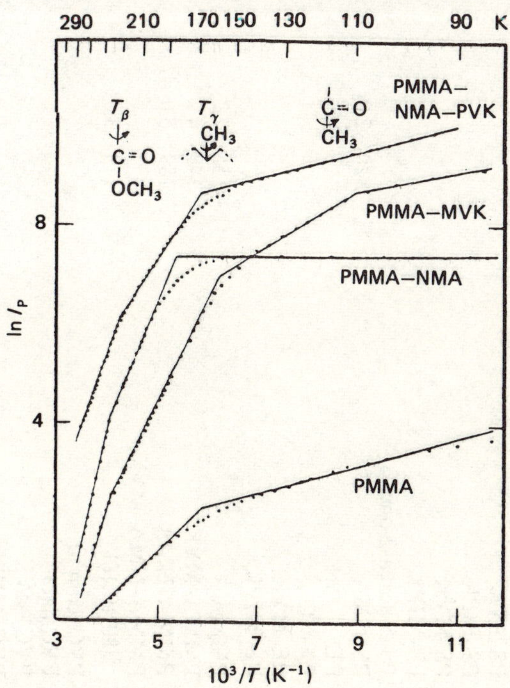

Table 8.1. *Transition temperatures* (T_{trans}(K)) *and activation energies* (E_a(*kcal/mol*)), from *Arrhenius plots of polymer phosphorescence intensity*

Polymers	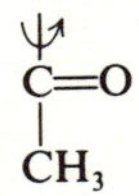		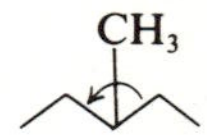		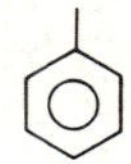		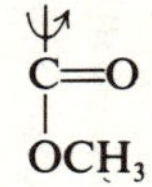		Chain segment rotation	
PS					178	5.2				
PS–PVK					182	7.0				
PS–PIPK					183	6.2				
PS–MVK					175	4.7				
PS–MIPK					183	4.9				
PS–5-hexen-2-one					175	5.4				
PS–NMA			120	2.0						
PE									163	4.3
PE–CO									163	4.5
PE–MVK									163	4.5
PE–MIPK									163	4.5
PMMA			~155	2.2						
PMMA–MVK			148	4.2			248	7.6		
PMMA–NMA			189	5.6			245	9.2		
PMMA–NMA–MVK (80:10:10)			173	3.0			236	6.0		
PVC (comm)									180	4.0
PVC (lab)									163	4.7
PVC–MVK									188	5.0
PVC–MIPK									163	4.2

PAN					~216	7.5
PAN–PVK					235	9.8
PAN–MVK	125	2.1			216	9.8
PAN–MA–MVK						
PAN–MA–MVK (75:20:5)	136	3.7			238	5.8
PMAN					234	7.6
PMAN–MVK					238	7.0
PMAN–MMA–MVK (75:20:5)					230	5.6
PMVK	145	5.8			230	12.2
PMIPK					233	7.2
PPVK			196	5.7		

the excited triplet and hence cause a reduction in phosphorescence. This work also demonstrated that the lifetime of phosphorescence emission decreased appropriately at the same temperature in a manner consistent with this hypothesis.

In the presence of oxygen one must consider also the effect of the second term in (8.26). This consists of two parts, one relating to the equilibrium solubility of the gas in the polymer denoted [Q] and the other related to the change in diffusion rate constant as expressed in Arrhenius form. If the concentration or the partial pressure of oxygen is very low, then this term will not become important until the diffusion rate becomes very high. That is, the effect will be shown at higher temperatures. Similarly, the temperature at which Arrhenius behavior will be observed, according to (8.28), will depend on the relative magnitudes of the two expressions and until the second expression is an order of

Figure 8.8. Arrhenius dependence of PMMA-NMA phosphorescence at various oxygen partial pressures. (From Andrews 1978.)

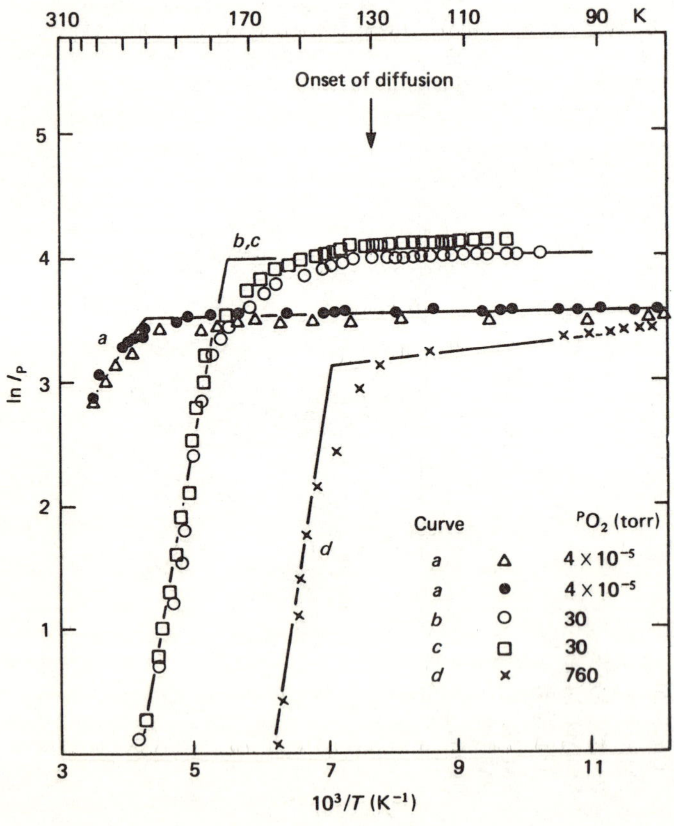

magnitude larger than the first, one should not expect to find a linear Arrhenius plot. This would then explain the observed shift in the linear portions of the high temperature region with oxygen partial pressure. The activation energies associated with Henry's law coefficients for gaseous solubility in polymers are typically 4–8 kJ mol^{-1}. This is small compared to E_D and hence the slope of the high temperature region of the Arrhenius curve should be primarily dependent on the activation energy for diffusion E_D, which explains why the slopes are relatively independent of experimental conditions and oxygen pressure.

Of more importance in identifying solid phase transitions is the fact that in the presence of oxygen all curves begin to deviate from their low temperature linearity at about the same temperature, 130 K. It is possible to identify this point as the temperature at which diffusion of oxygen becomes large enough so that the second term in (8.26) is now comparable in magnitude to the first term. If we assume a model for the polymeric solid phase in which diffusion of small molecules is assisted by small rotations or vibrations of specific groups, then one can use this point to identify the temperature at which such movements become detectable. It seems likely that this temperature should then correlate with certain of the solid phase transitions identified in polymers by other means. This method is analogous to that developed by Braun and Guillet (1976) for identifying the glass transition temperature in polymers from gas chromatographic data relating to the diffusibility of much larger organic molecules in the polymer phase.

8.8 Oxygen diffusion rates measured by phosphorescence quenching

The identification of a diffusive oxygen quenching mechanism of phosphorescence in polymer films permits an interesting application of the technique to the determination of diffusion rate constants in solid polymers. The procedure was first suggested by Oster *et al.* (1962) and has been used by Unterleitner and Hormats (1969) and by Shaw (1967) for measuring the diffusibility of oxygen in cylindrical samples of PMMA.

Jones (1968) reported measurements on the permeability of thin polymer films to oxygen by measurements of changes in the phosphorescence lifetime with oxygen pressure. Andrews (1978) reports an application of this procedure to the determination of diffusion coefficients for oxygen at very low temperatures in thin films of polymers containing copolymerized naphthyl groups. Numerous experiments have shown that diffusion of low molecular weight gases through polymer membranes obeys Fick's law of diffusion which can be given for the simple one-dimensional case by:

$$\frac{\partial c}{\partial t} = D\left(\frac{\partial^2 c}{\partial \xi^2}\right) \tag{8.29}$$

where c is the concentration of diffusant, D is the diffusion coefficient of gas moving in direction ξ and t is time. In deriving solutions for this equation it is usually assumed that Henry's law is valid so that c is proportional to the partial pressure of the gas surrounding the sample. General solutions to this type of equation are formally analogous to those obtained from the flow of heat in a solid bounded by parallel planes and developed by Carslaw and Yeager (1959). A detailed derivation of the equation is beyond the scope of this book and only the relevant results will be quoted here.

In the experiment described by Andrews, the films were coated on polished aluminum plates so that diffusion occurred in one direction from the front surface only. Phosphorescence was observed by reflection through the film. The amount of gas absorbed by the film at any time t is given by the expression:

$$c = lc_0 + l\left(\frac{c_2}{2} - c_0\right)\left(1 - \sum_{m=0}^{\infty} \frac{1}{(2m+1)^2} \exp\left[\frac{Dt(2m+1)^2\pi^2}{l^2}\right]\right) \tag{8.30}$$

Similarly, the intensity of phosphorescence is given by

$$I_P(t) = I_A B \phi_P \sum_{m=0}^{\infty} \frac{1}{(2m+1)^2} \exp\left[\frac{-Dt\pi^2(2m+1)^2}{l^2}\right] \tag{8.31}$$

This series converges rapidly for times only slightly greater than zero and hence only a small error is incurred by retaining the first term of the series which gives the simple expression of the form shown below:

$$I_P(t) = B' \exp\left[\frac{-Dt\pi^2}{l^2}\right] \tag{8.32}$$

Here B and B' are instrumental constants relating to the geometry of the sample and the optics of the phosphorimeter. The diffusion constant can be obtained readily from this equation by plotting the log of the phosphorescence intensity as a function of time.

In this experiment a thin film of polymer coated on a reflecting aluminum plate is evacuated in a cryostat inserted into a phosphorimeter. The intensity of phosphorescence is then monitored as the cryostat is exposed to a given partial pressure of oxygen. Typical data with films of differing thickness at various temperatures are shown in figure 8.9. The exponential nature of the decay is clearly shown from these experiments.

Diffusion constants obtained from these data are shown in table 8.2. A plot of these diffusion constants in Arrhenius form shows a linear relationship from $-110°C$ to room temperature and an activation energy of 28 kJ mol^{-1} for oxygen in polystyrene. This value agrees very well with those estimated from the data of Somersall *et al.* (1974) and is similar to the activation energy obtained by Stannet (1968) of 35 kJ mol^{-1} for oxygen in glassy polystyrene above 25°C. It should be pointed out that the advantage of this procedure for measuring diffusion constants is that it permits the determination of very low diffusivities which would be nearly undetectable by conventional procedures. Util-

Figure 8.9. Semi-logarithmic plot of phosphorescence intensity vs. time after exposure of PS-NMA films to oxygen. (From Andrews 1978.)

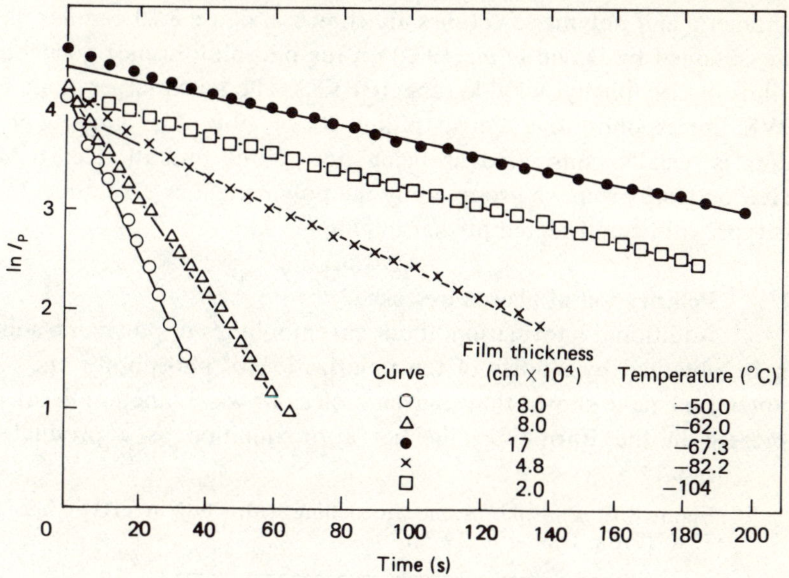

Table 8.2. *Diffusivity of oxygen in films of PS–NMA of various thickness at different temperatures*

Temperature (°C)	Film thickness (cm)	Diffusion constant (cm^2 s^{-1})
$+23.2 \pm 0.1$	2.8×10^{-2}	3.1×10^{-7}
$+3.0 \pm 1.0$	1.7×10^{-3}	1.2×10^{-7}
-50.0 ± 2.0	8.0×10^{-4}	4.3×10^{-9}
-62.0 ± 1.0	8.0×10^{-4}	3.2×10^{-9}
-67.3 ± 2.0	1.7×10^{-3}	2.1×10^{-9}
-82.2 ± 0.5	4.8×10^{-4}	3.3×10^{-10}
-104 ± 1.0	2.0×10^{-4}	3.6×10^{-11}

ization of this technique to measure diffusivities of polymers at cryogenic temperatures should provide much needed data on the correlation of gaseous diffusivities with molecular motion in solid polymers.

Another application of phosphorescence measurements in polymer systems relates to the quenching of phosphorescence from polymers containing sequential chromophores. When polymers are dissolved at high dilution in solvents capable of forming clear glasses at 75 K, strong phosphorescence is observed from ketone or naphthalene chromophores. In the presence of quenching molecules static quenching will occur by the exchange mechanism. Typical Perrin plots (1.46) for a number of both monomeric and polymeric ketones are shown in figure 8.10. Similar data were obtained by David *et al.* (1972) on the phosphorescence quenching of films of poly(phenyl vinyl ketone) (PPVK). The much larger slope for PPVK corresponds to a Perrin radius R_0 of 19 Å. The larger Perrin radius is usually considered as being due to the migration of triplet excitation from group to group along the polymer chain. Experiments of this type will be considered further in chapter 9.

8.9 Polarization of phosphorescence

Additional information about chromophores in polymeric solids can be obtained by studies of the polarization of phosphorescence. In chapter 1 we have shown that a given molecular wave function ψ can be expressed in the Born–Oppenheimer approximation as a product of

Figure 8.10. Phosphorescence quenching with COD in THF–ether at 77 K. (From Dan *et al.* 1973.)

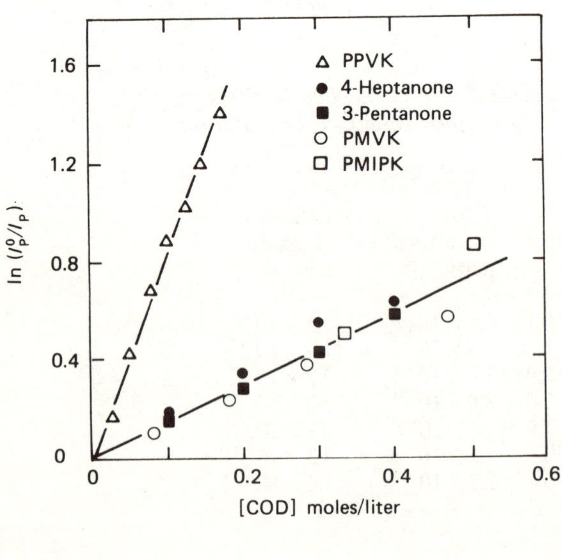

electronic, vibrational and spin components, respectively. The probability of a transition between states *m* and *n* is proportional to the square of the transition dipole moment $\mathbf{R}_{mn}$ which is a vector quantity possessing a definite orientation relation to the molecular skeleton. In an arbitrary coordinate system the vector $\mathbf{R}_{mn}$ can be resolved as the sum of three components:

$$\mathbf{R}_{mn} = \mathbf{R}_x + \mathbf{R}_y + \mathbf{R}_z \tag{8.33}$$

An allowed transition must have a non-zero transition dipole moment, implying that at least one of the components in (8.33) must be non-zero. The allowed transition is then polarized along a particular direction and the probability of absorption is highest for those quanta whose electric vectors vibrate parallel to $\mathbf{R}_{mn}$. An electronic transition which is symmetry forbidden can become partially allowed due to vibronic interactions resulting in coupling between the excited electronic state and a non-totally symmetric nuclear vibration. For example, in formaldehyde the transition to the n-π^* singlet is symmetry forbidden. However, coupling of this S_1 state with the rocking vibration of the O atom out of the CH_2 plane results in this transition becoming allowable. The forbidden transition can be thought of as borrowing intensity from an allowed transition to an upper state which has been added by vibration of suitable symmetry. Also, it can be shown that the component of intensification possesses the polarization direction of the allowed transition. In the case of fully allowed transitions vibronic interactions can actually cause a change in the direction of polarization.

Similarly, transitions which are spin forbidden can become partially allowed by means of spin-orbit coupling. The phosphorescence of formaldehyde, for example, occurs because of the mixing of the triplet n-π^* state with the second excited singlet π-π^* state and possesses a polarization direction characteristic of the latter. The theoretical basis between spin-orbit coupling and vibronic interactions has been developed by Albrecht (1963).

8.10 Theoretical treatment of phosphorescence polarization

The mathematical treatment of phosphorescence polarization is formally identical to that for fluorescence as outlined in chapter 6.

In order for an electronic transition to occur, the transition dipole moment must have a component in the direction of the oscillating field of the incoming photon. Consider a single molecule located at the origin of a coordinate system as shown in figure 6.2 and exposed to light that is plane polarized in a plane perpendicular to the *x* axis. Suppose also that its transition moment for absorption, **R**, makes an angle θ with this direction. If the molecule does not change its orientation between the time of absorption and emission, then the emitted light when viewed along YO, will have the two linearly polarized orthogonal components $I_\parallel$ and $I_\perp$ for two intensities along OZ and OX, respectively. The polarization P is then defined as:

$$P=\frac{I_\parallel - I_\perp}{I_\parallel + I_\perp} \tag{8.34}$$

In the case where absorption and emission transition moments are parallel, Weber (1966) has shown that polarization can be described by

$$\frac{1}{P}-\frac{1}{3}=\frac{2/3}{\frac{3}{2}\cos^2\theta-\frac{1}{2}} \tag{8.35}$$

where $\cos^2\theta$ is the average cosine square of the angle determined by the emission oscillator and the electric vector of the exciting light. If no motion occurs between absorption and emission, then $\cos^2\theta = 3/2$ and the substitution of this into (8.35) yields $P=0.5$.

When the absorption and emission oscillators are parallel but form an angle λ, P is given by:

$$\frac{1}{P}-\frac{1}{3}=\frac{5}{3}[2/(3\cos^2\lambda-1)] \tag{8.36}$$

When the absorption and emission oscillators are perpendicular, i.e., $\lambda=90°$, the value of P becomes -0.33. Values of P for other situations are summarized in table 8.3 where R_a and R_e denote the absorption and emission oscillators, respectively. The polarization values listed in this table are seldom achieved experimentally since depolarization can occur due to a number of extrinsic factors. For example, depolarization can be caused by unequal transmission of plane and vertically polarized light by the optical system and diffraction gratings of the spectrometer. This can

Table 8.3. *Theoretical maxima for emission polarization*

Orientation of		
R_a	R_e	P
Any	Parallel to R_a	1/2
Any	Perpendicular to R_a	−1/3
Perpendicular to an axis of symmetry C_n where $n>2$	Parallel to C_n	−1/3
Perpendicular to an axis of symmetry C_n where $n>2$	Perpendicular to C_n	1/7

usually be corrected for experimentally. However, the major and most interesting causes of extrinsic depolarization are random Brownian motion and energy migration. It is usually possible to determine which model of depolarization is occurring by varying experimental conditions. In rigid media or highly viscous solvents, Brownian motion is greatly reduced and hence depolarization is often the result of energy migration. The amount of depolarization by Brownian motion is usually determined in dilute fluid solutions where energy migration can often be neglected. The magnitude of the induced anisotropy or the rate at which it decreases can be used to estimate the kinetic parameters defining these two major processes.

8.11 Depolarization by Brownian motion

The extent of depolarization caused by the rotation of a chromophore in fluid solution due to Brownian motion during its excited state lifetime was originally described by Perrin (1929) for the case of spherical molecules

$$\frac{1}{P}-\frac{1}{3}=\left(\frac{1}{P_0}-\frac{1}{3}\right)\left(1+\frac{3\tau}{\rho_0}\right) \tag{8.37}$$

where

$$\rho_0=\eta V/kT \tag{8.38}$$

Here P is the measured degree of polarization, P_0 is the limiting degree of polarization in the absence of rotation (for example, in frozen solution), τ is the excited state lifetime, η is the viscosity of the solvent at temperature T and V is the effective molar volume of the chromophore. The quantity ρ_0 is usually called the rotational relaxation time. For spherical molecules the rotational relaxation time is defined as the time required for the value of cos θ to change from its value of 1 at $t=0$ to a value of $1/\varepsilon$ (see also §6.5).

8.12 Depolarization by energy migration

A number of theroetical treatments have been developed to explain concentration depolarization due to energy migration. All of these theories assume a random spatial distribution and orientation in the chromophores in the media. The models due to Jablonski, Forster-Öre and Weber all predict a linear relationship between $1/P$ and the luminescent solute concentration at low values of the latter. Details of these treatments and experiments relating to energy migration are given in §9.7.2.

8.13 Experimental studies of phosphorescence decay and polarization

Kilp and Guillet (1981) studied the time-dependence of phosphorescence decay for atactic and isotactic poly(acrylophenone). The simple relationship

$$I_P = I_0 \exp[-t/\tau_P] \tag{8.39}$$

should give rise to a linear first-order decay curve when $\ln I_P$ is plotted against t. However, the experimental curves could only be fitted accurately as a sum of two exponentials in the form of

$$I_P = I_1^0 \exp[-t/\tau_1] + I_2^0 \exp[-t/\tau_2] \tag{8.40}$$

An analysis of this form gives rise to two components in the phosphorescence, one having a lifetime τ_1 of about 2.5 ms, the other τ_2 of about 14.0 ms. The relative intensity of the two components I_1/I_2 is about 2.6. Similar curvature in decay functions was observed by David *et al.* (1977) in copolymers of styrene and vinyl benzophenone. It was proposed that these results might be explained on the basis of emission from two states with close lying triplet levels but different lifetimes. A similar explanation has been advanced to explain curvature observed in the decay of phosphorescence from small molecules entrapped in organic glasses. However, the generality of non-single exponential decay in organic glasses may suggest that the behavior is dependent on site distributions in the glass (see chapter 5).

Related to these studies are those of Miller and North (1975) on phosphorescence depolarization of small molecule probes such as anthrone and benzophenone in polymeric matrices. They noted that phosphorescence polarization of anthrone in SAIB (sucrose acetate isobutyrate) varied with temperature in a manner consistent with changes in the internal viscosity of the solution. Figure 8.11 shows their results along with data on the dielectric loss peak for SAIB. From calculations of the rotational lifetime ρ, they concluded that depolarization cannot arise from isotropic rotation of the chromophore in the solvent on this time scale and that therefore it probably occurs by way of either a rapid restricted vibrational motion comparable with the β process, or by an intramolecular butterfly or twisting motion observed in dielectric studies. This would explain the fact that considerable depolarization of anthrone occurs below the glass transition temperature of the solvent where rotational motion of the conventional kind is presumably precluded.

Horie and Mita (1978) have utilized the quenching of phosphorescence of benzyl by anthracene to study the rate constants for collision of end groups in polystyrene molecules. They determined the rate of quenching of phosphorescence of polystyrene terminated with benzyl groups with other polystyrene molecules terminated by anthracene. The system was used as a model for bimolecular termination·of radical polymerization in solution, and it was shown that the rate constants for triplet transfer were far larger than the chain termination constants determined from non-luminescent techniques. The higher values of the rate constants were attributed to the occurrence of some energy migration in these systems. The quenching constants decreased with degree of polymerization, P, as shown in figure 8.12. In benzene, a relatively normal relationship was observed in which the rate of quenching k_q decreased with $P^{-0.29}$. More complex behavior was observed in cyclohexane. A theoretical model based on the segmental diffusion of non-interpenetrating coils predicted the correct qualitative variation of k_q with degree of polymerization P except at very high degrees of polymerization.

8.14 Intramolecular phosphorescence quenching

An elegant application of phosphorescence quenching measurements has been developed by Winnik and co-workers (Winnik 1981) to study end-to-end cyclization probabilities in model compounds of interest for polymer studies. They studied intramolecular phosphorescence quenching in ω-alkenyl esters of benzophenone-4-carboxylic acid:

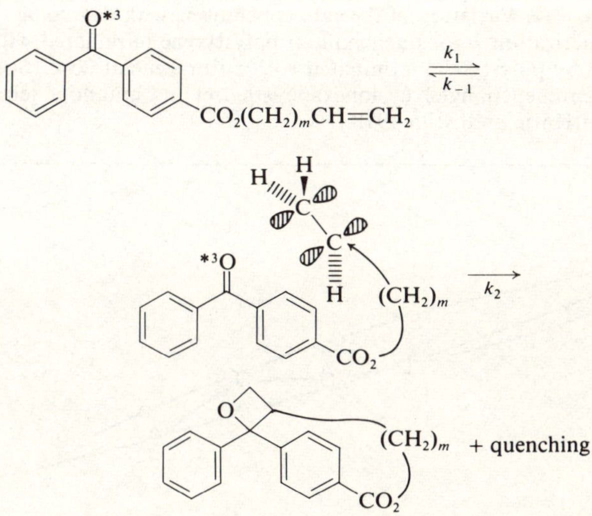

In these molecules the quenching group is a mono-substituted alkene attached to the end of the hydrocarbon chain. Since triplet quenching by such groups is 10^2–10^3 lower than k_{diff}, intramolecular reaction occurs after conformational equilibrium has been attained. Careful measurements of the phosphorescence lifetimes (τ^3) as a function of chain length m allow calculation of the relative rate constant for intramolecular quenching. Winnik's results are shown in figure 8.13 (Mar *et al.* 1981). The steep rise and maximum at $m = 12$ illustrate the steric requirements of the quenching reaction. Chains shorter than eight CH_2 groups cannot,

Figure 8.11. Degree of polarization, p, of phosphorescence plotted against temperature for, circle, 10^{-3} mol n^{-3} anthrone in SAIB λ_{ex} 350 nm, λ_{em} 440 nm: broken line, dielectric loss of SAIB (ε'') measured at 3.5 kHz. (From Miller and North 1975.)

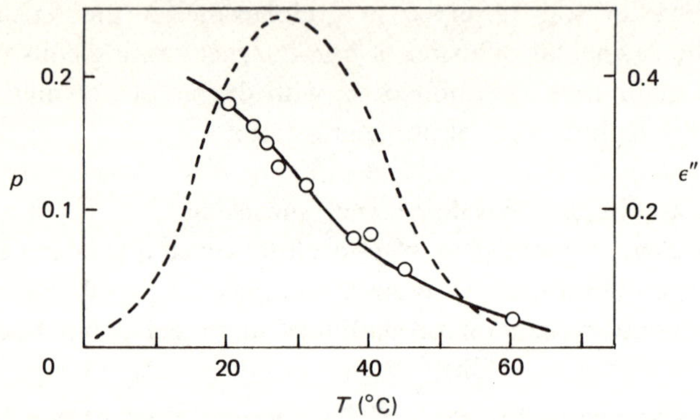

Figure 8.12. Variation of the rate constant k_q with degree of polymerization P for quenching of polystyrene terminated with triplet benzil by polystyrene terminated with anthracene at 20°C. Solvents are benzene (triangle), cyclohexane (square) and butanone (circle). (From Horie and Mita 1978.)

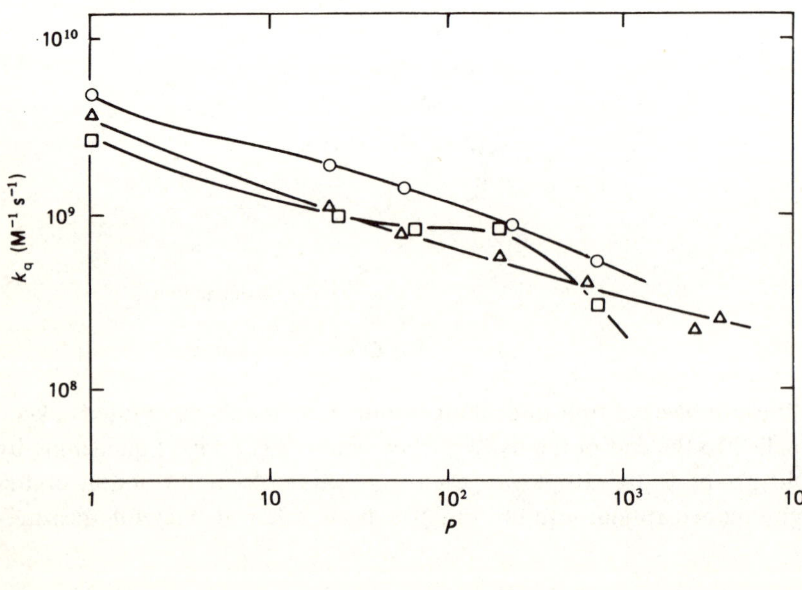

Figure 8.13. The rate of intramolecular phosphorescence quenching, by the terminal alkene group in the molecules benzophenone–4 CO_2–$(CH_2)_m$–CH=CH_2, as a function of chain length, *n* (CCl_4 solution at 5×10^{-4} M at 22°C). (From Mar *et al.* 1981.)

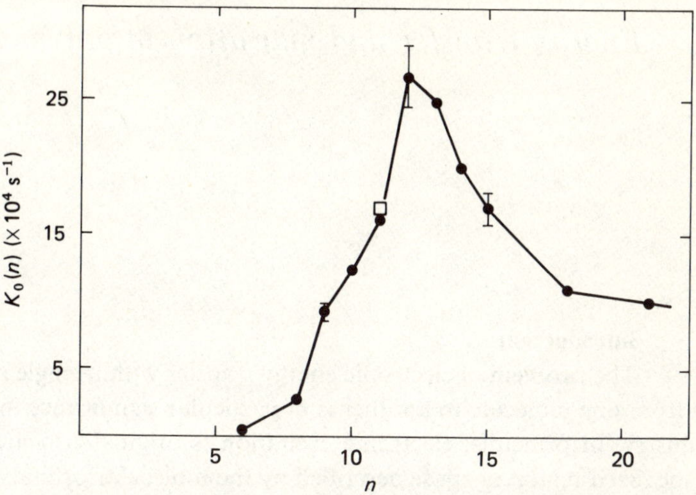

without serious distortion, achieve conformations where the π electrons of the double bond overlap the half-filled *n* orbital of the keto group. For chains longer than $m=8$, the activation energy is independent of chain length and the decrease above $m=11$ reflects the decrease in the number of 'reactive' conformations as the chain length increases. Winnik has shown good correlations between these experimental results and those predicted from Monte Carlo simulations of random walks on a tetrahedral diamond lattice for hydrocarbon chains. Extension of these end-to-end cyclization studies to actual polymer systems has been discussed in §7.4.6.

9

Energy transfer and migration in polymers

9.1 Introduction

The problem of electronic energy transfer within single molecules and from one molecule to another is of particular significance in polymer chemistry. In principle, electronic excitation is originally located within the localized region of space described by the molecular orbitals involved in the ground and excited states. Normally this localized excitation is restricted to the small number of atoms involved in the chromophore itself, however it is well known that the excitation energy can be transferred to other parts of the same molecule, or indeed from one molecule to another. The principles of these excitation transfer processes have been outlined briefly in chapter 1.

At least three types of energy transfer can be distinguished. The first of these is the transfer of the excitation from the locus of the absorbing group, herein called a chromophore, to bonds adjacent to that chromophore within the same molecule. This is illustrated for example in aliphatic ketones where the light is absorbed by the ketone carbonyl, but reaction occurs by the breaking of bonds adjacent to the carbonyl, such as the α–β or β–γ carbon–carbon bonds (see chapter 10). In these primary photochemical reactions the energy absorbed by one bond system has been transferred to another. As the size of the molecule and its complexity increases, one can expect to find quite extensive delocalization of the original excitation. This is particularly true if, for example, the chromophore consists of extended sequences of double bonds. The theory for this case has been well described in terms of exciton band mechanisms but there is still much to learn about the reason for the migration of energy in systems where delocalized excitons are not likely to occur.

A second type is collisional energy transfer, where the electronic excitation localized on one group is transferred to another group on the same or different molecule as a result of the overlap of electronic charge clouds which occur during a collision (§1.14).

The third type of energy transfer which can be distinguished is that which takes place, not during the collision, but by resonance transfer over extensive regions of space as a result of dipole–dipole and dipole–multipole interactions, of which the simplest example is that described by Förster (§1.13).

9.2 Types of energy transfer in polymers

Polymer molecules undergo similar energy transfer processes. However, we can distinguish at least two major types of energy transfer in a polymer system. The first of these involves the transfer of excitation from or to a large molecule from a small molecule, as shown in figure 9.1*a*. For example, the excitation originally localized on a small molecule can be transferred to a large polymer chain, causing a sensitized photochemical reaction. Alternatively, the excitation energy localized on a chromophore in a polymer chain can be transferred to a small molecule, thus quenching a photochemical or photophysical process. Examples of

Figure 9.1. (*a*) Energy transfer from small donor (D) to polymer acceptor (A). (*b*) Intramolecular energy transfer. (*c*) Energy migration.

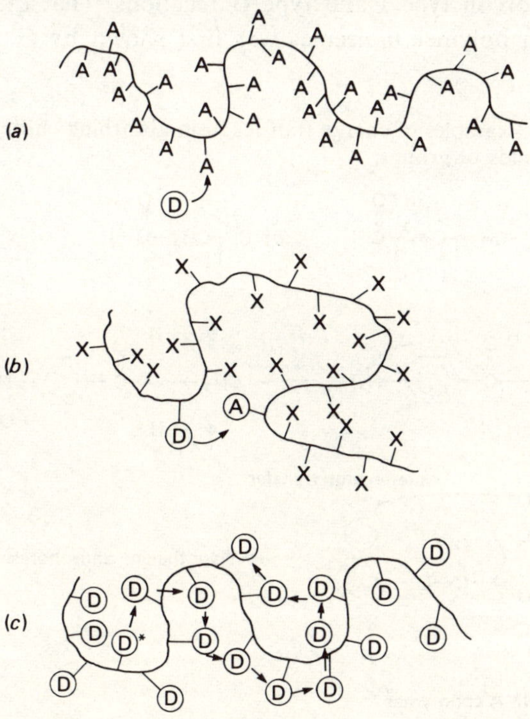

these types of reactions have already been discussed extensively in chapters 6, 7 and 8, and will be treated only briefly in the present chapter. Of special interest to polymer chemists, however, is energy transfer from one group to another in the same polymer chain. This intramolecular energy transfer can be exhibited in a number of ways. For example, a quenching group on one part of the chain may react with an excited group on another part of the chain to change the position of the excitation energy (figure 9.1*b*), or secondly, excitation localized on a sequence of chromophores may be transferred from one chromophore to the next by a hopping mechanism (figure 9.1*c*). This latter process is termed *energy migration.*

9.3 Energy delocalization in polymer chromophores

9.3.1 *Delocalization within chromophores*

There are many examples of energy delocalization in small molecule photochemistry. One of the most prominent is the delocalization of energy absorbed by the carbonyl groups in simple ketones. This is illustrated in figure 9.2*a*. In aliphatic ketones containing γ-hydrogens, the excitation originally localized on the carbonyl group ultimately results in the breaking of either the α, β or β–γ carbon bond by processes now called the Norrish type I and type II reactions. That these same processes occur in polymer molecules was first shown by Guillet and

Figure 9.2. Examples of energy transfer from absorbing chromophores to other bonds or groups.

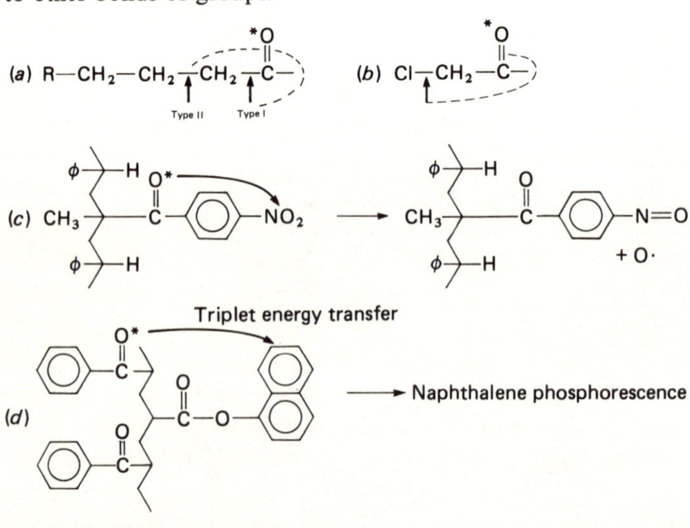

Norrish (1954, 1955*a*,*b*). A similar delocalization occurs in α-chloro ketones where the bond which reacts is the carbon–chlorine bond adjacent to the absorbing site (figure 9.2*b*).

A further example of such delocalization in polymer systems is the transfer of the excitation energy from the ketone carbonyl in *p*-nitrophenyl isopropenyl ketone copolymers with styrene to the nitro group on the aromatic ring, as described by Sugita *et al.* (1976) (figure 9.2*c*). In contrast the *p*-hydroxy and *p*-chloro derivatives showed the expected photochemistry of the ketone carbonyl instead of that of the nitro group.

9.3.2 *Delocalization to adjacent groups in copolymers*

A more interesting example, which is unique to polymers, is shown in figure 9.2*d*. When copolymers are prepared from monomers having two different functional groups (chromophores), it is possible for the energy absorbed by one chromophore to be transferred to the chromophore on the adjacent unit, either by Förster or exchange mechanisms. In this example, a 1:1 copolymer of phenyl vinyl ketone (PVK) and naphthyl methacrylate (NMA) was shown to emit exclusively naphthalene phosphorescence, whether the original excitation was localized on the phenyl ketone carbonyl or on the naphthalene (Somersall and Guillet 1973). The transfer in this case was triplet energy transfer from the triplet phenyl ketone to the naphthalene moiety. This is an important result because it is quite possible that by making copolymers one may induce photochemical reactions which will not occur in either of the two homopolymers of the corresponding monomers.

9.3.3 *Delocalization in poly conjugated systems*

Another interesting example of energy delocalization in polymers is shown by polymers and copolymers of the phenyl acetylenes. For example, studies of p-methoxyphenylacetylene (North *et al.* 1977) show that the energy in this alternating single and double bond polymer (figure 9.3*a*) is extensively delocalized. In solution, the rapid rotation of this flexible chain limits the number of sequences over which energy can migrate. The reason for this is that each chain of alternating double and single bonds is visualized as a sequence of conjugated units, but the conjugation extends only over essentially coplanar regions of the chain. When internal rotation of a single bond occurs, this results in destruction of the coplanar arrangement. Since the absorption and emission characteristics of conjugated sequences depend on the conjugation length, an ensemble of such chains functions as a mixture of chromophores. This results in broad absorption and emission envelopes.

In fluid solution, the rapid rotation of backbone segments maintains a dynamic distribution of segment lengths and provides a mechanism for rapid conversion of the excitation energy from the high energy short sequences to the lower energy long sequences. Thus, in solution, emission usually occurs from the longer conjugated sequences. Because of the complexity of the number of emitting species involved, decays of the fluorescence emission are non-exponential but can be fitted by the sum of two exponential decays (North *et al.* 1977). Theoretical treatment is difficult but attempts have been made by North and co-workers (North *et al.* 1974, North and Ross 1976).

Later a somewhat simpler system was studied (Guillet *et al.* 1978). This consisted of a copolymer of styrene and phenylacetylene (S-PA) in which the lengths of the conjugated double bond sequences were limited by the distribution of styrene units in the polymer, as indicated in figure 9.3*b*. In this case the amount of styrene in the copolymer was 20 wt-%, which would give an average sequence length of 5 for the conjugated double bonds, but because of the random nature of the copolymerization process, a distribution of sequences was obtained within each polymer molecule ranging from $n=1$ to $n>20$.

It is known that the absorption and emission spectra of sequences of conjugated double bonds are red-shifted as n increases up to about 10, after which no further effect is observed. The presence of energy migration in this polymer was deduced from the time-resolved fluorescence spec-

Figure 9.3. Structure of (*a*) poly(paramethanyl phenyl acetylene), (*b*) styrene-phenyl acetylene copolymer.

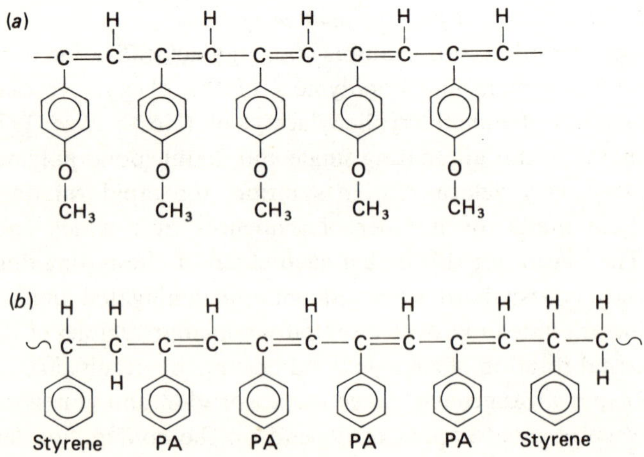

trum of the S-PA copolymer when dissolved at low concentration in polystyrene film where no rotational or conformational changes of the emitting polymer are expected. The results are shown in figure 9.4. In this experiment, the excitation wavelength was 313 nm, which would excite primarily the shorter sequences, i.e., $n=1$ to 3. However, even at the shortest time intervals one sees emission from the longer sequences, which was taken to indicate the presence of rapid energy migration or transfer from the shorter to the longer sequences within the same chain. As expected, as the time after the initiating pulse increases, the emission occurs more and more from longer sequences in the polymer as the result of the quenching of emission from the shorter sequences by energy transfer to the longer sequences. Further evidence for this energy transfer process was deduced from the fact that the emission viewed at 500 nm was depolarized, whereas that emitted at shorter wavelengths still retained a high degree of polarization, suggesting that it arose from the original absorbing site (Guillet *et al.* 1978).

The study of these polymers with a conjugated unsaturation is of particular interest in that they mimic certain biological photo systems such as those in chlorophyll micelles and chloroplasts.

9.3.4 *Delocalization by the exciton band mechanism*

When chromophores are arranged at precisely regular distances and with similar orientations, energy can be delocalized over a very large region of space, as is observed in molecular crystals. In this case, the energy absorbed can be treated as a delocalized exciton and theories have been developed by Partridge (1970*a*,*b*,*c*) and Philpott (1975). The carbon

Figure 9.4. Time resolved fluorescence spectra of SPA in polystyrene film. Exciting wavelength 313 nm, at intervals (ns) from the lamp maximum (*a*) 0–0.23, (*b*) 1.1–4.4, (*c*) 4.6–7.8, (*d*) 20.2–25.1. The spectra are adjusted to fit on the same scale. (From MacCallum *et al.* 1980.)

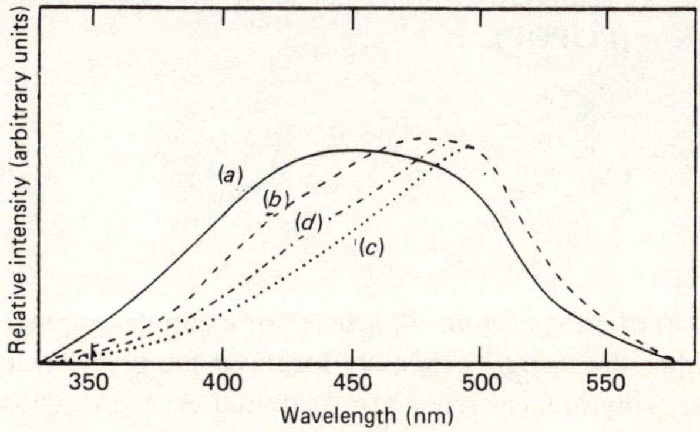

atoms in a vinyl polymer chain are arranged precisely at distances corresponding to the carbon–carbon bond length and so, in principle, the σ-bonds can be treated in this fashion. The theory developed by Partridge, however, shows that the exciton band for σ-bonds in polymers will only be excited at wavelengths below about 150 nm. Spectroscopic evidence on polymers such as polyethylene gives strong support for the Partridge theory. Further evidence for energy delocalization by an exciton band mechanism in polyethylene was obtained by Slivinskas and Guillet (1974) and is described further in chapter 14. Very few synthetic polymers have sufficient regularity between the side chain groups in amorphous polymers to meet the requirements of the exciton band theory of Philpott. However, as the techniques of organic polymer synthesis improve, it seems likely that such polymers will be synthesized and these should show energy delocalization with high efficiency.

9.3.5 *Delocalization of energy in crystalline polymer systems*

Both monomer and polymer crystals lend themselves to extreme delocalization of excitation energy. Essentially the entire crystalline region can be reached by a migrating exciton unless it is first trapped at crystal defects or at other traps in the crystal region. Depending on the purity of the crystal, there appears to be very little limit to the range of a migrating exciton in such systems. It is only recently that photochemical studies of monomers and polymers have been carried out in the crystalline phase, and these are dealt with in chapter 12.

9.4 Intermolecular transfer with small molecules

The classic example of energy transfer to small molecules from polymeric materials is exemplified by the practical use of scintillator films for the detection of atomic and nuclear radiation. In their simplest form, these detectors consist of a matrix of polystyrene containing low concentration of a scintillator molecule such as 1,4-bis[2-(5-phenyloxazoly)]benzene (POPOP).

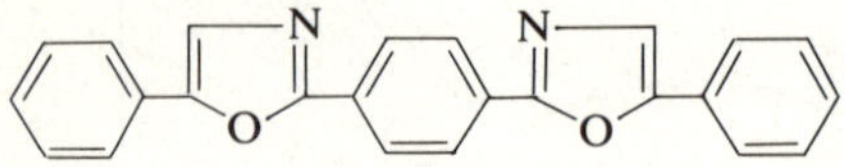

POPOP

I

When a photon of γ-rays, or an α-particle, for example, impinges on the polystyrene film, the energy originally localized along the track of the particle or the γ-ray induces secondary scattered electrons whose energy is ultimately transferred to the scintillator molecule, which then gives off a photon of fluorescent light. The exact mechanism by which the energy is deposited and transported to the scintillator is not yet completely understood (for a detailed review of early work, see Birks 1964), however it probably includes at least one long-range transfer by the Förster mechanism involving dipole–dipole interaction between the excited states induced in the solid polymer matrix and the scintillator molecule.

Until recently, it was not known that collisional energy transfer could also occur in solid polymers. One of the first examples is the work of Heskins and Guillet (1968, 1970) who showed that solid polymers of ethylene containing carbonyl groups introduced by the copolymerization of carbon monoxide would quench the fluorescence of naphthalene dissolved in the polymer, at close to collisional rates. The reaction could be followed both by the reduction in fluorescence of the naphthalene and by the increase in the rate of degradation of the polymer chain. Although the efficiency of quenching is reduced by the higher microviscosity of the polymer matrix ($\eta \doteq 1$ poise) this reduction is not as great as might have been expected (§3.4). This is an example of a sensitized photodegradation of a polymer initiated by energy transfer.

It was also possible to show the reverse effect, that is, introduction of a triplet quencher, cyclooctadiene, reduced the rate of chain scission in polymer. These two examples show the possibility of either sensitized degradation or the use of small molecule quenchers as stabilizers against the photodegradation of polymeric materials in the solid state. Because cyclooctadiene is a triplet quencher, only that part of the reaction occurring from the triplet manifold was quenched and it was therefore possible to estimate that approximately 60% of the degradation arose from the triplet excited state of the ketone.

Further examples of quenching of excitation of polymeric molecules are illustrated in the chapters dealing with phosphorescence, fluorescence and excimer processes (chapters 6, 7 and 8).

9.5 Intermolecular energy transfer in polymers

One of the more important applications of energy transfer in polymer systems involves the intermolecular transfer of energy between two or more polymer coils in solution or in the melt. This allows very precise measurements to be made on such important properties of the polymer coil as its diffusion constant, its size, and whether or not it can interpenetrate with other coils of either the same or different composition. Examples of studies of this kind are the work of Tazuke and Matsuyama (1975), Tazuke *et al.* (1975) and Tazuke and Banba (1976), who used intermolecular excimer and exciplex formation to study the possibility of chain interpenetration in solution. They used polyesters containing naphthalene or carbazole groups of the general structure:

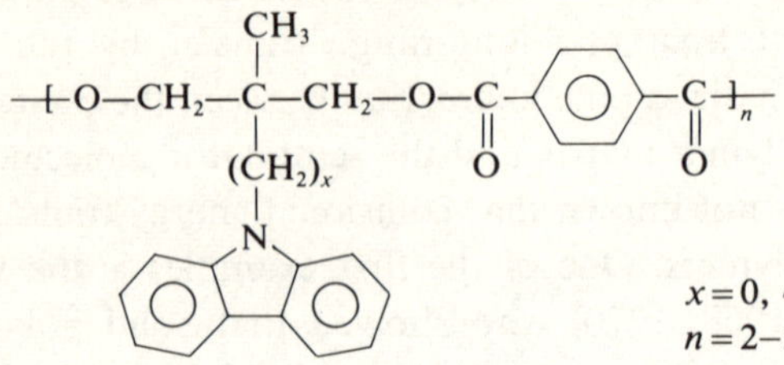

$x=0, 4, 10$
$n=2–5$

II

The variation in emission intensity from the excimer or exciplex with concentration was used to define the concentration at which chains began to interpenetrate. A typical example is shown in figure 9.5.

In a similar series of investigations, Horie and Mita (1978) studied the effect of molecular weight, solvent and concentration on the rate of

Figure 9.5. Exciplex emission of II as a function of concentration ($x=4$). Solvent (THF (hollow circle) and dioxane (solid circle)): (*a*) concentrated system, (*b*) dilute system. (From Tazuke and Matsuyama 1975.)

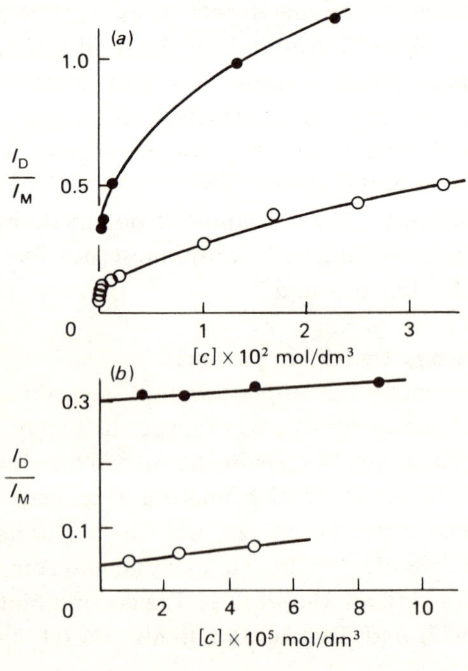

quenching of phosphorescence of polystyrene terminated by benzyl, mixed with a similar polymer terminated with anthracene (§8.14).

Similarly, Kiwi and Schnabel (1976) used triplet energy transfer from phenyl vinyl ketone units in a poly(methyl methacrylate) copolymer to small amounts of vinylnaphthalene, also copolymerized with methyl methacrylate as a probe of intermolecular contacts in polymer solutions. The quenching rate constants were of the order of $2.2 \times 10^8\ M^{-1}\ s^{-1}$ which were one-third as large as that for the quenching of poly(methyl methacrylate-*co*-phenyl vinyl ketone) by the small molecule naphthalene. This unusually high value of the quenching constant and the fact that it was independent of the amount of naphthalene in the copolymer suggested to the authors that the two coils had to interpenetrate in order to quench. This problem of interpenetration of macromolecules in solution is important to the better understanding of the solution properties of concentrated solutions of both homo- and copolymers and the luminescence quenching technique appears to be able to provide useful data in an area which is difficult to study by other methods.

Mechanical blends of two different polymers are frequently used to improve physical or chemical properties. There is a general rule that most polymers are incompatible with each other and therefore a physical blend usually consists of relatively small domains of the two components. Recently, Morawetz and Amrani (1978) have developed a promising new technique for studying the compatibility of polymers in the molten and solid phase. They labeled poly(methyl methacrylate-*co*-N-butyl methacrylate) with naphthalene and poly(methyl methacrylate) with anthracene. The variation in the ratio of donor to acceptor fluorescence was plotted as a function of the butyl ester BMA in the copolymer (figure 9.6). At high BMA content, phase separation of the two polymers occurred so that the ratio of donor to acceptor fluorescence decreased substantially. It was of particular interest that the value of $I_F(N)/I_F(A)$ increased gradually with increasing BMA content. This shows that there was no sharp transition between compatibility and incompatibility in these polymer blends. The two polymers were incompatible when the mole fraction of BMA was greater than 0.4, under which conditions very little energy transfer occurred from naphthalene to anthracene.

9.6 Intramolecular energy transfer in polymers

One of the more important theoretical problems in polymer chemistry relates to the calculation of the probability of interaction of any two groups in a single polymer chain. In particular, it is of interest to determine the rate at which two end groups on a polymer can interact as a function of the length of the polymer chain. This so-called end-to-end chain cyclization has been studied by several groups using luminescence spectroscopy by synthesizing polymers with chromophores attached to both ends of the chain. For example, Cuniberti and Perico (1977) used intermolecular excimer formation between pyrene end groups to study chain cyclization in a series of polyethylene oxides of known molecular weights. Haas *et al.* (1975) studied the variation in end-to-end distance with chain length in the polypeptide N,5-(2-hydroxyethyl)-L-glutamine. In this system, energy transfer from a terminal naphthalene group to a terminal dansyl group was used as the probe. From the viscosity dependence of the acceptor fluorescence decay curves, it was concluded that there was a dynamic contribution to energy transfer on the nanosecond time scale. The same conclusion was reached by Sisido *et al.* (1979) who performed energy transfer experiments on random coil poly(sarcosine) chains.

Winnik (1977) and Maharaj *et al.* (1979) have studied the reaction called 'tadpole cyclization', which is best described as the reaction of the end group of a polymer or small molecule with any accessible point along a hydrocarbon chain. They synthesized molecules of the structure:

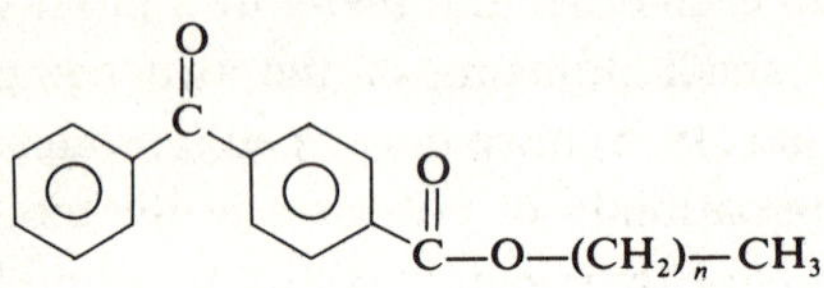

III

Figure 9.6. Ratio of naphthalene to anthracene emission in blends of labelled PMMA and MMA–BMA copolymers. (From Morawetz and Amrani 1978.)

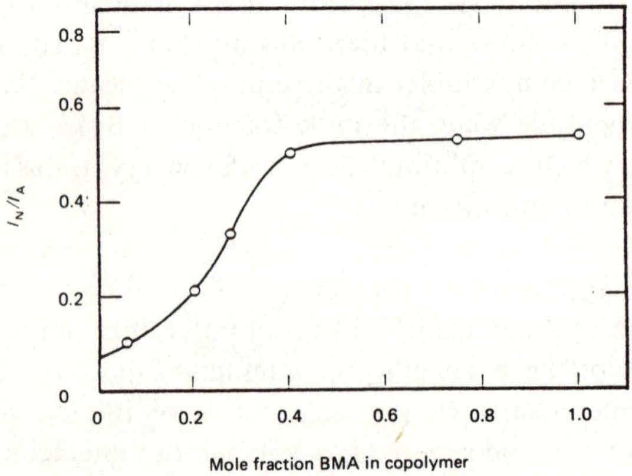

and excited the benzophenone moiety with a pulsed laser. The rate of quenching by the hydrocarbon tail could be used to probe the frequency of collision of the carbonyl group with the methylene groups of the side chain. The overall rate of intermolecular hydrogen abstraction was measured by determining the decrease in the phosphorescence lifetime of the triplet benzophenone groups. In this case, the use of a triplet probe whose lifetime is from 10 to 100 microseconds permitted the study of the cyclization rates on a time scale longer than that permitted by fluorescence methods using singlet probes.

In similar work, Tazuke and Matsuyama (1977) studied the cyclization of a side chain to the chain backbone of a polyester by observing the extent of exciplex formation in compounds of structure II. In all of these experiments it was shown that the fraction of chain conformations which can undergo end-to-end cyclization within the excited state lifetime decreases as the molecular weight increases. This effect is illustrated in figure 7.27 for poly(ethylene oxide)s terminated with pyrene. On the other hand, the fraction of chain middle conformations increases with molecular weight in the small molecule model system studied by Winnik (1977).

The importance of these new experimental methods is that they can be compared with both theory and mathematical models of chain cyclization carried out by Monte Carlo calculations on suitable lattices. Models must consider the effect of the rotational energy barriers as well as excluded volume and polymer–solvent interactions and thus the experimental results can help to refine both the model and the methods of treating such systems. It is expected that these methods will contribute a great deal to our knowledge of polymer chain dynamics in the near future.

9.7 Singlet energy migration in polymer systems

The examples given in the previous sections are relevant to simple energy transfer in polymer systems. Energy migration, however, is a much more interesting and more complex phenomenon. A polymer, consisting of repeating units of chromophores is illustrated in figure 9.7. No matter how far the polymer is diluted, the 'concentration' of groups of chromophores along the chain backbone remains relatively constant. Furthermore, their separation from adjacent units will be determined by the geometry of the chain and the flexibility of the bonds, and for the repeating groups in simple vinyl polymers, will be of the order of 3–8 Å. This is an important factor and permits the distinction between inter- and intramolecular transfer in polymer systems since intermolecular transfer would be expected to depend on the polymer concentration, whereas intramolecular processes would be relatively independent.

A number of possible types of intramolecular energy transfer are shown in figure 9.8. These include (1) the hopping of excitation from one group to another, often the adjacent group along the polymer chain, (2) movement of excitation energy along the carbon–carbon backbone of the chain by an exciton band mechanism, and (3) movement of excitation across loops in a single polymer chain. Loops could be formed either by

Figure 9.7. A conceptual comparison between dilute solutions of a polymer and a small molecule. (From Holden and Guillet 1980.)

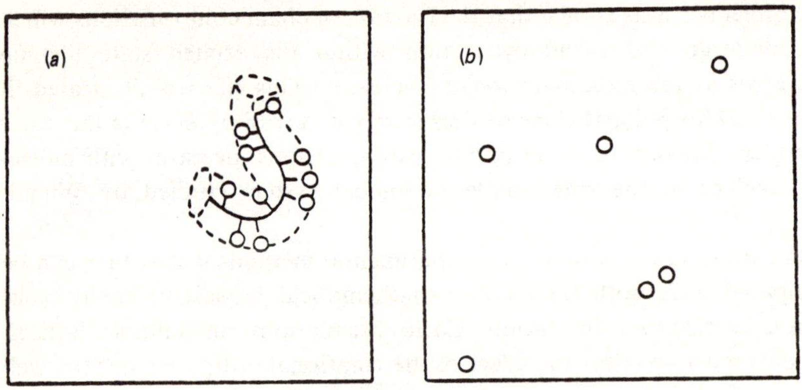

Figure 9.8. Intramolecular energy exchange in polymers. (From Guillet 1977.)

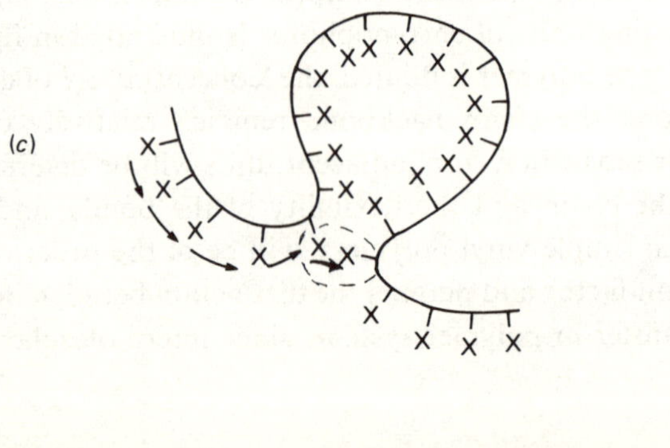

the packing of the molecule in a crystal, or as a temporary collision, i.e., the close approach of two groups on the same polymer chain.

The important feature of this type of migration is that, in principle, the excitation can be localized, at least for some finite period of time, on a particular group before it moves to another group in the chain. We will refer to this type of energy delocalization as *energy migration.* In some of the earlier literature concerning intramolecular singlet energy migration in polymer molecules, some confusion has been introduced with regard to the exact meaning of the term 'energy migration'. Some authors seem to have assumed that energy migration can occur only between nearest neighbor chromophores in a polymer, as defined by the monomer sequence in a copolymer, for example. While it is possible that this does represent the major mechanism in certain polymers it certainly is not true for others, such as in the alternating copolymers of vinyl naphthalene and maleic anhydride, and in poly(acenaphthalene). In this and succeeding work we define 'intramolecular energy migration' as any process involving more than one exchange of excitation energy between spectroscopically identical chromophores attached by covalent bonds to a polymer chain. Thus a single step between identical units would be termed 'energy transfer', while more than one such step in sequence constitute 'energy migration'.

As a consequence of this definition we can identify three distinct types of energy transfer steps which can contribute to energy migration in a polymer chain. These are illustrated in figure 9.9. The first of these, 'nearest neighbor' transfer, is defined as transfer between chromophores where n, the number of monomer units between those bearing the transfer sites, is equal to zero. This will be important in polymers containing small chromophores attached to flexible chains which permit the proximity of nearest neighbor groups, as in polystyrene.

The second type, 'non-nearest neighbor' transfer, where $n = 1, 2, 3$, will occur with chromophores which are prohibited by steric or structural effects from approaching an adjacent chromophore by facile bond rotation. The most stable conformation of flexible polymer chains usually places large substituent groups as far apart as possible and thus may preclude nearest neighbor interactions, particularly if the lifetime of the excited state is short relative to the rotational relaxation time of the polymer chain and its side groups. In such polymers the interposition of a single monomer unit ($n = 1$) may give rise to the closest approach of the two chromophores. The observation of excimer emission in alternating copolymers such as poly(vinyl naphthalene-maleic anhydride) is strong confirmation of this type of interaction.

The crystal structure of isotactic polystyrene has been shown to be a 1:3 helix in which each phenyl ring in the chain is superimposed on the unit three units down the chain. Thus energy migration in this system would be expected to occur largely through non-nearest neighbors where $n=2$ in our scheme. There is strong evidence that such helical structures can persist in solution for isotactic sequences of vinyl polymers, at least over short periods of time, and thus could contribute to this type of energy transfer and migration. As the size of the chromophore group is increased, the periodicity of the helix may increase to 1:4, and make possible energy transfer steps where $n=3$. Because of the relatively rigid geometry of the aromatic rings in such helical arrangements, the depolarization of fluorescence arising from energy migration in these systems may well be much less than is the case for randomly oriented chromophores.

The third type of process, 'loop' transfer ($n>3$), was first proposed by Somersall and Guillet (1973) to explain the strong solvent effect on excimer emission from poly(1-naphthyl methacrylate). Although the concept has been criticized by Morawetz on theoretical grounds, there is much experimental evidence that it can take place. Its importance lies in the fact that the occurrence of only one such event can result in the transfer of excitation across very many units along the contour length of

Figure 9.9. Possible energy transfer steps in intramolecular migration. 1. Nearest neighbor transfer ($n=0$). 2. Non-nearest neighbor transfer ($n=1$, 2, 3). 3. Loop transfer ($n>3$). The interger n represents the number of 'repeating' or 'monomer' units between the units bearing the chromophores involved in the transfer step.

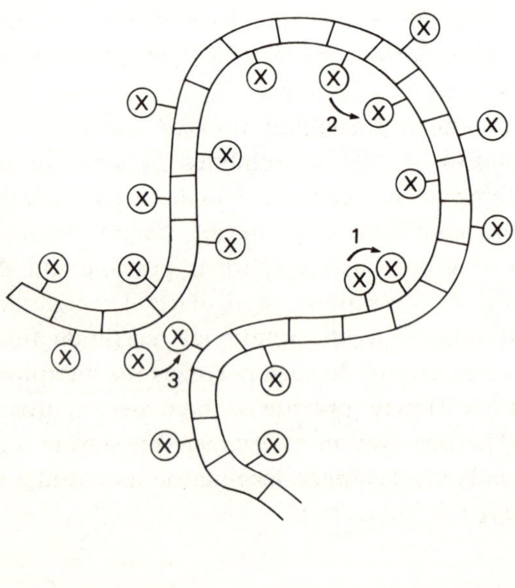

the macromolecular chain. This can be very significant when energy migration is required to excite a small number of traps localized in a long polymer chain.

In all of these steps, energy migration will take place by a random walk process, and may be treated theoretically by one-, two- or three-dimensional random walk theory, depending on the circumstances. It is worth noting that in all such walks, the most probable location of the excitation remains at the absorption site, and the maximum distance attained $\bar{l}$ is proportional to the square root of the number of steps or hops.

9.7.1 *Detection of energy migration*

In spite of extensive research over the past decade, the quantification of the extent, and indeed the existence, of singlet energy migration remains a difficult task. The usual methods to test for singlet energy migration fall into three categories. The first of these is the quenching of polymer fluorescence by small molecule quenchers in rigid media or in solution. If the quenching rate is larger than expected, this has been taken to mean that the sphere of action of either the quencher or the polymer is increased by energy migration. In rigid glasses the Perrin model is usually used (1.46). An increase in the radius R of the quenching sphere from that in systems involving the same chromophores where no migration is possible is considered to be strong evidence for energy migration.

The extent of energy migration is usually described in terms of a time-independent diffusion coefficient Λ. This is related to the root-mean-square distance $\bar{l}$ travelled by the excitation during the donor lifetime τ_0:

$$\bar{l}=(2(D+\Lambda)\tau_0)^{1/2} \tag{9.1}$$

where D is the diffusion coefficient of any diffusing species involved in the quenching process and Λ is the migration coefficient for the excitation. If one assumes that there is no material diffusion in a solid glass, then this relationship reduces to

$$\bar{l}=(2\Lambda\tau_0)^{1/2} \tag{9.2}$$

In fluid solutions, the Stern–Volmer relationship (1.41) is usually employed and the value of k_q is related to the sum of D and Λ, thus

$$k_q=D+\Lambda \tag{9.3}$$

for many random copolymers containing aromatic groups which have been studied, values of Λ are typically from 1 to 10×10^{-5} cm^2 s^{-1} which is of the same order of magnitude as mass diffusion constants for small molecules in fluid solution.

A second general procedure is to analyze the decay curves of quenched and unquenched systems with either external or internal quenchers based on some theoretical model of the transport processes. Such methods, however, should be used with caution because many factors other than energy migration can affect the shape of emission curves from polymer systems. Furthermore, the characteristics of Perrin, Stern–Volmer and Förster theories do not give large differences in the shape but only in the position of decay curves.

A third method involves the depolarization of fluorescence emission, which will be treated theoretically in the next section. A fourth method, which has been used primarily in a qualitative way, is the observation of excimer emission from aromatic polymers in dilute fluid solution. However, because of the complexity of excimer kinetics, it is difficult to extract a reliable measure of the degree of energy migration, or even its existence from the ratio of excimer to monomer emission in polymer solutions.

9.7.2 *Fluorescence depolarization*

The importance of fluorescence depolarization is that measurements can, in principle, distinguish the number of hops occurring, provided the number is less than about five or ten. If emission occurs from a chromophore before it can rotate or relax its geometry, that emission will be polarized with a specific direction relative to the exciting light. If, however, a number of transfer steps occur, then each will involve some change in the polarization of the light and ultimately after five to ten steps the emission will be completely depolarized. The theory of these processes has been developed using several models.

Jablonski model. In this model (Jablonski 1955, 1958) the medium consists of excited centres surrounded by an active sphere containing a variable number of ground-state chromophores. If energy transfer to each of the latter is equally probable and no orientation factors are involved, then the following relationship is derived:

$$\frac{1}{P}-\tfrac{1}{3}=\left(\frac{1}{P_0}-\tfrac{1}{3}\right)(1+\tfrac{1}{3}v) \tag{9.4}$$

where v is the average number of molecules within the active sphere. This only holds in dilute solutions where concentration quenching can be neglected. Emission from those chromophores excited directly is considered to be completely polarized while that from molecules excited by energy migration is completely depolarized.

Förster–Öre model. The Förster (1948) model is similar to the Jablonski model but energy transfer is considered to be probable only with the nearest neighbor, rather than with all molecules in the active sphere. This introduced a radial dependence to the depolarization. Öre (1959) modified Förster's model to include the possibility of back transfer of energy to the originally excited molecule. The fraction of the observed fluorescence n_1/n coming from the molecules excited only by absorption is given by

$$\frac{n_1}{n}=\int_0^{\infty}\frac{\xi^2(\xi^2+\gamma^2)}{(\xi^2+\gamma^2)^2-\gamma^4\exp(-11\xi/16)}+\exp(-\xi)\mathrm{d}\xi \tag{9.5}$$

where ξ is the average number of molecules in a sphere of radius R for a given concentration. The quantity γ is the ratio of this concentration to the critical concentration $c_0=(3/4)\pi NR_0^3$ where R_0 is the critical transfer distance at which transfer and spontaneous deactivation occur with equal probability. In this case the polarization is given by

$$\frac{P}{P_0}=\frac{6n_1/n}{5+n_1/n} \tag{9.6}$$

Weber model. Weber (1966) attempted to overcome the neglect of orientation effects in the energy transfer step, as well as contributions to the observed polarization from molecules excited by the energy transfer process. He considered the situation in which an emission oscillator is subjected to a series of small changes in orientation, each characterized by an angle θ and occurring an average of $\bar{n}$ times during the lifetime of the excited state. He derived the following relationship:

$$\frac{1}{P}-\tfrac{1}{3}=\left(\frac{1}{P_0}-\tfrac{1}{3}\right)(1+\tfrac{3}{2}\bar{n}\overline{\sin^2\theta}) \tag{9.7}$$

This equation can be applied equally well to situations involving both random Brownian motion or energy migration. While the radial dependence of energy transfer by an exchange mechanism is well established, its angular dependence remains uncertain. Adamczyk and Phillips (1974) have shown that if an orientation effect does exist in triplet energy transfer, it is extremely weak and not detectable by current experimental techniques.

9.7.3 *Experimental studies of singlet migration*

Excimer formation in aromatic polymers. Among the earliest work postulating singlet energy migration were studies of excimer formation in aromatic polymers, both in the solid phase and in solution. Many of these have been dealt with in chapter 7. Unfortunately, because of the difficulties in establishing the exact concentration of excimer sites in a polymer chain, and in the absence of a good theory, it is difficult to quantify the amount of energy migration in such systems. There is even some question as to whether migration plays an important role in polymers such as polystyrene, which obey the Hirayama rule, but in which there is little overlap of absorption and emission spectra of the donor chromophores. This leads to small values of R_0, the critical distance for Förster transfer between the polystyrene units (see MacCallum and Rudkin 1977).

Quenching by small molecules. A large number of studies have invoked singlet energy migration based on the quencing of polymer fluorescence by added small molecules (Klöpffer 1969*a*, *b*; Powell 1971; David *et al.* 1972; Samedova *et al.* 1972*a*, *b*; North and Treadaway 1973; North *et al.* 1974; North and Ross 1976; Nowakowska *et al.* 1976; Ishii *et al.* 1977; Ueno *et al.* 1977; Ishii *et al.* 1978; Ueno and Osa 1978). North and Treadaway (1973) estimated the values of Λ, the energy migration coefficient, from spectral overlap of the homopolymers using the Förster equation and

$$2\Lambda/r^2 = (1/\tau_D)(R_0/r)^6 \tag{9.8}$$

where r is an estimate of the average distance between adjacent chromophores on the polymer. These calculations were then compared with quenching experiments in solid polymer films where D for the quenching molecules was assumed to be negligibly small. The experimental Λ values were those required to make the integrated form of the Yokata–Tanimoto equation (Yokata and Tanimoto 1967) fit plots of quenching efficiency versus the concentration of a small molecule acceptor. Typical Λ values were found to be of the order of 7 to 14×10^{-5} cm^2 s^{-1} for poly(1-vinylnaphthalene) and poly(vinylcarbazole). As expected, these migration coefficients are many orders of magnitude lower than those observed for organic crystals where the high degree of order leads to migration coefficients as high as 10^{10}.

In a similar series of experiments using solid polymers of poly(vinylcarbazole) Klöpffer (1969*a*,*b*) showed that the efficiency of quenching of excimer emission by perylene and hexachloroxylene were similar, in spite of the fact that the former quenches by the Förster mechanism, whereas the latter quenches by electron transfer. From the variation of the quenching efficiency with trap concentration, Klöpffer estimated that the singlet energy migrated an average of 20 nm before it was trapped. In a later series of experiments, Johnson (1980) studied singlet energy transfer to perylene and dimethyl terephthalate from *n*-isopropylcarbazole dispersed in polystyrene films. An interesting transition from the Förster relationship to nearly Stern–Volmer kinetics was observed on increasing the donor concentration. From this, Johnson concluded that singlet energy migration functioned similarly to material diffusion at high donor concentration. Extrapolation of the donor concentration to that corresponding to pure poly(vinylcarbazole) gave an energy migration distance of 26 nm, comparable to that measured by Klöpffer. Similar experiments on amorphous films containing 1,3-dicarbazoyl propane, and films of poly(2,9-dicarbazoyl ethyl vinylether) (Itaya *et al.* 1979) also confirmed the existence of energy migration.

The use of small molecule quenching to quantify the extent of energy migration is open to some question, particularly when the experiments are carried out in the solid phase where variations in thermal history and indeed a lack of a suitable model of the amorphous glassy solid makes exact treatment difficult. The treatment of North and Treadaway (1973), for example, uses Λ as an adjustable parameter. It is often difficult to quantify the emission spectrum of many aromatic polymers because of the contribution of both monomer and excimer, thus the calculation of Λ directly from spectroscopic data is also imprecise. Furthermore, the Förster theory treats donor and acceptor as point dipoles, an approximation which may well be suitable when the Förster radius R_0 exceeds several molecular diameters, but is certainly inappropriate for the distances of 3–5 Å normally encountered between adjacent chromophores on vinyl polymer chains.

In solution, further problems arise from the effect of local concentration. For example, if the concentration of a quencher inside the macromolecular coil differs from the bulk concentration, erroneous quenching constants will result from the use of the Stern–Volmer equation. Studies by Keyanpour-Rad *et al.* (1978) involving a variety of carbazole-containing polymers showed that the migration coefficient was the same for a variety of quenchers. Since it was unlikely that each quencher would show the same local concentration effect, this suggested that the values in this case were not dependent on such factors. However, in aqueous systems, for example, the very high efficiency of energy transfer between ionic dyes in the presence of polar polymers probably arises because the dye concentrations are highly inhomogeneous (Cundall *et al.* 1970, Taha and Morawetz 1971).

Additional errors associated with the Stern–Volmer kinetic analysis can be introduced by the assumption of single exponential decay, a requirement for the Stern–Volmer analysis to be valid. Since most aromatic polymers studied to date have been excimer-forming, the single exponential decay required for the Stern–Volmer analysis is not observed experimentally, and thus the conclusions are open to some question.

9.7.4 *Fluorescence depolarization*

Perhaps the strongest evidence for the existence of energy migration between like chromophores involves studies of the depolarization of monomer fluorescence which occurs during the hopping mechanism. For example, if energy migration occurs by the Förster mechanism, the Weber equation takes the form

$$\frac{1}{P} \mp \frac{1}{3} = \left(\frac{1}{P_0} \mp\right)\left(1 + \frac{4\pi N_A C R_0^6}{15000(2r)^3}\right) \tag{9.9}$$

where P_0 is the polarization of the chromophore in the limit of the lowest concentration, C is the concentration in mole/liter and r represents the chromophore collision radius. When applied to dispersions of organic dyes in viscous solutions, Weber (1954) calculated R_0 for like chromophores using this equation and literature data for fluorescein, eosin and rhodamine B. Using this equation Weber estimated experimental values for R_0 of about 30 Å, in close agreement with values calculated directly from the use of the Förster equation. These are the first experimental results which suggested that the Förster theory can be applied to energy transfer between identical chromophores.

It is important to recall that experiments of this type have to be done in rigid glasses, usually at 77 K, because in solution at room temperature the fluorescence is completely depolarized by the normal molecular motion in the chromophore. A number of studies have shown that monomer fluorescence of a variety of aromatic homopolymers was completely depolarized in rigid solutions at 77 K (Yokoyama *et al.* 1975; David *et al.* 1976, 1977; Reid and Soutar 1977, 1978; McInally *et al.* 1979; Schneider 1979). For example, Reid and Soutar (1977, 1978) showed a linear relationship between P^{-1} and l_m the mean sequence length of fluorescent monomer units in copolymers of styrene or 1-vinylnaphthalene with methyl methacrylate. These authors proposed that depolarization occurred by singlet energy migration between consecutive styrene or 1-vinylnaphthalene units.

There are a number of experimental problems with the use of fluorescence depolarization in solid films which may affect the absolute value of migration coefficients. The measured values of P can be seriously affected by scattered light and are also wavelength-dependent. Both phosphorescence and delayed fluorescence emission can also complicate the analysis since this emission is usually completely depolarized. MacCallum and Rudkin (1977) have shown that room-temperature emission from polystyrene films retained a high degree of polarization, a result which is difficult to reconcile with the idea of extensive energy migration in this polymer at room temperature.

Perhaps the most direct evidence for energy migration is obtained from experiments in which aromatic polymers are prepared with chemically bound traps. These experiments will be discussed next.

9.8 The antenna effect in polymers

Early studies by Schneider and Springer (1971) and by Fox and coworkers (1969, 1972) showed that efficient fluorescence emission occurred from small amounts of copolymerized chemically bound fluorescent traps in polymer chains. Schneider and Springer used styrene-acenaphthalene copolymers and Fox and Cozzens (1969) used styrene-vinylnaphthalene polymers. They showed that emission of naphthalene fluorescence from such copolymers was very much higher than in a solution of equivalent amounts of the two homopolymers. Both groups proposed that this phenomenon was due to energy migration between styrene sequences to the naphthalene moieties in the polymer chain. Guillet and coworkers have made quantitative studies with a variety of polymers containing naphthalene and phenanthrene donors with traps containing the anthracene moiety (Hoyle and Guillet 1978; Aspler *et al.* 1978; Holden and Guillet, 1980, Holden *et al.*, 1981; Ng and Guillet 1982 *a,b*). Because this effect mimics that observed in the ordered chlorophyll regions of green plant chloroplasts, i.e., the antenna chlorophyll pigments, this has been termed the *antenna effect*.

Other studies (Guillet 1979) have demonstrated that the effect is not exclusively due to energy migration among the chromophores making up the antenna but to a combination of migration and direct Förster transfer to the trap. The mechanism of energy transport in such systems is illustrated in figure 9.10 in the limiting case where only one trap occurs in a long single antenna chain. A theory may be derived from the assumption that on the short time scale available from the lifetime of the singlet state in typical aromatic chromophores, i.e., $\tau < 100$ ns, collisional energy transfer is a relatively minor factor. Energy migration and transfer in such systems are therefore assumed to be primarily due to long-range Förster transfer by the dipole–dipole mechanism outlined previously. Combining (1.34) with (1.35) one obtains for the rate of exchange between donor and acceptor in the chain

$$k_{DA}=(1/\tau_D)(R_0/R)^6 \quad (9.10)$$

where R is now the distance between the donor and acceptor chromophore in the polymer. If we visualize the migration process as occurring between donor units on the chain but not necessarily adjacent units, the rate constant k_{DD} for the migration step is

$$k_{DD}=(1/\tau_D)(R'_0/R_D)^6 \quad (9.11)$$

where R'_0 is the Förster radius for transfer between the identical donor chromophores and R_D is the distance between them. The number of hops n is given by

$$n=2k_{DD}\tau_D=2(R'_0/R_D)^6 \quad (9.12)$$

where R_D is now the average distance between chromophores along the polymer chain and the factor 2 is introduced because there will be two chromophores approximately equally spaced on either side of each excited chromophore in the chain. In the absence of any trap in the polymer, this hopping will follow a random walk, primarily along the backbone of the polymer chain until the energy is deactivated by other processes. The average displacement of the excitation in this one-dimensional random hopping mechanism will be given by

$$\bar{l}=(2nR_D)^{1/2}$$

However, in the presence of a singlet energy trap, the lifetime of the excitation will be reduced and the number of hops as well as the distance diffused by energy migration will also be reduced. The probability of an energy hop from donor to donor versus one directly to the acceptor trap

Figure 9.10. Mechanism of the 'antenna effect' in a naphthalene-substituted polymer, containing an anthracene trap.

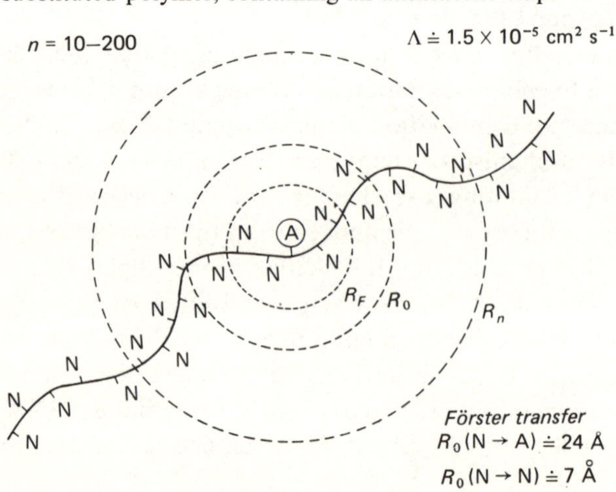

is given by the ratio of the two rate constants:

$$k_{DD}/k_{DA} = \frac{1}{2}(R'_0/R_D)^6(R/R_0)^6 \qquad (9.13)$$

Since R_0 and R'_0 can be calculated from spectroscopic data and R_D can be estimated from the polymer dimensions, this equation permits an evaluation of the relative sphere of influence of the two types of energy transfer.

This situation is shown in figure 9.10 using naphthalene chromophores as the donors and anthracene as the acceptor. Such a polymer is similar to those synthesized by Guillet and coworkers where the anthracene is the terminal unit on a polymer chain. In this system, the best estimates for R_0 and R'_0 are 24 and 7 Å, respectively. The only other unknown is the value of R_D, the average distance between the naphthalene chromophores. Since most naphthalene polymers exhibit excimer formation, which requires an approach of close to 3 Å between naphthalene groups, a value of approximately 4 Å is plausible for the average distance in simple copolymers such as 1- and 2-vinylnaphthalene and possibly even for the naphthyl esters, such as 1- and 2-NMMA. With these assumptions, one can calculate the values of R_1, R_{10} and R_{100} as 12.2, 17.9 and 26.2 Å, respectively, using the relation

$$R_N = 0.89(R_D/R'_0)R_0 n^{1/6} \qquad (9.14)$$

These correspond to the radii of spheres centered on the anthracene trap in which, for example, at R_1 the possibility of a naphthalene to naphthalene transfer is equally probable to the transfer from naphthalene to anthracene. At R_{10} there will be, on the average, ten naphthalene-to-naphthalene transfers for every naphthalene–anthracene transfer and at R_{100} there will be 100 naphthalene–naphthalene transfers for each transfer from naphthalene to anthracene.

The radius R_F defines a sphere around the acceptor chromophore inside of which direct Förster transfer from the absorbing chromophore to the trap is favored. Outside of R_F at least one energy migration step will occur before the energy is trapped by the acceptor.

These spheres are illustrated in a two-dimensional projection in figure 9.10 which also includes the standard Förster radius R_0 of 24 Å. From this diagram it can be seen that although a substantial number of naphthalene-to-naphthalene transfers may occur at distances well within the Förster radius, they will not be expected to contribute to efficiency of energy collection, since the energy would be transferred to anthracene in any case as long as the radius was less than the Förster radius. The only increase in efficiency of energy collection due to the energy migration process will occur for naphthalene chromophores outside the Förster radius R_0 where migration will occur at a rate corresponding to that in the absence of the anthracene trap. The number of migration steps can be calculated from (9.11) and (9.12), substituting the value of the natural lifetime of the naphthalene chromophores and an estimate of their separation distance in the polymer. This gives an average number of hops of 57, and a migration distance $\bar{l}$ of 21.3 Å.

The contribution of this energy migration to the increased efficiency of collection of photon energy is difficult to calculate without further knowledge of the structure and conformation of the chain in solution. Since this migration essentially follows the backbone of the chain, it will correspond to a random walk in one dimension. For a homogeneous distribution of chromophores in the sphere, one can calculate the additional radius contribution to be approximately 9 Å outside the Förster radius. Although this would contribute a substantial increase in the total volume of the sphere of action, the random coil does not consist of a homogeneous distribution of naphthalene chromophore units. When rapid segment motions can occur, the density of polymer segments follows a gaussian distribution about the centre of mass of the polymer. Consequently, if the anthracene is close to the centre of mass, as in the diagram, the concentration of naphthalene units in the outer shells of the hydrodynamic volume will be considerably reduced.

On the other hand, if instead of a random coil polymer, one has a conformation more nearly approximating the rigid rod, energy migration would be expected to play a much larger role in improving the efficiency of energy delocalization to the anthracene trap. Furthermore, the lifetime of the typical excited singlet (10–20 ns) is too short to permit sufficient segmental motion to make the gaussian distribution of segment density a reasonable approximation.

For the reasons discussed above, the detection and quantification of the role of energy migration in antenna polymer systems requires very careful experimentation and analysis. Clearly, for systems in which the average distance of migration $\bar{l}$ is comparable with the Förster radius for direct transfer from donor to acceptor, most of the prompt emission from the trap will come as a result of direct single-step Förster transfer from donor to acceptor, and only at longer times after the initiating pulse will the effects of energy migration be observed. These conclusions have been confirmed by experimental studies of such systems.

The distinction between this type of energy transfer and that observed in scintillators and other solid aromatic polymers is that the photon energy is collected within a single polymer molecule, and all transfer processes are intramolecular. The antenna effect thus permits the collection of photon energy from the entire region of space representing the hydrodynamic volume of the polymer, and its transmission to traps located on the polymer chain. Unlike the process in scintillator systems, the efficiency is relatively independent of concentration, and can be very efficient even in very dilute solutions. The practical results of this phenomenon remain to be exploited.

9.8.1 *Experimental studies of the antenna effect in polymer solutions*

In early studies of transfer within single polymer molecules in solution, Aspler *et al.* (1978) showed that anthracene attached to the ends of poly(naphthyl methacrylate) chains gave strong fluorescence emission from anthracene when the naphthalene chromophores were excited. Time-resolved fluorescence studies showed that emission from anthracene increased while that from the naphthalene monomer and excimer decreased with times of the order of 1 to 20 ns. In more detailed studies, Holden and Guillet (1980) showed that the efficiency of energy transfer in the naphthyl methacrylate system was low but much higher efficiencies could be obtained in polymers of 1-naphthylmethyl methacrylate terminated with methyl anthracene or with copolymerized 9-anthrylmethyl methacrylate monomer units. The energy efficiency χ can be calculated for such systems from the relations:

$$\phi_F = \chi\phi_A - (1-\chi)\phi_N \tag{9.15}$$

$$\chi/(1-\chi) = \phi_N I_A/\phi_A I_N \tag{9.16}$$

where

$$\chi = \frac{\text{number of photons transferred to anthracene}}{\text{number of photons absorbed by naphthalene}}$$

and ϕ_F, ϕ_N and ϕ_A are the fluorescence quantum yields of the total copolymer, the corresponding naphthalene homopolymer, and the anthracene trap on direct excitation, respectively. Usually the naphthalene excitation was at 280 nm and that for direct excitation of anthracene was at 340 nm.

The fluorescence quantum yields were obtained by comparing the integrated fluorescence intensities of samples of known optical density with standard solutions of compounds with known fluorescent quantum yields such as 1-naphthol ($\phi_F = 0.21$) and 9-methyl anthracene ($\phi_F = 0.35$) in cyclohexane. It was necessary to correct for inner filter effects also, using the relation

$$\phi_2/\phi_1 = I_2 A_1 (\mathrm{IF}_2) n_2^2 / I_1 A_2 (\mathrm{IF}_1) n_1^2 \tag{9.17}$$

where subscripts 1 and 2 denote standard and sample, respectively, I is the fluorescence intensity, A is the absorbance at excitation wavelength and IF is the inner filter correction factor given by

$$\mathrm{IF} = 10^{Ad} \tag{9.18}$$

where d is a parameter used to fit the dependence of fluorescent intensity on solution absorbance for 1-naphthol solutions of varying concentrations. Typical experimental fluorescence ($\lambda_{ex} = 280$ nm) curves for anthracene containing poly(naphthylmethyl methacrylate) in THF at 25°C and in MTHF glass at 77 K are shown in figure 9.11*a* and *c*.

At room temperature there is still a correction for excimer emission

Figure 9.11. Steady-state fluorescence spectra in deoxygenated THF at 25.0°C of (*a*) poly(NMMA-co-0.67% AMMA), $\lambda_{ex} = 280$ nm, (*b*) 9-anthrylmethyl pivalate (5×10^{-5} M), $\lambda_{ex} = 340$ nm, and (*c*) poly(NMMA-co-6.7% AMMA) in 2-methyl THF at 77 K, $\lambda_{ex} = 280$ nm. ((*a*), (*b*) From Holden *et al. 1981;* (*c*) from Holden and Guillet 1980.)

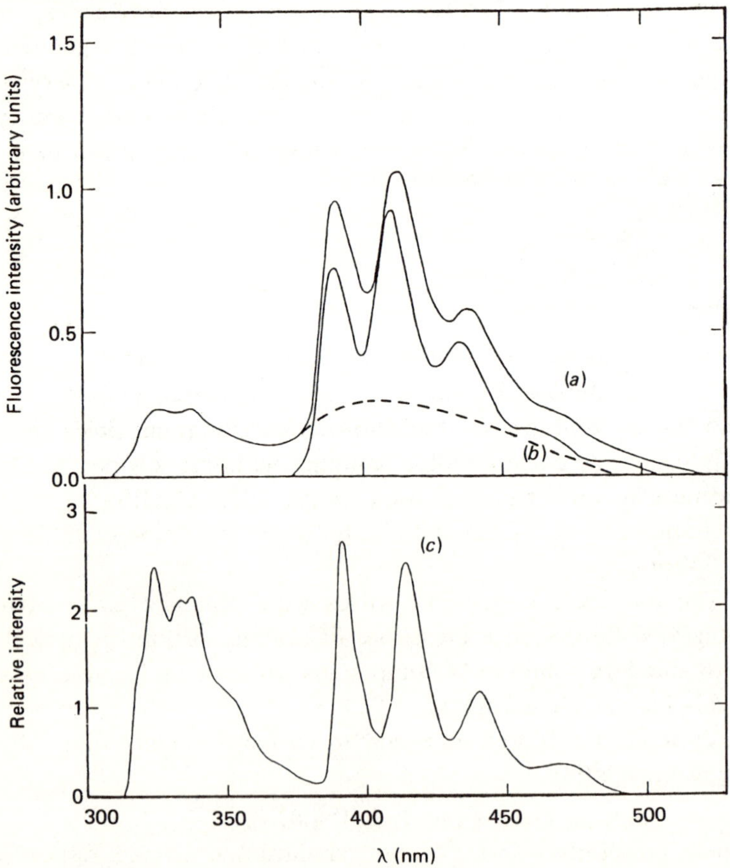

(dotted line in figure 9.11), while at 77 K the emission from N and A units is completely resolved. In each case the concentration of anthracene is less than 1 mol-%, but the emission intensity of A is substantially greater than that of the absorbing N group, illustrating the substantial delocalization of the excitation energy.

The χ values for a series of copolymers with increasing amounts of anthracene traps are shown in table 9.1. The efficiencies in excess of 70% are comparable to those estimated for the antenna chlorophyll in the chloroplasts in green plants. Furthermore, the relatively high efficiency in MTHF glass at 77 K suggests that collisional processes and molecular mobility are not required for efficient energy migration and transfer.

Time-resolved emission spectroscopy (figure 9.12) and emission decay curves (figure 9.13) show that the anthracene emission is observable even at the shortest time scale, and that in good solvents and expanded coils, energy migration and transfer require 5–20 ns or longer to occur. The rapid quenching of the naphthalene units close to anthracene is demonstrated by the very rapid decay of naphthalene monomer fluorescence in the first few ns, as shown in figure 9.14.

Kinetic analysis of the fluorescence decay curves of the donor and acceptor are consistent with a mechanism involving rapid one-step Förster transfer from the N chromophores nearest the trap, followed by a slower contribution from energy migration through more distant N donors.

Ng and Guillet (1982*a*,*b*) have synthesized similar polymers using phenanthrylmethyl methacrylate (PhMMA) as the antenna monomer.

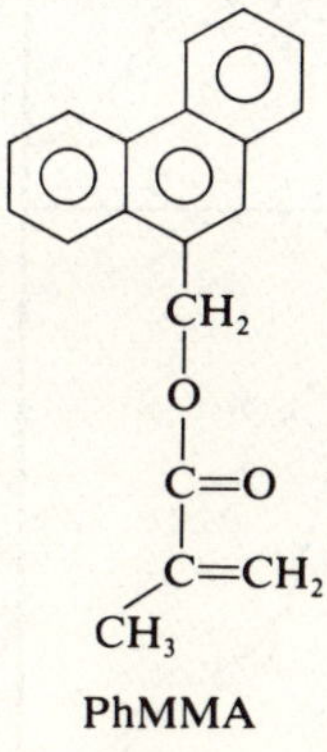

PhMMA
IV

This polymer has the advantage that no excimer emission is observed for the phenanthrene moities, even at room temperature, hence removing a complication of the kinetic analysis. Based on a detailed study of

Table 9.1. *Variation of the efficiency χ of energy transfer from naphthalene to anthracene in copolymers of NMMA with AMMA*[a]

Mol% AMMA	χ (%)
0.074	3.7
0.174	9.9
0.348	18
0.665	30
1.42	69
2.06	84

[a] Solutions in THF at 25 °C.

Figure 9.12. Time-resolved fluroescence spectra of anthracene-terminated PNMMA in toluene at 25.0°C on excitation at 292 nm. Lower and upper time limits in ns from maximum of the exciting pulse are (*a*) 0.0–2.2, (*b*) 11.5–13.8, (*c*) 21.0–24.0, (*d*) 44.3–48.6 and (*e*) 179.2–228.5. The curves are normalized to the same intensity at 390 nm. (From Holden and Guillet 1980.)

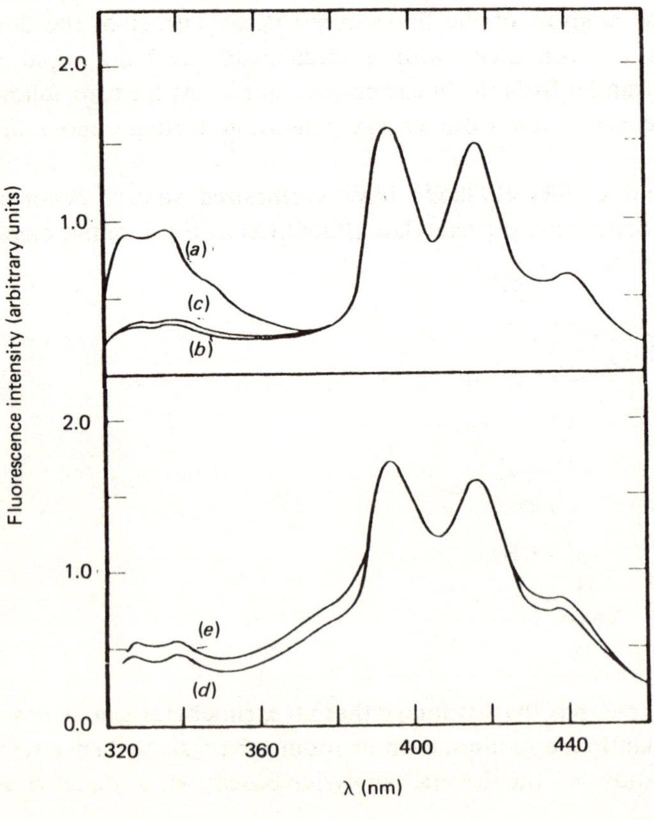

emission kinetics and depolarization studies of copolymers of poly(naphthylmethyl methacrylate) (PNMMA) with anthrylmethyl methacrylate and methyl methacrylate spacer groups, Ng and Guillet (1982*b*) calculated an average distance for energy migration $\bar{l} = 120$ Å and a value of $\Lambda = 1.5 \times 10^{-5}$ cm^2 s^{-1} which is comparable to the bulk diffusion constant of small molecule diffusants.

In both the naphthylmethyl methacrylate (NMMA) and PhMMA copolymers, the efficiency χ depends on the hydrodynamic volume (which can be estimated by the intrinsic viscosity in solution) being higher in poor solvents where the coil is compressed to smaller dimensions. Typical results for PNMMA are shown in table 9.2. As the non-solvent, methanol, is added to the solution, the polymer coil contracts, thus reducing the average distance between donor chromophores and the anthracene trap, thereby increasing both the rate and efficiency of energy transfer.

In aqueous solution (at pH 14) copolymers of NMMA with acrylic acid

Figure 9.13. Fluroescence decays at 420 nm on 280 nm excitation of PNMMA–AMMA samples in nitrogenated THF at 25°C. The mol % AMMA in the copolymer samples is (*a*) 0.348, (*b*) 0.665, (*c*) 1.42, (*d*) 2.06. Curve (*e*) is the decay of the anthracene emission on direct excitation at 366 nm. (From Holden *et al.* 1981.)

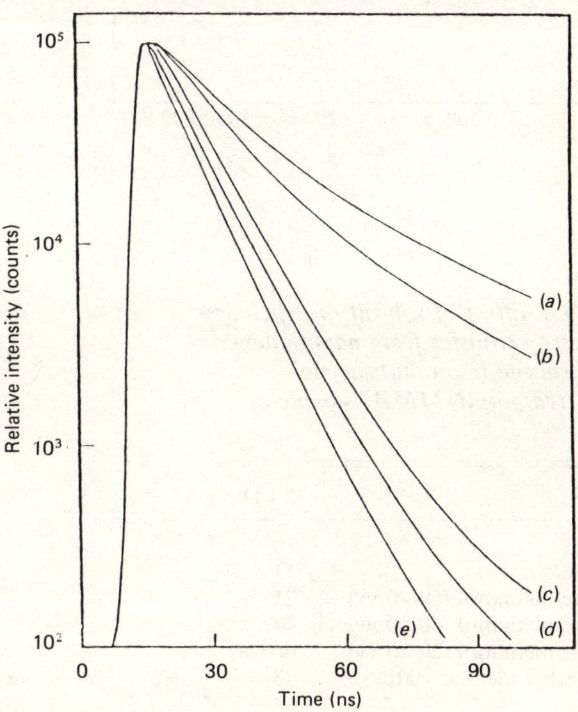

Figure 9.14. Decay of the fluorescence of the free naphthalene chromophore at 340 nm on 280 nm excitation of PNMMA–AMMA copolymers in 2-methyl THF at 77 K. The mol % AMMA in the copolymer samples is (*a*) 0.174, (*b*) 0.348, (*c*) 1.42, (*d*) 2.06. (From Holden 1980.)

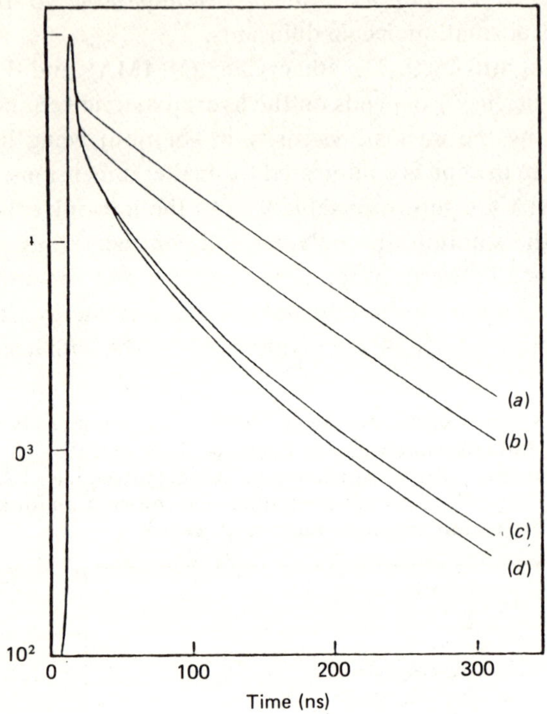

Table 9.2. *Effect of solvent on efficiency χ of energy transfer from naphthalene to anthracene in an anthracene-terminated poly(NMMA) sample at 25 °C*

Solvent	χ (%)
THF	10
Dioxane	11
Dioxane/methanol (60/40 vv)	21
Dioxane/methanol (40/60 vv)	47
Dioxane/methanol (20/80 vv)	66
Poly(methyl methacrylate)	43

showed higher than expected energy transfer efficiencies (Holden *et al.* 1981), presumably because of the clustering of the hydrophobic naphthalene moieties with equally hydrophobic anthracene traps.

Merle-Aubry *et al.* (1980) have identified a phenomenon they termed 'cyclic' energy transfer. If traps such as phenyl ketones or anthraquinone units are included in naphthalene-containing antennas like PNMMA there is a marked increase in the intensity of naphthalene phosphorescence, as well as delayed fluorescence. The phenomenon can be explained by the energy level diagram shown in figure 9.15.

The photon energy absorbed excites the naphthalene singlet. S_1^N ($E \doteq 89.2$ kcal) and is transferred to the trap via the antenna effect, giving rise to the excited singlet state S_1^K of the trap. In the case of both phenyl ketone and anthraquinone traps there is a rapid intersystem crossing to the excited triplet T_1^K which, being higher in energy than the triplet of the adjacent naphthalene units (T_1^N) undergoes triplet transfer back to the antenna naphthalene groups which subsequently phosphoresce or undergo delayed fluorescence (§1.11.3). This apparently provides a more facile route to the naphthalene triplet than intersystem crossing in the naphthalene trap itself. Furthermore, it provides a specific locus for the initiation of triplet migration processes in the antenna chain, which will be dealt with in the next section.

Figure 9.15. Energy level diagram for poly(1-NMA–PIPK). Also shown are the principal deactivation pathways on excitation of naphthalene at 313 nm (full line) and on excitation of aromatic ketone at 366 nm (broken line). (From Merle-Aubry *et al.* 1980.)

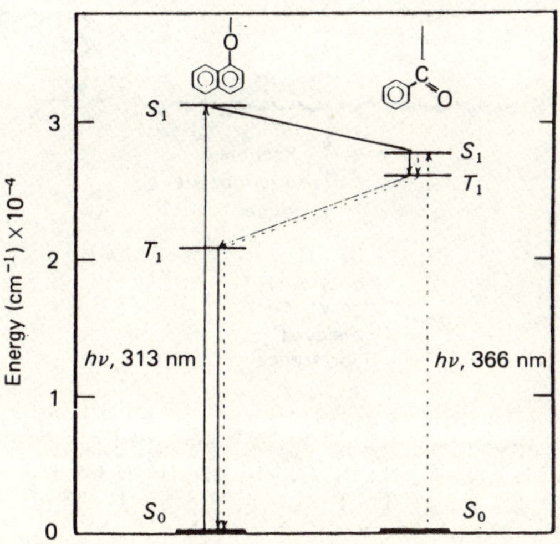

9.9 Triplet energy migration

The possibility of energy transfer from one group to an adjacent chromophore in the polymer chain will be dependent to a large extent upon the lifetime of the excitation and its alternative modes of deactivation. For this reason the most readily observed form of energy migration is that occurring through the mechanism of the triplet state. The first important work in this field was done by Cozzens and Fox (1969*a*,*b*) who showed the occurrence of delayed fluorescence in polymers of poly(1-vinylnaphthalene) when dissolved at high dilution in a low temperature glass.

The mechanism of delayed fluorescence has been dealt with in §1.11.3. In a long polymer chain of perhaps as many as a thousand units with a relatively long excited state lifetime, of the order of fractions of seconds, there is a finite possibility of having two excitations on the same chain at the same time. The process of delayed fluorescence is illustrated in figure 9.16 where absorption of photons at two different locations in the chain gives rise, first of all, to two naphthalene singlet states which intersystem cross into the triplet manifold. If energy migration can take place along the polymer chain, there is a finite chance that two naphthalene triplets will arise on adjacent chromophores and under these conditions the two triplets can combine to give an excited singlet state of the naphthalene and a thermally excited ground state.

Emission from the excited singlet will give emission with a wavelength corresponding to that of normal fluorescence, but because the emission is

Figure 9.16. Triplet–triplet annihilation scheme. (From Somersall and Guillet 1973.)

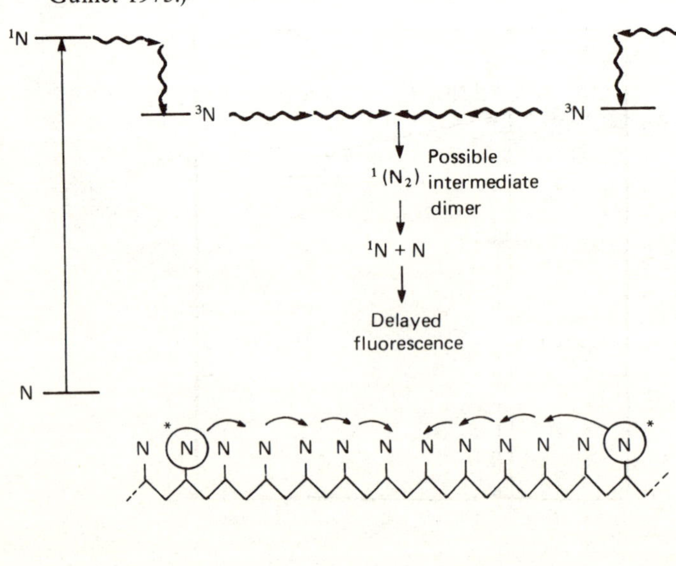

the result of triplet processes, the lifetime of this emission will be of the same order as that of phosphorescence. Furthermore, because this is a biphotonic process, the intensity of the delayed emission will be proportional to the square of the exciting light intensity. Although Fox's early work was subject to considerable controversy, the experiment has been repeated a number of times with different polymers and the results are qualitatively similar. There seems to be no reason now to question the mechanism proposed by Cozzens and Fox. Examples of delayed fluorescence in other aromatic polymers have been given by David *et al.* (1970), Somersall and Guillet (1973), Cozzens and Fox (1969*b*), and Longworth and Battista (1970).

Since phosphorescence intensities from typical chromophores in polymeric systems are usually very low at room temperature as a result of deactivation of the precursor triplet states, it is necessary to cool the system and often to use low temperature glasses in order to study these photophysical processes. A fundamental problem related to these studies is whether or not one can correlate the behavior of a polymer in an organic glass at 77 K with various processes occurring at room temperature or higher. Rauh and Leermakers (1968) consider that if the solute–solvent interactions are not drastically different, then one can assume the emitting states will be similar at the two temperatures. Thus a qualitative if not quantitative correlation may be made. Further problems are related to the limited solubility that polymers have in many solvents, which restricts the number of solvents available for studies of this type. One of the best seems to be α-methyl tetrahydrofuran.

9.9.1 *Experimental studies of triplet migration in vinyl aromatic polymers*

The existence of triplet excitons in synthetic polymers was first postulated by Lashkov and Ermolaev (1967) based on experimental studies of the quenching of phosphorescence in poly(vinyl phthalimide). However, the earliest studies of delayed emission from polymers were reported by Fox and coworkers (Cozzens and Fox 1969*a*,*b*, Fox *et al.* 1971, Fox 1972) who demonstrated delayed emission in homopolymers of 1- and 2-vinyl naphthalene (P1VN, P2VN) and their copolymers with methyl methacrylate and styrene. In solid films at 75 K, the characteristics of the emission seemed to depend on the presence of various adventitious traps introduced either by air oxidation or from catalyst residues from the polymerization. In solid films, triplet migration leading to delayed emission by the mechanism described in §1.11.3 and 9.9 can occur both inter- and intramolecularly. However, when the polymer is dissolved at high dilution in an organic glass, migration appears to occur efficiently within a single chain. The intensity and spectral emission may, however, be altered by traps included in the chain, such as end groups derived from polymerization initiators. Furthermore, quenching impurities in the solvent can also be very much more efficient because of energy migration. Triplet migration was concluded to be as effective in an alternating copolymer of 2-vinyl naphthalene and methyl methacrylate as in P2VN itself, but random copolymers had substantially less migration, presumably because the triplet exciton cannot span the distance interposed by more than one methyl methacrylate unit in sequence along the chain (Fox *et al.* 1974).

Early measurements of the spectral distribution of delayed emission in polymers was complicated by impurity traps. More recently, careful purification of the monomers and polymers has provided better data. Typical delayed emission spectra from solid films of P1VN are shown in figure 9.17 (Burkhart *et al.* 1981). The lower intensities observed at higher temperatures are probably caused by greater triplet quenching by impurities or oxygen.

Recently, elegant experiments by Webber and his coworkers have quantified many aspects of triplet energy migration in poly(vinyl naphthalene)s and other aromatic polymers by pulsed laser experiments using various quenchers.

Webber and Avots-Avotins (1980) studied quenching of the triplet excitons in P2VN and polyacenaphthalene by biacetyl. Biacetyl emission was observed over much longer periods than its normal lifetime, which was related to the rate of triplet exciton migration in each polymer. The authors

Figure 9.17. Delayed-luminescence spectra of P1VN solid films at various temperatures, using a 4-ms delay between excitation and emission. (From Burkhart *et al.* 1981.)

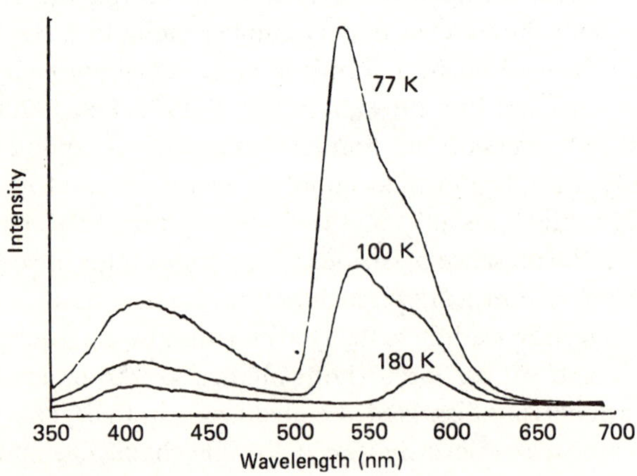

concluded that migration distances were much longer in P2VN than in polyacenaphthalene. Sensitization persisted for longer periods in high molecular weight samples of P2VN, presumably because of the longer migration distances possible. The transient absorption spectrum characteristic of the lowest triplet in P2VN is shown in figure 9.18 (Bensasson *et al.* 1979). Lifetimes of the triplet migration process at 77 K are of the order of seconds, and there is some evidence that at low light intensities the migration extends along the entire length of the chain. At higher intensities the concentration will be reduced by triplet–triplet annihilation. At moderate intensities the relative intensity of delayed fluorescence will be proportional to the square of the light intensity (Somersall and Guillet 1973) (figure 9.19), whereas the phosphorescence is directly proportional to I_A.

9.9.2 *Triplet migration in aromatic ketone polymers*

Kilp and Guillet (1981) have reported quenching and polarization data on isotactic and atactic poly(phenyl vinyl ketone) (PPVK) polymers in which triplet energy transfer can occur. A typical phosphorescence emission spectrum for PPVK is shown in figure 9.20. The spectrum obtained by irradiation at 340 nm into the ketone n-π^* singlet state shows a series of vibronic bands representing the 00, 01 and 20 transitions which occur at 396, 424 and 456 nm, respectively. Their spacing is uniformly 1660 cm^{-1} which corresponds to the carbonyl stretching frequency of 1675 cm^{-1}. The spectrum is independent of the

Figure 9.18. Transient absorption changes, immediately (open circle) and 20 μs (solid circle) after pulse radiolysis of P2VN in benzene solution ($\bar{P}=318$, $8.8+10^{-3}$ M in naphthalene groups). (From Bensasson *et al.* 1979.)

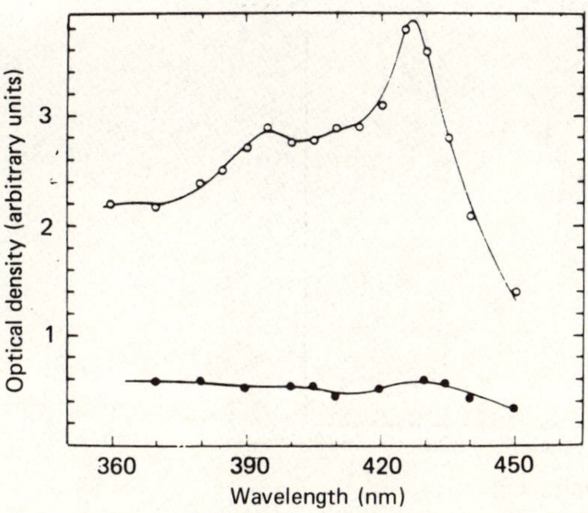

Figure 9.19. Log–log plot for the delayed fluorescence (DF) and phosphorescence (P) of poly(1-naphthyl methacrylate) in THF/ether at 77 K with varied excitation slit widths. (From Somersall and Guillet 1973.)

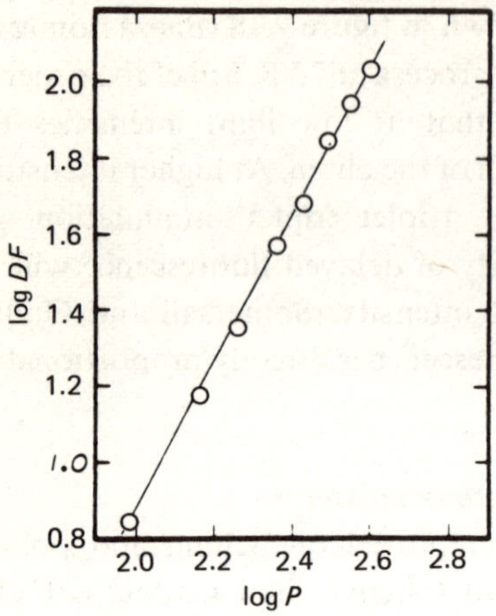

Figure 9.20. Phosphorescence emission of PPVK in 77 K MTHF glass, $\lambda_{ex} = 340$, 10^{-3} M in ketone. (From Kilp 1979.)

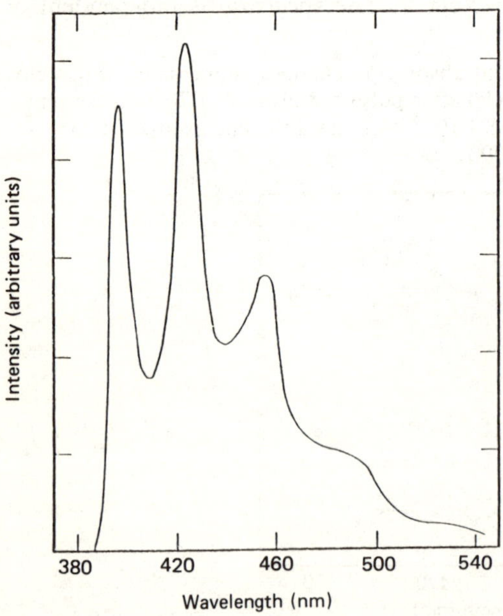

molecular weight of the polymer or of its tacticity. Earlier studies of these and other systems involving aromatic chromophores have confirmed that energy migration can occur along the chains in polymers of this type (David *et al.* 1970, Dan *et al.* 1973). In this case, two additional equations are required to derive a kinetic relationship for the quenching process:

$$D_3^* + D \rightarrow D + D_3^* \quad k_{mig} \tag{9.19}$$

$$D_3^* + A \rightarrow D + A_3^* \quad k_q \tag{9.20}$$

In solution in rigid glasses at 77 K the chromophores can be considered to be rigidly held in space. The possibility of migrating excitation gives rise to a kinetic equation of the same form as the Stern–Volmer relationship in which excitation diffuses with an apparent diffusion constant Λ.

A linear relation of the Stern–Volmer type was observed by David *et al.* (1975, 1977) in studies of energy transfer migration in films and glassy solutions of poly(styrene-*co*-vinyl benzophenone) and by Somersall and Guillet (1973) in the phosphorescence quenching of poly(naphthyl methacrylate) in THF-ether glasses. However, Kilp and Guillet obtained curved Stern–Volmer plots for PPVK systems and pointed out that the Stern–Volmer relation is valid only in the case where energy migration is sufficiently rapid so that a statistical mixing of donor and acceptor groups is achieved during the excited state lifetime of the donor. When this is not the case, the active sphere model of Perrin provides a better representation of the experimental data, as shown in figure 9.21.

Since Förster transfer in this system is unlikely because the overlap between donor emission and acceptor absorption is negligible, energy transfer in the PVK naphthalene system must occur by an exchange mechanism. On this basis the radius of the active sphere R_A should be about 11 Å, as indicated in the dotted line in figure 9.21. The much larger slopes which are obtained experimentally correspond to active quenching spheres of radius 22.5 and 20.3 Å for isotactic and atactic PPVK, respectively. It should be noted that it is necessary to correct for volume contraction in these calculations. Failure to do so increases these values by about 2 Å. The increased magnitude of the active sphere of quenching is explained by the occurrence of intramolecular triplet energy migration in PPVK.

Excitation energy initially located on chromophores not within the active sphere of the quencher migrates down the polymer chain and enters the interaction volume of a quenching molecule and is quenched. This process results in an effective active sphere of quenching which is much larger than that calculated on the basis of molecular diameters. The lack of a molecular weight effect which was observed in this work was taken to imply that the exciton migration length was shorter than the contour length of the polymer, and the higher value of R_A in the isotactic polymer indicates a greater extent of intramolecular energy migration. This is also consistent with the greater tendency for internal quenching of chain scission in the photolysis of isotactic and atactic PPVK (Kilp and Guillet 1981).

Dan *et al.* (1973) have reported a similar effect in phosphorescence quenching of atactic PPVK with cis-cis-1,3-cyclooctadiene as the quencher, in which case the value of R_A is 19 Å, and David *et al.* (1970) have reported a value of 26 Å for atactic PPVK quenched by naphthalene. These conclusions are also supported by the results of Ishi *et al.* (1977) who showed higher rates of energy migration in isotactic, as compared to atactic polystyrene in fluid solutions.

Kilp *et al.* (1982) also studied the phosphorescence depolarization of a number of copolymers of methyl methacrylate and methacrylophenone at high dilution in solid organic glasses. They showed that the phosphorescence depolarization as expressed by $1/P$ was a linear function of the ketone composition in the polymer, as shown in figure 9.22. Since phosphorescence emission from the homopolymer methacrylophenone is

Figure 9.21. Perrin plots for phosphorescence quenching. (From Kilp and Guillet 1981.)

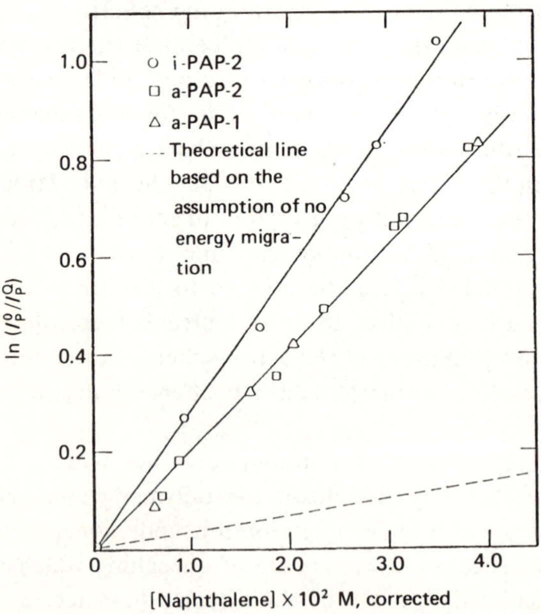

completely depolarized, this behavior is in good accord with that predicted by the Jablonski, Förster–Öre and Weber models of emission depolarization as a function of concentration, except that in this case, the concentration changes are within the individual polymer molecule rather than uniformly dispersed throughout the solution. David *et al.* (1976) observed similar behaviour for copolymers of styrene and vinyl benzophenone. From the kinetics of the copolymerization reaction it is possible to calculate the sequence distribution in the copolymers of

Figure 9.22. Phosphorescence polarization as a function of PMMA–MAP composition λ_{ex} was 350 nm and all solutions were $(1\text{–}3) \times 10^{-3}$ M in ketone units. (From Kilp *et al.* 1982.)

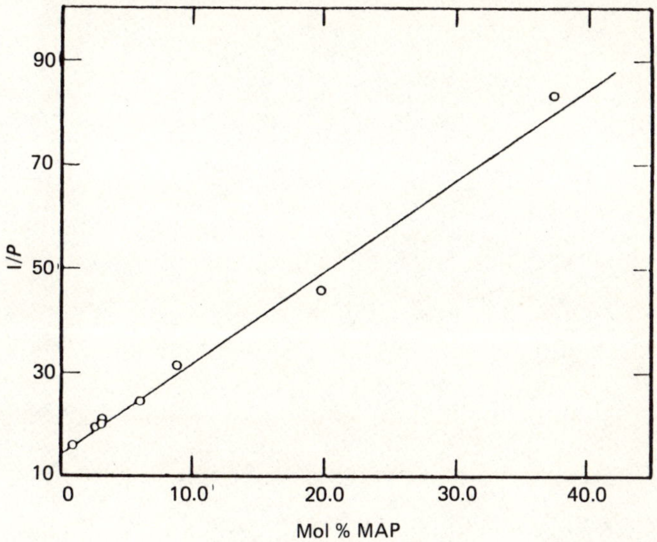

Table 9.3. *Sequence distribution of PMMA–MAP copolymers*[a]

Polymer	Mol% MAP	F_{ABA}	F_{BBA}	F_{BBB}	$P \pm 0.002^b$
PMMA–MAP-1	1.00	0.9869	0.0131	0.0000	0.068
PMMA–MAP-3	3.15	0.9667	0.0633	0.0003	0.050
PMMA–MAP-5	6.20	0.9252	0.0733	0.0015	0.040
PMMA–MAP-6	8.80	0.9043	0.0933	0.0024	0.031
PMMA–MAP-7	20.06	0.7511	0.2311	0.0178	0.017
PMMA–MAP-8	37.70	0.5211	0.4015	0.0776	0.010
PMAP	100.00	0.0000	0.0000	1.0000	0.006

[a] A and B refer to MMA and MAP units, respectively.
[b] Polarization of 0–0 band following 340 nm excitation.

methyl methacrylate and methacrylophenone. Table 9.3 shows the phosphorescence depolarization along with the mole fraction of various types of sequences in the polymer chain. It is clear that substantial amounts of depolarization occur, even when 90% of the carbonyl groups are isolated in the polymer chain ($F_{ABA}=0.90$). This phenomenon strongly suggests that triplet energy migration can take place through non-nearest neighbors in the polymer chain. Similar effects were observed in the phosphorescence depolarization of copolymers of styrene and methacrylophenone and for polystyrene with PVK and methyl methacrylate.

10

Photochemistry of carbonyl-containing polymers

10.1 Ketone-containing polymers

10.1.1 *Excited states of keto and aldehyde carbonyls*

The ketone group C=O is one of the most important chromophores from a photochemical point of view. Historically, the earliest work is that of Bodenstein and Lind but real progress followed the classic studies of Bamford and Norrish (1935, 1938*a,b*).

A particular advantage of the ketone and aldehyde carbonyls is that although they are photochemically labile, they are thermally stable. Furthermore, they absorb in the near ultraviolet and hence can be excited selectively in the presence of large numbers of other chemical groups. In synthetic polymers they are particularly important because they are known to play an essential role in initiating and propagating the photo-oxidation processes which occur in hydrocarbon polymers on exposure to terrestrial radiation. Before discussing the photochemistry of polymeric ketones, it is useful to review our present understanding of the photophysics of small molecule ketones.

In most simple ketones and aldehydes the highest energy occupied molecular orbital (MO) is non-bonding (n) and the lowest unoccupied MO is a π^* orbital. The transitions which are of particular interest to photochemists are then the n-π^* and π-π^* transitions which occur upon irradiation with uv radiation as shown schematically in figure 10.1. The weak absorption usually observed between 280 and 350 nm in aliphatic and aromatic ketones corresponds to the n-π^* transition to the $S_1(n\text{-}\pi^*)$ state. This common designation of excited states identifies the multiplicity of the state S_1 (first excited singlet) and the singly occupied orbitals (n-π^*). T_1, T_2 and T_3 would designate the first, second and third excited triplets, respectively. Since the n and π^* orbitals are orthogonal, the n-π^* transition violates both symmetry and orbital selection rules. This explains the low intensity of these absorptions, usually ε_{max}, from 10 to 100, in contrast to completely allowed transitions which have ε_{max} of the order of 10^5, such as the π-π^* transition in carbonyls which are usually observed around 180 nm. In aromatic ketones the n-π^* absorption maximum is red-shifted to longer wavelengths (~320 nm) with a corresponding increase in ε to 100–150 mol^{-1} cm^{-1}. This red shift is a consequence of the inductive effect of the π system of the phenyl ring

Figure 10.1. Molecular orbitals and their approxmate energy levels in formaldehyde. (Adapted from Kasha 1960.)

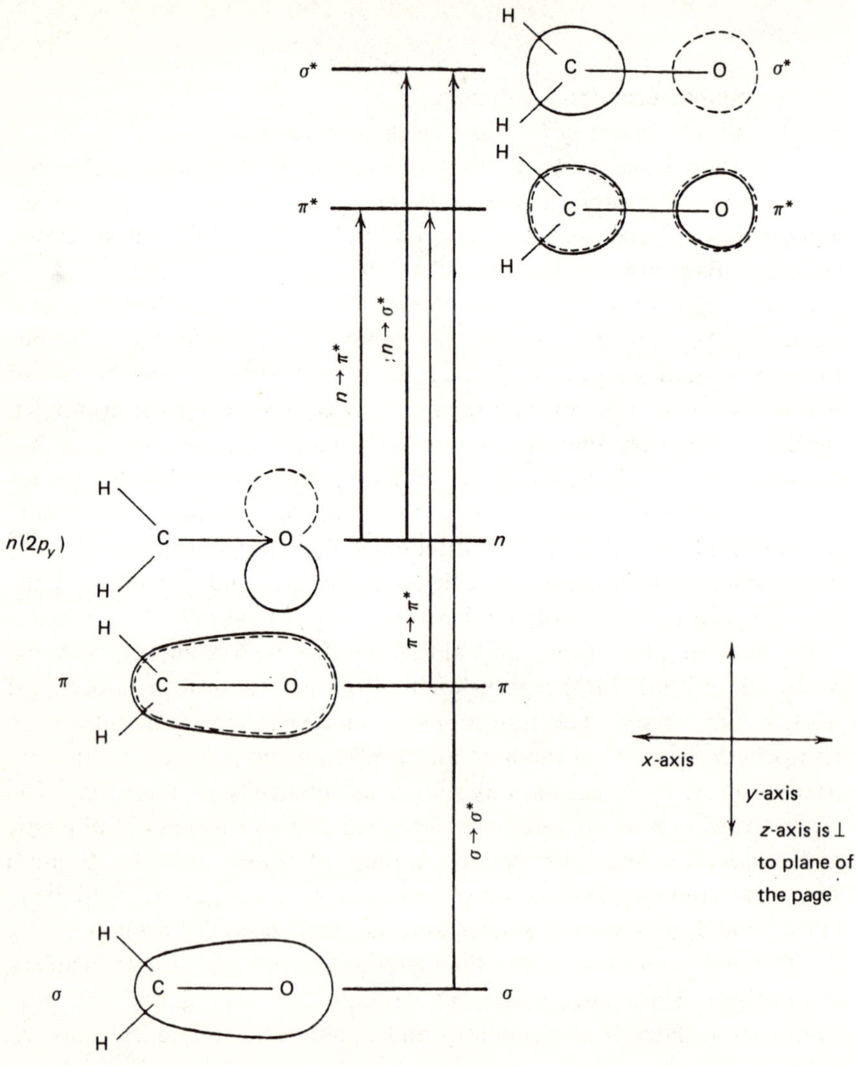

which lowers the π^* energy level of the ketone group and mixes the n-π^* state of the ketone with the π-π^* state of the phenyl system.

Figure 10.2 shows a sketch of the ketone carbonyl orbitals which are important in photochemistry. The n orbital is localized on the carbonyl oxygen atom, while the π^* orbital is delocalized over the entire carbonyl function. Thus, moving an electron from an n orbital to a π^* orbital will tend to remove an electron from the immediate vicinity of the oxygen, giving a product which has many of the properties of a diradical. For this reason n-π^* states are (1) electrophilic and radical-like in the vicinity of the oxygen atom as a consequence of the half-vacant n orbital in the plane of the carbonyl function, and (2) nucleophilic above and below the molecular plane of the carbonyl because of the π^* electron. For this reason also, carbonyl triplets react very much like alkoxy radicals.

Aliphatic ketones fluoresce very weakly with a maximum around 400 nm. This has been assigned to emission from the $S_1(n\text{-}\pi^*)$ state and the quantum yields and lifetimes are typically of the order of 0.01 and 10^{-9} s, respectively. Phosphorescence from aliphatic ketones is usually too weak to be measured in solution at room temperature but may be observed in glassy matrices at low temperatures and occasionally in polymer matrices at close to room temperature. More recent determinations of the natural triplet lifetime $^3\tau_0$ for a variety of aliphatic ketones has given a value of the order of $\sim 10^{-5}$ s.

In contrast to aliphatic ketones, aromatic ketones do not fluoresce because intersystem crossing (ISC) from the singlet to the triplet state is very efficient. In most aromatic ketones Φ_{ISC} is usually assumed to be equal to unity and this is considered to be due to the presence of a close-

Figure 10.2. Charge distributions associated with ground and $(n-\pi^*)$ excited states in a carbonyl group.

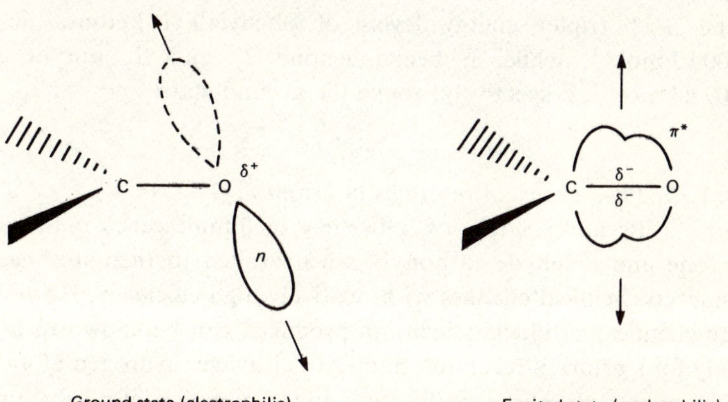

lying π-π^* triplet state which provides the necessary spin-orbit coupling interaction with the π-π^* singlet. El-Sayed (1963) has shown that spin-orbit coupling between states of different electronic configuration are about 100 times greater than those between states of the same electronic configuration. Thus aromatic ketones which have a π-π^* triplet energetically intermediate between the n-π^* singlet and triplet states should show a higher rate of intersystem crossing than would aliphatic ketones where the π-π^* triplets lie above the n-π^* singlet states. For example, acetone has a value of k_{ISC} equal to $5 \times 10^7\ s^{-1}$ compared to that for benzophenone of approximately $10^{11}\ s^{-1}$.

A further enhancement of k_{ISC} in aromatic ketones is caused by the smaller energy separation between S_1 and T_1 which increases the Franck–Condon overlap factor. In benzophenone the energy gap is 25 kJ mol^{-1}, whereas in acetone it is 42 kJ. Phenyl ketones also have a second excited π-π^* triplet, T_2 which lies less than 310 kJ mol^{-1} above the ground state. This can affect the reactivity of the excited molecule depending on the contribution of this π-π^* state to the close-lying n-π^* triplet. This explains the large changes in photochemistry which can be observed by varying the substituents on the ketone, or changing the solvent, which can often invert the energy levels of the two states. For example, in acetophenone, the lowest triplet T_1 is n-π^* ($E_T = 310$ kJ mol^{-1}) while for p-methoxy acetophenone it is π-π^* in character ($E_T = 301$ kJ mol^{-1}). Furthermore, n-π^* transitions show a blue shift in the presence of hydroxylic solvents which is due to hydrogen bonding and is partially responsible for the inversion of the π-π^* states which can occur in solvents of strong polarity.

Because of the lack of vibrational structure in both the absorption and emission spectra of aliphatic carbonyls, a 0–0 transition energy is not usually assignable. By other methods, approximate values of S_1 and T_1 energy levels has been determined to be 351 and 326 kJ mol^{-1}, respectively. The n-π^* triplet energy levels of phenylalkyl ketones lie around 300 kJ mol^{-1}, while in benzophenone T_1 and S_1 are at 284 and 309 kJ mol^{-1}, respectively, above the ground state.

10.1.2 *Photochemical reactions in ketones*

Because of the low efficiency of luminescence outlined earlier, ketone and aldehyde carbonyls, when excited to their n-π^* states, can undergo chemical changes with relatively high efficiency. However, all of the chemical products which are produced can be shown to arise from only four primary reactions, namely α-cleavage, hydrogen atom abstraction, charge-transfer complexation and elimination of α substituents.

10.1.3 *Reactions involving α-cleavage*

In acyclic ketones, α-cleavage is usually designated the Norrish type I reaction:

$$R{-}\overset{O}{\overset{\|}{C}}{-}R' \xrightarrow{h\nu} R{-}\overset{O}{\overset{\|}{C}}\cdot + R'\cdot \xrightarrow{\Delta} R\cdot + R'\cdot + CO \tag{10.1}$$

$$R{-}\overset{O}{\overset{\|}{C}}\cdot + SH \rightarrow R{-}\overset{O}{\overset{\|}{C}}{-}H + S\cdot \tag{10.2}$$

$$R'\cdot + SH \rightarrow R'H + S\cdot \tag{10.3}$$

The fate of the primary radicals produced depends on the experimental conditions, in particular the nature of the solvent medium (SH), but a hydrogen atom is usually extracted from the solvent, leading to an aldehyde and a saturated hydrocarbon. At elevated temperatures the acyl group will eliminate carbon monoxide to give a second alkyl radical. In cyclic ketones α-cleavage results in a photo-rearrangement, for example

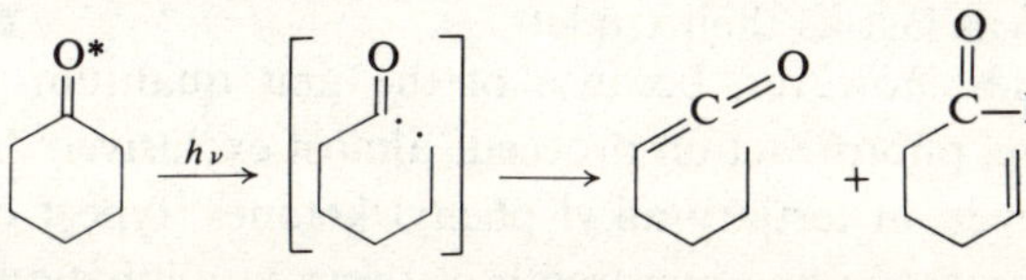

(10.4)

The efficiency of α-cleavage depends on both the structure of the carbonyl group and the viscosity of the medium and the rate constant is strongly dependent on the stability of the radicals formed. For example, the rate constants for cleavage of progressively more stable radicals are as shown in table 10.1 where we see that increasing the substitution on the α carbon greatly increases the primary quantum yield for type I processes.

Table 10.1. *Rate constants* k_c *for α-cleavage of* $R{-}\overset{O^*}{\overset{\|}{C}}{-}R$

Radical, *R*	k_c (s^{-1})
CH_3	10^3–10^4
CH_3CH_2	10^6–10^7
$(CH_3)_2CH$	10^8
$(CH_3)_3C$	2×10^9
 $-CH_2$	10^{10}

Source: Wagner (1966).

Rapid vibrational deactivation and 'cage' recombination can reduce the efficiency of the type I process in solution, but not in the gas phase at high dilution. In solution the free radicals initially formed are held in proximity by a so-called cage of solvent molecules which increases the possibility of recombination of the primary radicals to reform the original ketone. If the ketone is unsymmetric, the primary cleavage is that which produces the more stable alkyl radical, although at shorter wavelengths a greater degree of randomness in the product distribution is observed. In the gas phase, vibrationally excited acyl radicals resulting from the initial α-cleavage frequently undergo decarbonylation to produce carbon monoxide. In solution, decarbonylation occurs at room temperature only if the acyl group can decompose to form a very stable radical such as tertiary alkyl or benzyl.

The type I reaction in alkyl ketones can proceed either from the singlet or the triplet state. Excited singlets of tertiary butyl ketones undergo α-cleavage only 10^{-2} as fast as their triplets.

In phenyl ketones, however, because of the unit quantum yield of intersystem crossing, photoreaction proceeds almost exclusively from the triplet state. However, in tertiary alkyl phenyl ketones, type I cleavage occurs with rates only one-one hundredth of those in aliphatic ketones. Furthermore, studies on substituted phenyl ketones show that the rate of α-cleavage from π-π^* triplets is much lower than that from n-π^* triplets, indicating that n-π^* excitation favors α-cleavage because it weakens the carbonyl-α-carbon bond.

There is some evidence that vibrational excitation is required for the type I reaction. In the first place, ϕ_I is usually much greater in the gas phase than in solution, where vibrational deactivation is rapid, and, secondly there is usually a strong increase in ϕ_I with increase in temperature. Furthermore, Berger and Steel (1975) found that acetophenone excited into the S_1 state decomposed with a low type I quantum yield, but ϕ_I increased dramatically when it was excited to the S_2 state. In this case these authors postulated a vibrationally excited triplet state as the reaction intermediate although there is little experimental evidence for this. The relatively large activation energies for the triplet type I process in 4-methyl-2-pentanone and in 2-butanone, which are 60.2 and 53.1 kJ mol^{-1}, respectively, also argue for the participation of a vibrationally excited state. These activation energies are considerably greater than would be predicted from simple treatments involving microviscosity and the cage effect.

Theoretical calculations on carbonyl compounds have shown that the n orbital and C–C σ orbitals are mixed. For this reason, the n-π^*

excitation intrinsically weakens the carbonyl-α-carbon bond and predisposes that bond to rupture. The potential surfaces involved may well be analogous to those followed by alkoxy radicals in their β-cleavage reactions and in recent studies this concept has been increasingly helpful in explaining and predicting the reaction of n-π^* triplet states.

10.1.4 *Hydrogen atom abstraction*

Perhaps the most important reaction of excited carbonyl compounds is the abstraction of a proton which may occur either inter- or intramolecularly, the subsequent course of the reaction being dependent upon structural features in the molecule. It also depends to a certain extent upon the multiplicity and electronic configuration of the excited state. For example, it is generally agreed that the triplet n-π^* state is much more effective in hydrogen abstraction than is the triplet π-π^* state. This is because the n-π^* state is generally more electrophilic, and its subsequent reactions are the type which can replace the electron in the now vacant non-bonding orbital. However, in the triplet π-π^* state, there is an increasing negative charge on the oxygen, thus reducing its electrophilic nature and its tendency to undergo hydrogen abstraction. However, some ketones which possess a lowest π-π^* state triplet are not necessarily totally unreactive towards hydrogen abstraction. For example, Wagner (1976) has pointed out that if an upper n-π triplet lies sufficiently close to the lower π-π^* triplet, vibronic mixing of the two states can provide some reactivity towards hydrogen abstraction.

The most important hydrogen abstraction process in excited ketones is known as the Norrish type II photoelimination. This is an intermolecular hydrogen abstraction on the carbon γ to the carbonyl group, leading to cleavage of the α-β C–C bond and resulting in the formation of an enol and a terminal olefin. The enol then tautomerizes to the more stable ketone:

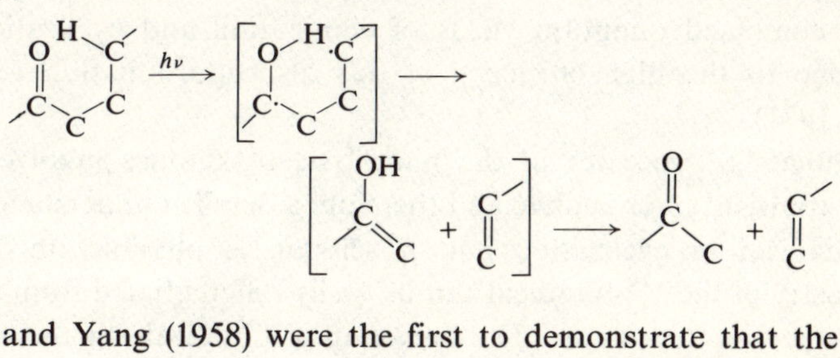

(10.5)

Yang and Yang (1958) were the first to demonstrate that the reaction proceeds via the intermediacy of a 1,4-biradical and the evidence for the biradical intermediate now seems so unequivocal that the possibility of a concerted process leading to C–C bond cleavage is no longer accepted.

Both the singlet and the triplet states of aliphatic ketones undergo this type of reaction but only the triplet state is reactive in phenyl ketones. In small molecule ketones the quantum yield for type II scission is usually independent of ketone concentration, solution viscosity, temperature and the presence of oxygen. In *n*-alkyl ketones, where the alkyl chain length is increased, there is a reduction in quantum yield with increasing chain length except for phenyl ketones (Golemba and Guillet 1972) and this reflects the greater difficulty in the formation of the six-membered cyclic intermediate which is required for the scission process. Increasing substitution on the γ-carbon also increases the efficiency of the reaction by improving the extractability of the proton.

10.1.5 *Photocyclization*

The 1,4-biradical has, however, other competitive alternatives, the first of which is cyclization to form cyclobutanols:

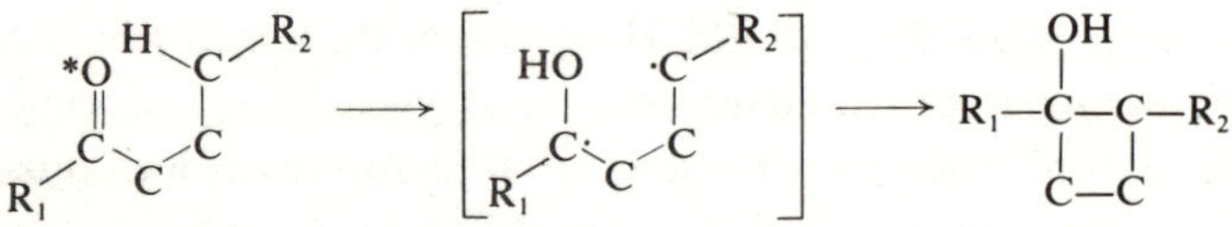

(10.6)

This reaction proceeds with high stereo specificity to yield predominantly the more stable stereoisomer (Wagner *et al.* 1972). Another possibility is internal disproportionation which results in a rearrangement back to the starting ketone

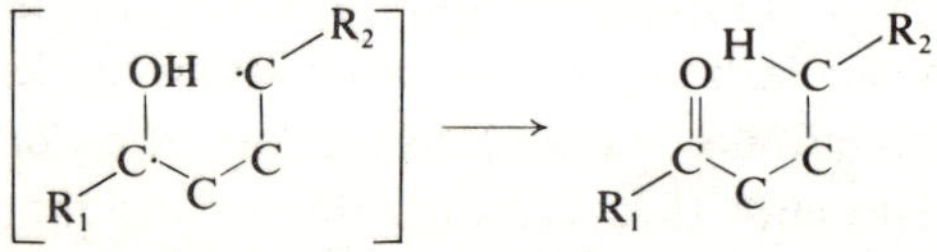

(10.7)

The low combined quantum yields of elimination and cyclization are usually due to the high efficiency of this disproportionation reaction (Wagner 1971).

Unfortunately, processes in the photolysis of ketones involve more than one transient intermediate and therefore a detailed understanding of the mechanism of cyclization and β scission is possible only when the chemistry of the 1,4-biradical can be easily differentiated from that of the precursor excited states. The behavior of biradicals in the type II reaction has received considerable attention in the last few years and several substrates have been shown to trap these biradicals in reactions involving hydrogen extraction from thiols, addition to double bonds, electron transfer, and addition of molecular oxygen to form bicyclic peroxides.

In phenyl alkyl ketones the type II quantum yields approach unity when the reaction is carried out in polar solvents such as alcohols. The formation of hydrogen bonds apparently stabilizes the 1,4-hydroxy biradicals in a conformation unsuitable for undergoing the disproportionation reaction back to the starting ketone. However, aliphatic ketones do not reach unit quantum yields, even in pure alcohol, possibly because of the shorter lifetime of the aliphatic 1,4-biradical.

The 1,4-biradical could conceivably be either singlet or triplet in nature, depending on the multiplicity of the original excited state and the efficiency of intersystem crossing. It has been suggested that the triplet biradical has a longer lifetime than has the singlet biradical, and that the singlet biradical undergoes α-cleavage to olefin and enol 10 to 20 times faster than rotation about $^{\beta}C$–$^{\gamma}C$. Since substitution on the carbon atoms would be expected to affect rotational rates around this bond, they could also be important in determining the reactivity of the 1,4-biradical. In phenyl alkyl ketones, the lifetimes of the biradicals produced has been estimated to be in the neighborhood of 100 ns, and are rather insensitive to changes in temperature, substitution and solvent. However, oxygen was found to be an efficient radical scavenger with a rate constant of $7 \times 10^9\ \mathrm{l\ mol^{-1}\ s^{-1}}$ in methanol, which is close to the diffusion limit.

The intermediacy of the enol has been well demonstrated experimentally. For example, the enol from the type II cleavage of 2-pentanone has been detected by NMR and lifetimes of more than 5000 s at 200 K have been estimated. In certain ortho-substituted alkyl aryl ketones, photoenolization appears to be the major reaction. When the substituent is a hydroxy group, this reaction can be proved, viz.

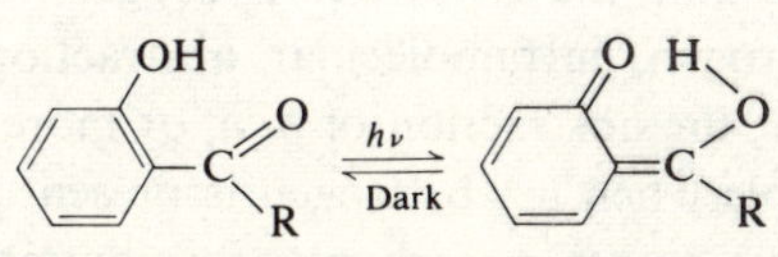

(10.8)

This type of process has been attributed by Baum *et al.* (1966) to be the cause of the lack of any type II cleavage in substituted butyro phenones where the substituent was either ortho-hydroxy or ortho-methyl.

As indicated earlier, the excited singlet and triplet have some characteristics of a biradical intermediate due to the change in distribution of the electronic charge in the excited state. In the presence of strong hydrogen donors, photoreduction of the ketone can take place by means of a ketyl radical intermediate:

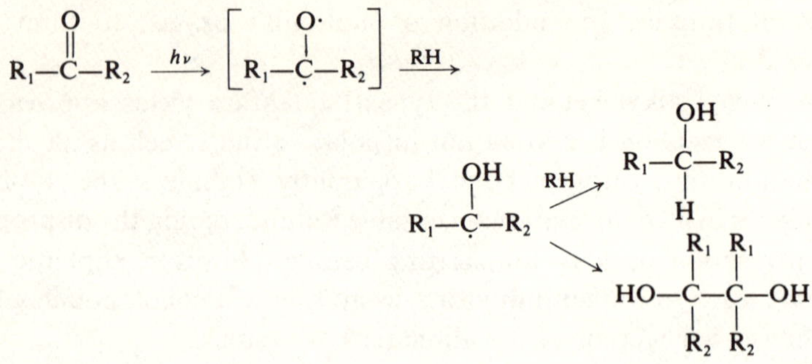

(10.9)

If the keto radicals are present at high enough concentrations or are stable enough to have long lifetimes, they may combine to form pinacols. The rate constant for this intermolecular hydrogen abstraction is typically about 10^4–10^6 mol^{-1} s^{-1}, about one-hundred to one-thousand times less than that for the intramolecular process in type II reactions. For example, the rate constant for hydrogen abstraction by acetophenone from cyclohexane was found to be 3.7×10^5 mol^{-1} s^{-1}, whereas the unimolecular rate constant for butyrophenone is 1.2×10^8 s^{-1}. This large difference between inter- and intramolecular rate constants indicates that the intermolecular process is unimportant in ketones capable of undergoing γ-hydrogen abstraction. Furthermore, intramolecular reactions do not require diffusion of reagents to form an encounter pair of separation of products by diffusion.

The proximity of the ketone chromophore to the γ-hydrogen produces a much more favorable entropic factor for hydrogen abstraction. Apart from this entropic factor, the rates of inter- and intramolecular hydrogen abstraction are similar, being primarily affected by the nature of the substituent on the carbon containing the abstracted hydrogen.

In the absence of a γ-hydrogen, intramolecular abstraction of δ-hydrogens is possible. However, the abstraction of β, δ, or more remote hydrogens is much less favorable when a γ-hydrogen is present because tortionally strained rings of five or seven carbons are required in the intermediate. Thus, for example, the ratio of γ to δ-hydrogen abstraction is 20:1 in *n*-hexyl ketone. In cyclic ketones such as cyclodecanone where the abstraction of a γ-hydrogen is sterically impossible, the principal photochemical reaction is the intramolecular abstraction of transannular hydrogen to give both cis- and trans-9-hydroxydecalin.

Rates of hydrogen abstraction are strongly dependent on factors such as inductive substituent effects, bond strengths and steric strain. Electron releasing groups attached to the γ-carbon atom greatly increase the abstractability of protons since they make it more susceptible to attack by the electron-deficient oxygen of the n-π^* state. Alternatively, strong electron hyphen withdrawing groups are deactivating. The reactivity is also strongly dependent upon the strength of the γ C–H bond. For example, the relative reactivities of tertiary to secondary to primary C–H bonds for the benzoyl triplet are 180:24:1, which is similar to the selectivity shown by alkoxy radicals.

The conformational effects in polymers which limit the accessibility of the abstractable hydrogen in the type II reaction were first recognized by Hartley and Guillet (1968*a*,*b*). They observed a decrease in the type II quantum yield with an increase in chain length of the carbonyl compound. Furthermore, there was a drastic decrease in the quantum yield of the type II process in polymers below their glass transition temperature where the molecular motion required for conformational changes is substantially reduced. The transition state for the intramolecular γ-hydrogen abstraction is a torsional strain-free chain conformation:

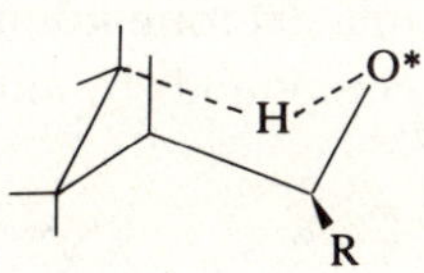

If there is not enough free volume available to achieve this conformation within the lifetime of the excited state, reaction will be restricted to the relatively small number of conformations which have this configuration at the time of excitation. In solid polyethylene-CO copolymers the reduction in ϕ_{II} with temperature below $-40°C$ could be attributed to a slowing down of the frequency of the crankshaft-type motion necessary to form this intermediate during the lifetime of the excited state.

10.1.6 *Reactions involving charge-transfer complexes*

In the excited state of ketones, an electron occupies a high energy anti-bonding orbital which is more easily oxidized or reduced. It has been shown that nearly all excited carbonyl compounds form exciplexes with electron donors and with many electron acceptors which can either revert to ground state reactants or lead to efficient chemical reactions. Two examples of reactions of carbonyl compounds which involve exciplexes are the photocycloaddition of carbonyl compounds to olefins and their photoreduction by amines.

Photocycloaddition (*Paterno–Buchi reaction*). The general mechanism of this reaction is shown below:

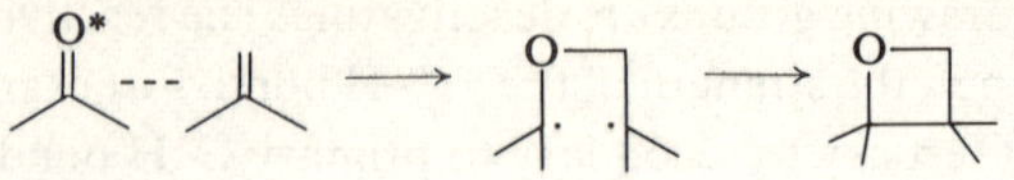

(10.10)

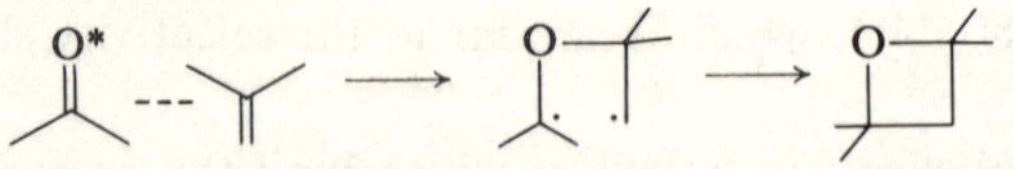

(10.11)

Recent evidence indicates that a charge-transfer complex precedes the formation of biradicals. Competing chemical and physical decay processes of the n-π^* species together with the non-radiative decay of the biradical intermediate to ground state reactants can cause the oxetane reaction to be inefficient. Oxetane formation proceeds from both the singlet and the triplet state and is observed for olefins such as 1-methoxy-1-butene. The singlet n-π^* state of acetone adds stereospecifically to electron-deficient olefins such as cis-dicyano ethylene, while its triplet state is quenched by these olefins without chemical reaction.

Photoreduction by amines. Kinetic studies and determination of the quantum yields for photoreduction of carbonyl compounds by amines are used to support an exciplex intermediacy:

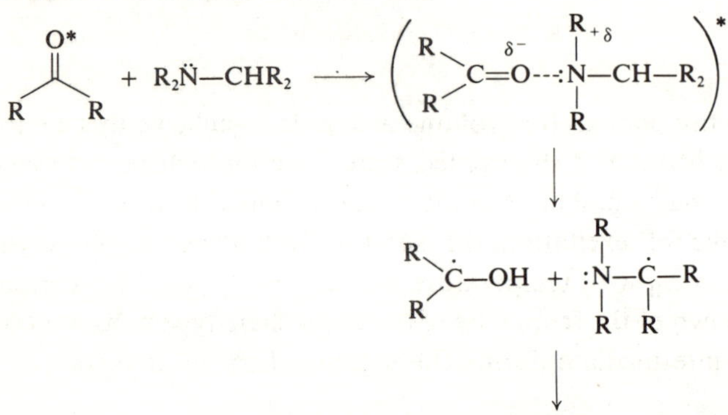

(10.12)

This mechanism explains the special features of this photoreduction process. For example, the rates of quenching are much faster than are those observed in intermolecular photoreduction while the quantum yields are lower than normal, and π-π^* states which are inefficient in normal hydrogen abstraction are efficiently reduced by amines. Furthermore, higher polarity of the solvent enhances the rates and quantum yields for reaction and, in some cases, transient species with absorption maxima greater than 600 nm have actually been observed and attributed to the absorption of an exciplex species. Other molecules containing hetero atoms such as sulfides, phosphenes and phosphites have been found to quench carbonyls efficiently. However, details of the relevant exciplex mechanisms are still subject to considerable controversy.

Pavlik *et al.* (1973) report that photolysis of aliphatic and aromatic ketones in carbon tetrachloride gives, in place of the expected ketone products, very high yields of chlorine atoms and tricholormethyl radicals. They also observed a quenching of ketone fluorescence and postulated that energy transfer from the excited ketone to the CCl_4 occurred via an exciplex mechanism.

10.1.7 *Elimination of α-substituents*

In carbonyl compounds which contain hetero atom substituents in the α position, such as α-halo or α-amino ketones, the energy localized on the carbonyl group can apparently be delocalized through the molecule, resulting in the elimination of a radical or an ion. For example, in a α-chloroacetone, the reactions are

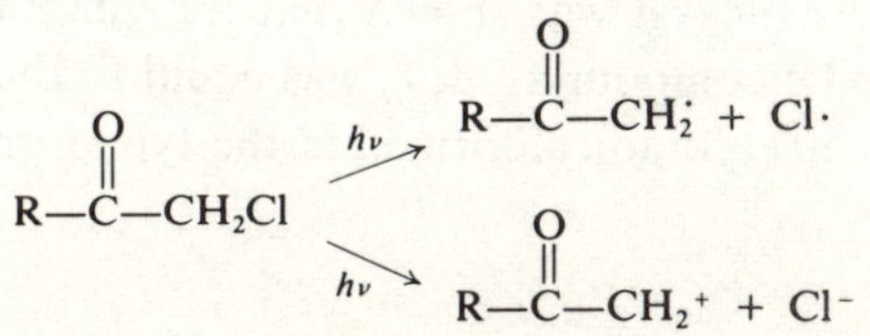

(10.13)

(10.14)

These processes are apparently sufficiently rapid that, when they occur, there is no evidence of competition from either the Norrish type I or type II processes.

10.1.9 *Effects of molecular weight and ketone structure on quantum yields*

The importance of the ketone chromophore in understanding the mechanism of photo-oxidation of hydrocarbon polymers such as polyethylene and polypropylene has prompted a series of studies on model aliphatic ketones in solution to determine the effects to be expected as the chain length is extended to polymeric dimensions. Some of these results are summarized in table 10.2. The last row of this table represents an extrapolation of these trends to infinite (i.e., polymeric) chain length.

The main conclusion from these results is that after the ketone chain length exceeds five or six carbon atoms, there is very little further change in quantum yield (not more than a factor of two) with chain length. Methyl ketones will have about double the type II quantum yields of the corresponding dialkyl ketones, but both, at 25°C, have very low type I (radical) yield, of the order of 0.002. The apparent activation energy for this process usually is from 20–35 kJ mol^{-1}, so that even above 100°C it will only correspond to about 10% of the type II, which is almost independent of temperature. However, this conclusion is only correct for linear alkanones. If the carbonyl group is attached to a secondary or tertiary carbon atom, as in structure III in table 10.2 there is a twenty- to thirty-fold increase in the type I quantum yield (to ~0.05, while ϕ_{II} drops to a similar value. This has important consequences in photo-oxidation, since branched chain carbonyls will produce many more initiating free radicals than will linear ketones.

The availability of improved infrared instrumentation has permitted estimates of photoreduction in model ketones of structure III, the determined quantum yields being about 0.03–0.04. Furthermore, none of the quantum yields were reduced when the model compounds were irradiated in an amorphous ethylene-propylene copolymer matrix.

When polymer films containing polymerized ketone groups are irradiated, the ir carbonyl absorption is observed to decrease. In work published by Li and Guillet (1984) it was shown that the quantum yield for carbonyl loss in the model compounds, ϕ_{CO}, was equal to the sum of ϕ_I and ϕ_{OH}. Apparently the acetyl radical formed in the type I process in

Table 10.2. *Quantum yields of photolysis of model aliphatic ketones in hydrocarbon solution*

Model	I. CH_3—C(=O)—R[a]		II. R—C(=O)—R[b]		III. R—CH(—C(=O)—CH_3)—$C_{18}H_{37}$[c]			
R	ϕ_I	ϕ_{II}	ϕ_I	ϕ_{II}	ϕ_I	ϕ_{II}	ϕ_{OH}	ϕ_{total}
1					0.028	0.066	0.034	0.128
2					0.045	0.063	0.041	0.149
3					0.030	0.063	0.040	0.133
4		0.25		0.11				
5		0.20						
6		0.20		0.09				
7	0.003	0.20	0.002	0.08				
11				0.07				
17		0.15						
18					0.056	0.049	0.031	0.136
19	0.002	0.124		0.06				
∞	~0.002	0.13	0.002	0.06	0.05	0.05	0.03	0.13

[a] Golemba and Guillet (1972); [b] Hartley and Guillet (1968); [c] Li and Guillet (1977).

compound III (table 10.2) and polymeric analogs abstracts a proton to give acetaldehyde, which subsequently is lost by vaporization. The formation of cyclobutanols appears to be favored in the secondary and tertiary ketones, since it is rarely observed in linear alkanones in aliphatic solvents.

10.2 Polymeric ketones

10.2.1 *Poly(methyl vinyl ketone)*

Most of the primary reactions of small molecule ketone carbonyls have been shown to occur in polymeric compounds containing the same chromophores in similar structural arrangements. The earliest work in this area was that of Guillet and Norrish (1954, 1955) who studied the photolysis of poly(methyl vinyl ketone) (PMVK) in solution. They showed that PMVK underwent the classic Norrish type I and type II reactions by the following scheme:

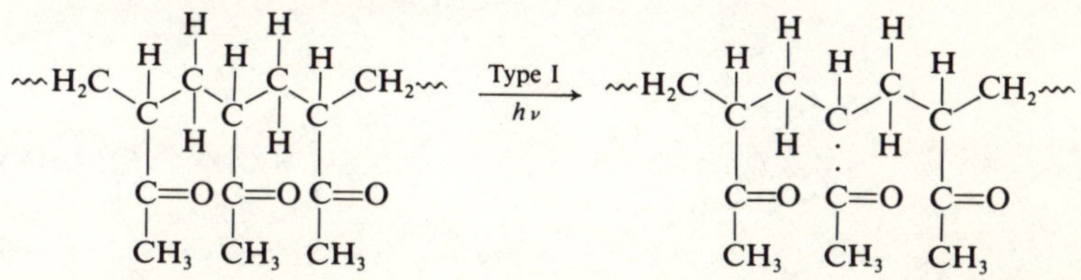

(10.15)

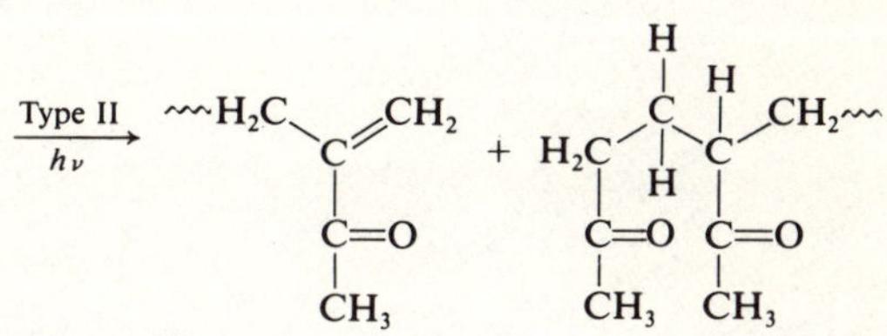

(10.16)

$$\cdot\overset{O}{\overset{\|}{C}}-CH_3 + SH \longrightarrow H-\overset{O}{\overset{\|}{C}}-CH_3 + S\cdot \qquad (10.17)$$

The type II reaction caused main chain scission with concomitant reduction in the molecular weight of the polymer, while the type I process involved the scission of the C–C bond adjacent to the carbonyl, resulting in the formation of an acetyl radical which subsequently abstracted a proton to give acetaldehyde. In dioxane solution and $\lambda = 313$ nm the quantum yields for these processes were 0.04 and 0.025, respectively. The quantum yield for chain scission decreased with the extent of degradation and was attributed to a repolymerization initiated by radicals produced by the type I process, by addition to the terminal double bonds created by the type II scission.

Further confirmation of the presence of free radicals resulting from the

Norrish type I scission was demonstrated by the formation of graft polymers by photolyzing PMVK in the presence of a vinyl monomer which did not absorb at the wavelength used for the photolytic reaction (Guillet and Norrish 1955):

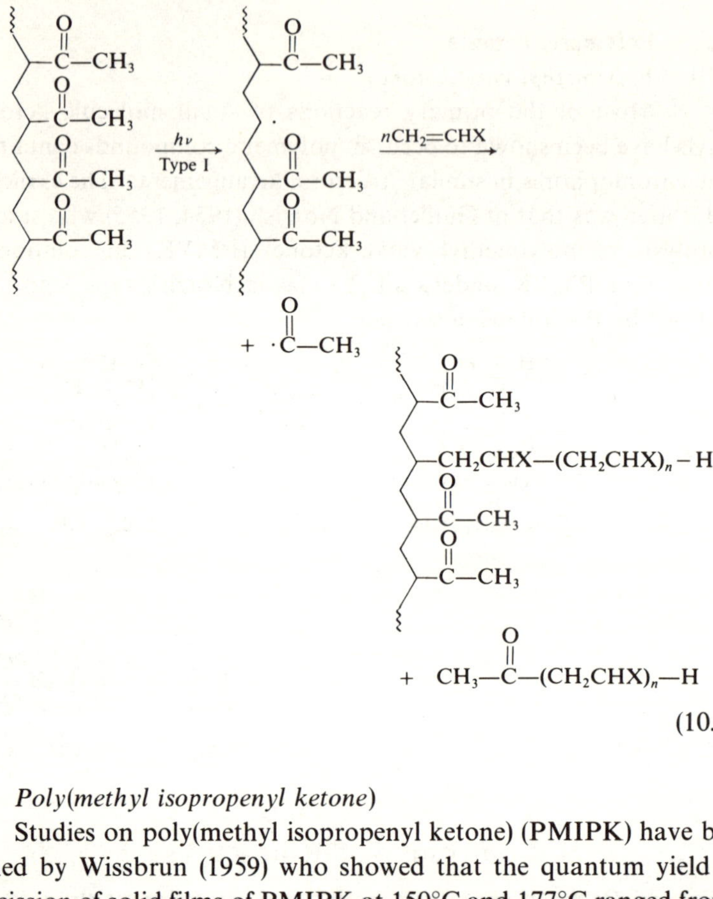

(10.18)

10.2.2 *Poly(methyl isopropenyl ketone)*

Studies on poly(methyl isopropenyl ketone) (PMIPK) have been published by Wissbrun (1959) who showed that the quantum yield for chain scission of solid films of PMIPK at 150°C and 177°C ranged from 1 to 7. At this high temperature the polymer degraded primarily by a chain unzipping mechanism to give the monomer. The high quantum yields were due to the fact that a large number of monomer units were produced from each radical formed in the primary photolysis. In later work, Shultz (1960) reported a value of $\phi_s = 0.22$ for irradiation of PMIPK films in air at 25°C. This relatively high efficiency for chain scission in the solid state has led to the use of PMIPK as a uv-sensitive photoresist. These photoresists are an essential part of the manufacture of microchip systems for computers and other electronic devices. This appears to be one of the more interesting practical applications of ketone photochemistry.

10.2.3 *Ethylene–carbon monoxide polymers*

Subsequent to the early work on PMVK, Guillet and coworkers published an extensive series of studies on the photochemistry of polymers containing simple ketone groups. Hartley and Guillet (1968*a*) studied copolymers of ethylene and carbon monoxide which have the general structure

$$\sim CH_2CH_2{-}CH_2CH_2{-}CH_2{-}CH_2{-}\overset{O}{\overset{\|}{C}}{-}CH_2CH_2CH_2CH_2\overset{C}{\overset{\|}{C}}{-}CH_2CH_2CH_2CH_2 \sim$$

and showed that the Norrish type I and II processes occurred by the following mechanism:

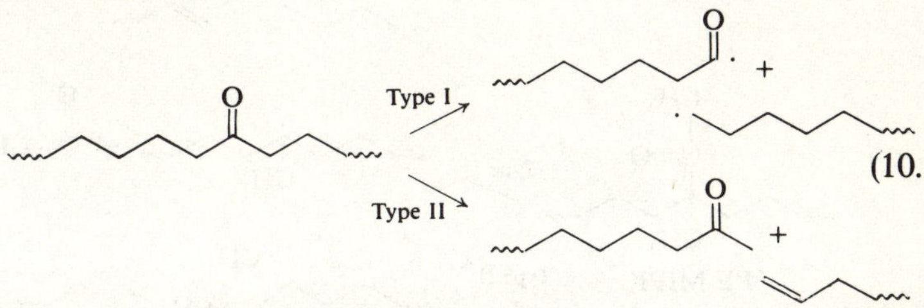

(10.19)

(10.20)

They showed that the quantum yield for the type I process was quite low, on the order of 0.001, but the type II process had a quantum yield of 0.02, which was independent of whether or not the polymer was in solution or in the solid phase. The low quantum yield for the type I reaction was rationalized on the basis that the cage effect in the solid and in solution would prevent the escape of the two primary polymeric radicals formed and consequently they would usually recombine. Studies of linear aliphatic ketones (§10.1.9) in which the length of the alkyl chain was increased showed that the quantum yields for the type I and II processes decreased with chain lengths up to about C20, after which no further change was observed. The observed quantum yields for these small aliphatic model ketones were similar to those obtained in ethylene–carbon monoxide polymers (Hartley and Guillet 1968*b*), thus demonstrating that studies of model compounds in solution were useful in predicting the photochemistry of solid amorphous polymers above their glass transitions.

10.2.4 *Copolymers of ethylene and vinyl ketones*

When the carbonyl group is included in a side chain adjacent to the main polymer backbone, the mobility of the keto group is substantially greater than when it is in the polymer backbone. For example, Sitek *et al.* (1976) and Li and Guillet (1980) have studied copolymers of ethylene with methyl vinyl ketone (MVK) and methyl isopropenyl ketone (MIPK). Again one obtains primarily the two main photochemical reactions, the Norrish type I and II, but in contrast to the ethylene–carbon monoxide copolymer, only the type II process leads to scission of the main chain:

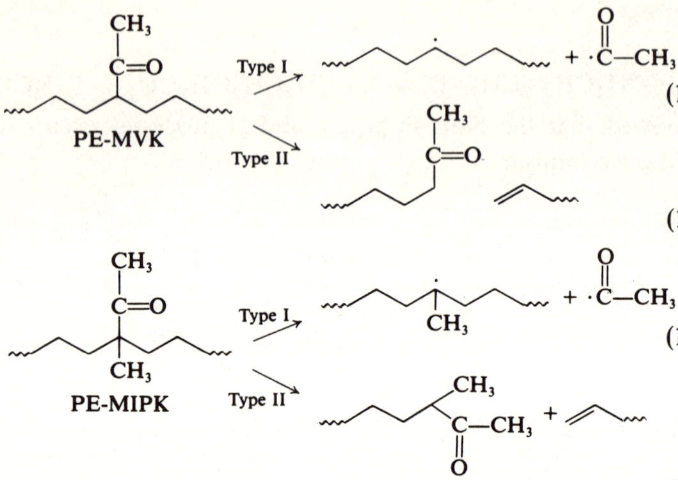

(10.21)

(10.22)

(10.23)

(10.24)

Since only one process gives rise to chain scission, it becomes very easy to follow the kinetics of the type II process by measuring the change in molecular weight of the polymer by the procedures outlined in §4.7.

The quantum yields are significantly higher than for the ethylene–carbon monoxide copolymer, as shown in figure 10.3 and table 10.3 (Sitek

Figure 10.3. Plot of ln $(\Delta K_0/\Delta K)$ for the carbonyl absorbance with time for irradiation of ketone copolymers in nitrogen. (From Sitek *et al.* 1976.)

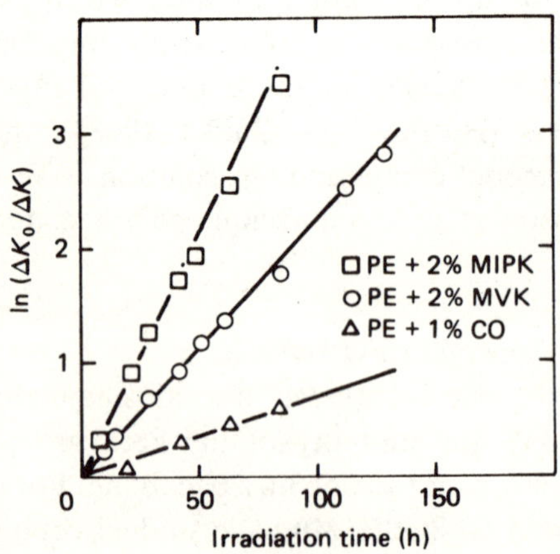

et al. 1976, Li and Guillet 1980). The loss of carbonyl via the type I reaction is greatest when the keto group is attached to a tertiary carbon as in polyethylene-MIPK. Typical rate curves for PE-MIPK are shown in figure 10.4*a*. In the presence of oxygen the type I radicals initiate a rapid photo-oxidation which causes the build-up of a new carbonyl absorption at 1713 cm^{-1} (figure 10.4*b*).

Table 10.3. *Relative quantum yields for loss of carbonyl groups in ethylene/ketone copolymers*

Ethylene copolymer	$\phi\varepsilon \times 10^3$ (cm^2 einstein^{-1})	Relative ϕ_s
1% CO	0.51	1
2% MVK	2.08	4.1
2% MIPK	4.25	8.3

Figure 10.4. Changes in infrared spectrum for a copolymer of ethylene and 2% methyl isopropenyl ketone: (*a*) in N_2, (*b*) in O_2. (From Sitek *et al.* 1976.)

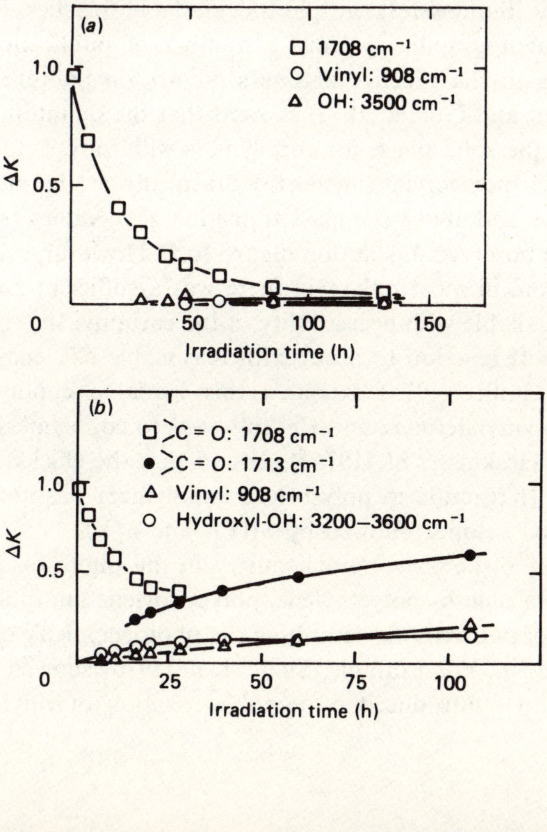

10.2.5 *Other copolymers with vinyl ketones*

Vinyl ketones such as methyl vinyl ketone (MVK) and methyl isopropenyl ketone (MIPK) copolymerize readily by free-radical processes with a wide variety of vinyl monomers including ethylene. The facility with which the Norrish type II process occurs, even in the solid phase, has led to a number of interesting applications with polymers of this type. For example, Guillet and coworkers have patented the synthesis of a wide variety of photodegradable polymer compositions based on the inclusion of ketone groups in conventional plastic resins, such as polyethylene, polypropylene, polystyrene, polyesters, and polyacrylonitrile. These polymers degrade directly on exposure to outdoor sunlight, but are relatively stable to indoor illumination because of the narrow band of radiation absorbed by the ketone carbonyl. They are now in commercial production and are used primarily for mulching film for agricultural purposes. A relatively small amount is used for control of litter in packaging applications. Also, the ethylene-carbon monoxide copolymers, first studied by Hartley and Guillet (1968*a*), are now commercially available for use in highly litter-prone packaging. Details of these processes are beyond the scope of this book, however the interested reader is referred to the review by Heskins and Guillet (1976).

As a result of the interest in photodegradable plastics for both packaging and photographic systems, a number of publications have appeared relating to the effective ketone structure on photosensitivity. For example, Dan and Guillet (1973) showed that the quantum yield of chain scission in the solid phase for copolymers with vinyl ketones with styrene or methyl methacrylate increased drastically at the glass transition temperature and above the glass transition gave values essentially identical to those observed in solution (figure 10.5). However, even below the glass transition, in most polymers there was a sufficient number of conformations available with accessibility of the carbonyl to γ-hydrogen to cause the type II reaction to occur with reasonable efficiency ($\sim$0.1). Alexandru and Guillet (1975) extended this work to copolymers of acrylonitrile with vinyl ketones, and Heskins *et al.* to copolymers of vinyl chloride (1976*a*). Heskins *et al.* (1976*b*) showed that the efficiency of the solid phase type II reaction in polystyrene copolymers was greatest for tertiary butyl vinyl ketone, followed by MIPK and MVK.

The importance of ketone photochemistry in the photo-oxidation of aliphatic polymers such as polyethylene, polypropylene and polystyrene led to a number of publications describing the photochemistry of natural weathering processes. For example, Sitek *et al.* (1976) showed that the side chain carbonyls introduced by copolymerization of vinyl ketones with ethylene could cause photodegradation by direct photolysis, via the Norrish type II process. Furthermore, photosensitization of natural oxidation was initiated by radicals formed in the type I process. In the ethylene–carbon monoxide polymers, on the other hand, where the type I process is drastically reduced, the polymer degraded primarily by direct photolysis of the carbonyl groups through the Norrish type II. In a natural weathering process, if no sensitizing impurities are added to the polymer, photo-oxidation occurs through a complex interaction of hydroperoxides and ketone carbonyl groups. The chemistry is complex and involves a very large number of reaction steps. Many reviews have summarized the latest information on this subject (Wiles 1973, Guillet 1980, McKellar and Allen 1979).

10.2.6 *Photochemistry of poly(acrylophenone)s*

Acrylophenone (AP) and its derivatives are among the most light-sensitive monomers which can be readily polymerized or copolymerized. The simplest member of the series, poly(acrylophenone) (PAP), often referred to as poly(phenyl vinyl ketone) (PPVK) has been studied for a number of years (Lukac *et al.* 1971, Golemba and Guillet 1972, Somersall *et al.* 1974, Dan *et al.* 1973, Dan and Guillet 1973). This is a particularly useful model polymer for photochemical studies because very rapid intersystem crossing occurs in the aromatic ketones so that nearly

Figure 10.5. Quantum yield of Norrish type II reaction as a function of temperature for 0.5-mm PS–PVK films irradiated with monochromatic radiation of 313 nm wavelength. Error bars represent errors ($\pm 20\%$) in reproducibility of results. (From Dan and Guillet 1973.)

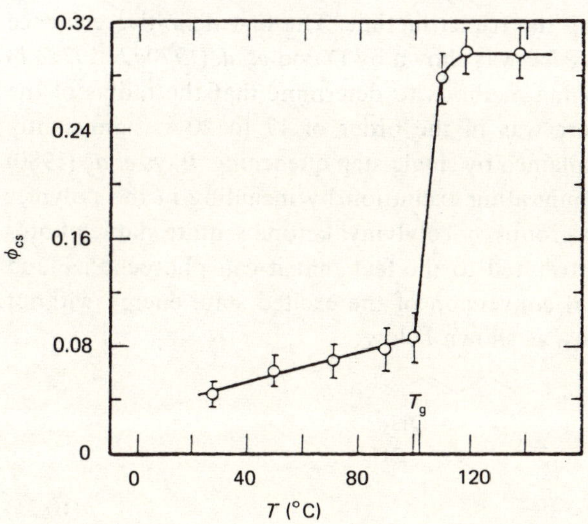

all of the photochemical processes, physical and chemical, occur from the triplet state. However, some care is required interpreting data on the homopolymer because of the very efficient triplet energy transfer which can occur along polymer chains. For example, Bays *et al.* (1980) have estimated that the lifetime of the triplet excitation on any given chromophore is about 30 ps, although the total triplet lifetime is of the order of 55 ns. They were also able to estimate the lifetime of the triplet biradical as 65 ns. The simplest photochemistry is observed in the case where acrylophenone is copolymerized with an excess of styrene or other comonomer so that the ketone groups can be considered as isolated. In this case, chain scission occurs efficiently with a quantum yield approaching 0.4 in solution. In early studies, Golemba and Guillet (1972) showed that the quantum yields were higher for isolated AP units as compared with those in sequence. They attributed this to the occurrence of energy transfer in long sequences of AP units which might be expected to reduce the quantum yield through delocalization of the original excitation. The quantum yields observed were in the range of 0.2 to 0.3 but this was later demonstrated to be due to the fact that the measurements were made at relatively high conversions. As pointed out by Golemba and Guillet (1972), the Norrish type II process leads to a double bond on the end of the chain which can act as a quencher for any migrating excitation. As the degradation process proceeds, more and more of these quenching groups will build up in the polymer chain itself and excitation can be quenched both internally and from intra- and intermolecular contacts in solution.

A more detailed theory of this effect was published by Encinas *et al.* (1979) who estimated that as many as 750 monomer units can be visited by the excitation during the triplet lifetime. The first definitive evidence for triplet migration in PAP was shown by David *et al.* (1970*a*,*b*, 1972*a*,*b*) who used static quenching methods to determine that the radius of the Perrin quenching sphere was of the order of 17 to 20 Å, significantly larger than could be explained by single-step quenching. Bays *et al.* (1980) attempted to trap the migrating excitation by including in the polymer ortho-tolylvinyl ketone groups. *o*-Tolylvinyl ketone is quite stable photochemically, which is attributed to the fact that it can photoenolize and thereby lead to thermal conversion of the excited state energy without photochemical reactions, as shown below:

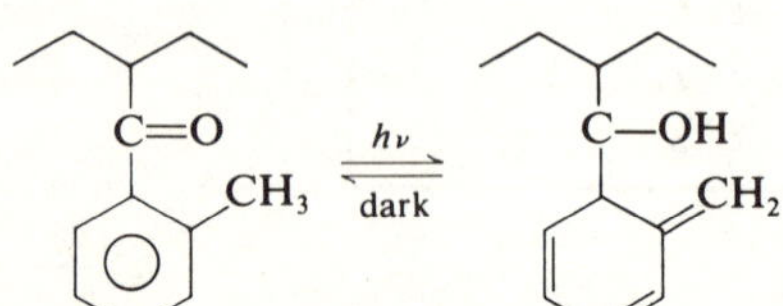

(10.25)

Kilp and Guillet (1981) showed that significant changes in the energy migration efficiency exist between the isotactic and atactic forms of PAP, although the quantum yields for photolysis were nearly identical. They attributed the differences to the fact that some helical configurations of the isotactic polymer are retained in solution and the overlap of the aromatic groups so obtained assists in the energy migration process. Kilp and Guillet (1977) also used copolymers with AP and other vinyl ketones as a means of preparing a series of samples of differing molecular weight by direct photolysis which could then be used for a rapid determination of *K* and *a* values relating the viscosity to the molecular weight of the polymer in solution.

Hrdlovic and Lukac (1982) have prepared a variety of substituted AP polymers and determined their quantum yields for chain scission, their triplet energies from phosphorescence measurements at low temperatures, and their quenching efficiencies with diene triplet quenchers. Their results are summarized in table 10.4. It appears that alkyl substitution in the para position on the ring can increase the quantum yield, both in nitrogen and air, halogen substitution gives relatively high quantum yields for fluorine, chlorine and bromine, but the iodine-substituted molecule apparently decomposes in a different manner and the polymer does not degrade by chain scission. Presumably iodine atoms are formed which may possibly recombine with the original polymer. Electron-withdrawing groups such as *p*-methoxy, reduce the quantum yield and the polymer of the 2,4-dimethoxy derivative is relatively stable. *p*-Phenyl substitution also reduces the quantum yield of chain scission to very low values.

An important application of PAP photochemistry is due to the pioneering work of Schnabel and his coworkers who have used laser flash photolysis of PAP and its copolymers with styrene to determine the rate of disentanglement of two macromolecules. After a main-chain scission of a polymer, two chains are formed from the original one, and these must diffuse apart. Schnabel and coworkers (Kiwi and Schnabel 1975, Schnabel 1979) follow this diffusion by light scattering measurements. Other intermediates in the process can be followed by transient flash spectroscopy. Schnabel (1979) distinguished two cases which can be related to the observed change in the light scattering intensities. In the first case, an intermediate is formed which has a lifetime much shorter than the time required for separation of the fragments in the macromolecules. In this case, the observed rate of decrease of light scattering intensity is due to the rate of separation of the fragments. In the second case, the lifetime of the intermediate causing chain rupture is longer than the time of

Table 10.4. *Quantum yields for chain scission and relative triplet lifetimes* (τ/τ_0) *for polymers of substituted acrylophenones* ($R{-}\overset{O}{\overset{\|}{C}}{-}CH{=}CH_2$)

Ketone, *R*	ϕ_s, N_2	ϕ_s, air	τ/τ_0	E_T (kcal)
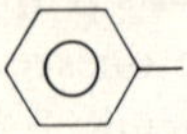	0.22	0.22	1.0 (τ_0)	69.8
CH_3– 	0.26	0.21	6.6	66.7
CH_3CH_2– 	0.25	0.22	5.2	67.7
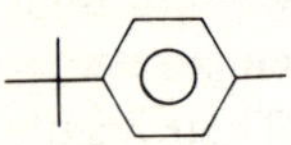	0.37	0.32	5.1	69.1
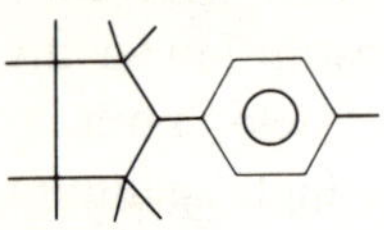	0.33	0.30	5.8	67.9
F– 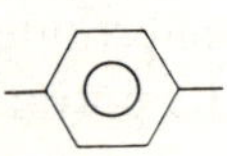		0.31	0.9	71.3
Cl– 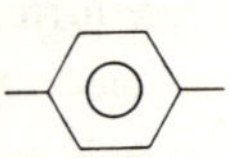		0.39	3.9	66.7
Br– 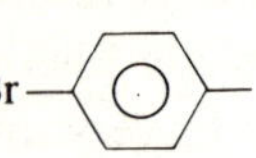		0.36	1.7	
I– 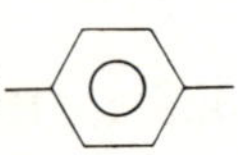		0		63.9
CH_3O– 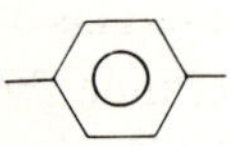	0.10		26	66.5
CH_3O– 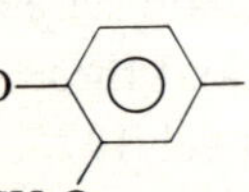 CH_3O	1×10^{-4}			
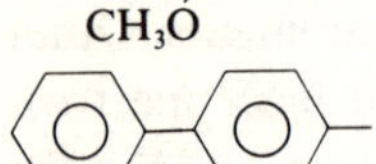	2.6×10^{-4}			62.2
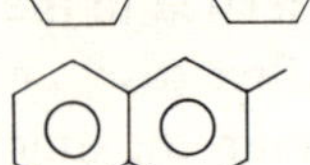	0			
CH_3–C(=O)– 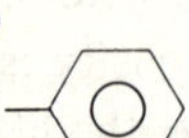		0.22	5.8	

separation of the fragment.

The observed rate of change of the scattering intensity can be correlated to the lifetime of the intermediate. In this way, time-resolved light scattering measurements can help to measure rate constants of reaction of large macroradicals undergoing main-chain scission or reaction with scavengers, etc. Schnabel (1979) has extended his earlier studies to look at problems relating to preferential solvation and the effect of solvent mixtures on disentanglement diffusion.

Salvin *et al.* (1976) have reported the determination of quantum yields of chain scission for oligomers, dimers and trimers of PAP irradiated in benzene solution under nitrogen at 365 nm. They showed that in the model compounds the carbonyl situated in the middle of the trimer had the same quantum yield for chain scission as those at the ends. The fact that epimerization took place in the photochemical reaction of the diastereomers of 2,4-dibenzoylpentane supported the evidence for a biradical intermediate in the Norrish type II reaction. It was shown that the quantum yields for the formation of various photoproducts were independent of the configuration of the oligomer. Salvin *et al.* also showed that the rate constant for a triplet energy transfer for PAP, which was determined from static quenching of the type II cleavage, was similar to that for valerophenone which indicated that the triplet–triplet energy transfer rate constant is independent of the carbonyl position whether it is free or bound in a polymer chain.

A further report of the detection of the biradical intermediate in the photolysis of PAP is given by Small and Scaiano (1978) who showed that the biradicals generated in the photodegradation of PAP are highly reactive and capable of undergoing intermolecular reactions. The absorbance due to these biradicals was monitored at 450 nm and the spectrum was similar to that observed for small aromatic ketones. The decay was first order with a lifetime of 76 ns in benzene at room temperature. The addition of radical scavengers such as atomic oxygen NO, SO_2 or tertiary butyl nitrite to the solution increased the rate of biradical decay. The authors were also able to initiate the polymerization of methyl methacrylate by the biradical and their efficiency was between that of the radicals of valerophenone and γ-methyl valerophenone.

10.2.7 *Chemistry of poly(methacrylophenone)s*

Poly(methacrylophenone) (PMAP) also represents an interesting photochemically active polymer. The homopolymer has not been prepared by free radical polymerization but can be prepared by anionic polymerization using lithium butyl as catalyst. Since the γ-hydrogen is blocked by the α-methyl group, the chain scission efficiency is expected to be lower than in the corresponding acrylophenones. Copolymers of this monomer with styrene can be readily prepared by free-radical mechanisms and these indeed appear to be very photosensitive. Among the earliest work with this system was that of Sugita *et al.* (1976) who synthesized copolymers of methacrylophenone and its para-chloro, para-nitro, para-amino and para-hydroxy analogues. The copolymers with styrene contained about 10% of the methacrylophenone and so in this case γ-hydrogens would be available from adjacent styrene groups. The quantum yields for Norrish type II reaction at $\lambda = 364$ nm are summarized in table 10.5 in comparison with those of model compounds. Both the p-chloro and unsubstituted methacrylophenones had quantum yields of the order of 0.25, while *p*-amino and *p*-hydroxy had much lower quantum yields. The *p*-nitro derivative apparently decomposed by way of a different mechanism since no chain scission was observed, although the nitro group apparently decomposed according to the mechanisms shown in §9.3.1. The very low quantum yields for type II in the *p*-amino compound was explained by the authors by the fact that *p*-amino phenones are known to have lowest triplets which are π-π^* in character instead of n-π^*. On the other hand, when MAP is copolymerized with styrene, an α position is available and it seems clear in this case that the major mechanism for decativation of the triplet is the Norrish type II photoscission (Ng and Guillet 1984).

Naito *et al.* (1980) reported the quantum yield for main-chain scission in MAP of ~0.05. They attributed this to a β-scission process. They also measured the quantum yield for α-scission of 0.3. Based on transient spectroscopy, Schnabel and coworkers report a carbonyl triplet decay

Table 10.5. *Quantum yields for type II reaction* ($\lambda = 364$ *nm*)

	Quantum yield		
Substituent, X	PS-PIPK-X copolymer[a]	X-Valerophenone[b]	X-Butyrophenone[c]
H	0.26	0.33	0.42
p-Cl	0.24	0.29	0.43
p-NH_2	0.002	—	0.00
p-OH	0.012	0.001	0.00
p-NO_2	0.000	—	—

[a] At 364 nm, 1,2-dichloroethane, 30 °C.
[b] At 313 nm, benzene; data of Wagner *et al.* (1972).
[c] At 313 nm, benzene, 25 °C; data of Baum *et al.* (1966).

constant of $6 \times 10^6\,s^{-1}$, which was independent of copolymer composition but lower than for the homopolymer which was $1 \times 10^7\,s^{-1}$. Using transient spectroscopy the formation of benzyl radicals was observed during photolysis. Furthermore, the yield of benzaldehyde recovered indicates a quantum yield of 0.3 to 0.4. The results show that when MAP is copolymerized with a monomer such as methyl methacrylate, which also has no γ-hydrogen, the decomposition is primarily by the type I process and the relatively low yield of chain scission is caused by β-scission processes.

10.2.8 *Poly(vinylbenzophenone)*

Another important aromatic carbonyl polymer is poly(vinylbenzophenone). This polymer can be made either directly by the polymerization of vinylbenzophenone or by the Friedel–Crafts reaction with benzoylchloride on polystyrene to give substitution on the benzene ring. The latter reaction is generally not 100% efficient and one usually obtains a copolymer containing a major amount of vinyl benzophenone units and a minor portion of unreacted styrene units. The

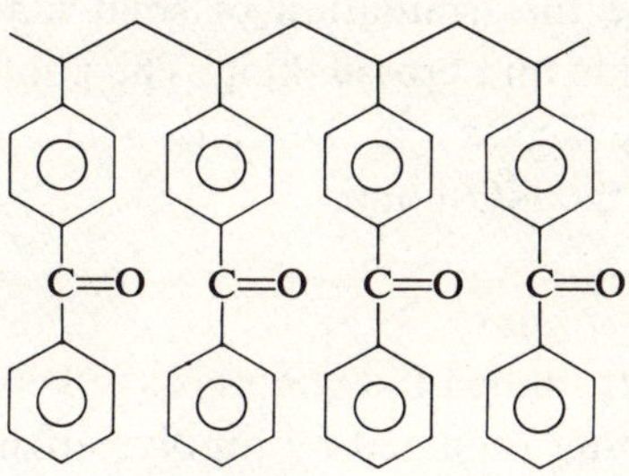

IV

polymer reacts primarily by photoreduction and it has been used as a catalyst for photoreduction processes.

10.2.9 *Other ketone-containing polymers*

Bonnans-Plaisance *et al.* (1978) have prepared copolymers with a combination of ketone and cyclic rings. The rate of photolysis of these polymers varied considerably although the carbonyl content was nearly the same. The higher degradation rate of structure VII results from the fact that the cleavage by the Norrish II can occur and lead to clean cleavage. On the other hand, this is not possible for structures V and VI. The rapid degradation of structure V is due to the fact that the Norrish type I process is more efficient for cyclopentanones than for cyclohexanones.

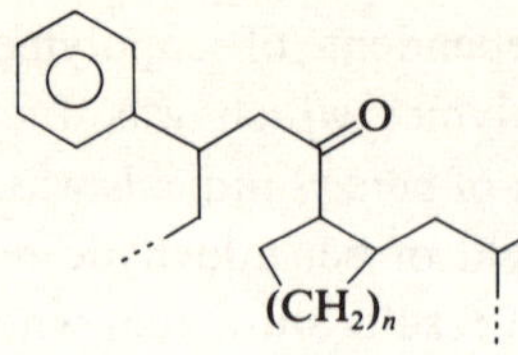

V $n=2$
VI $n=3$

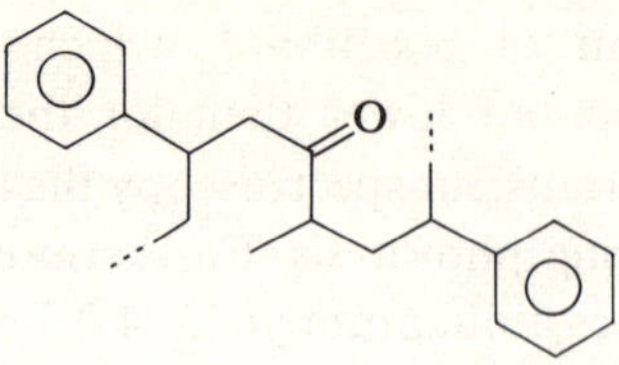

VII

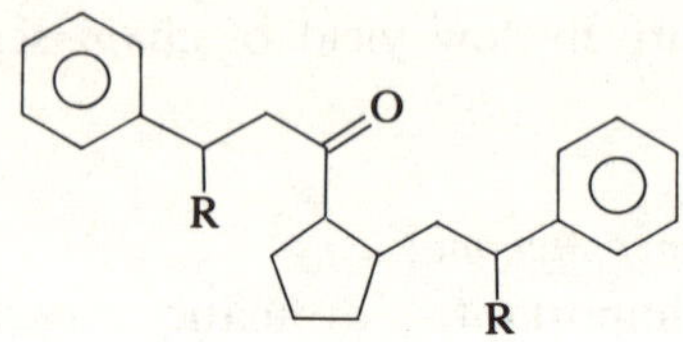

VIII

Tsunooka *et al.* (1977) prepared polymers containing *o*-acyloxime groups. It was shown that the photochemistry of the *o*-acyloxime polymers depended on the irradiation conditions. The irradiation of *o*-acyloxime polymers in benzene solutions led to degradation in air and crosslinking under nitrogen, while the irradiation of solid films in air resulted in simultaneous degradation and crosslinking. The photolysis of methyl β-naphthyl ketone oxime acetate in benzene solution under nitrogen resulted in a cleavage of the NO bond.

10.2.10 *Polyesters containing keto groups*

Plooard and Guillet (1972) studied the photolysis of keto groups located in the backbone of a polyester prepared by copolymerizing a diol with a dibasic acid containing a ketone group, in this case δ-ketoacelaic acid. Although the corresponding dimethyl ester of δ-ketoacelaic acid decomposes with high quantum efficiency to give the type II products, when the same ketone group was included in a long polyester chain, the quantum yield was reduced by two orders of magnitude. This appears to be a general rule with polymer systems. When chromophoric groups are contained in the backbone C–C chain one can expect decreases in quantum yields of processes which produce pairs of polymer radicals, due to the difficulty of their escape from the solvent cage. However, as we will see later, this restriction does not apply to other types of polymeric reactions, or to radical processes which produce a small mobile free radical along with a polymer radical. Dhanraj and Guillet (1968) extended this concept of the synthesis of block copolymers, by photolysis of copolyesters of γ-keto pimelic acid in the presence of vinyl monomers:

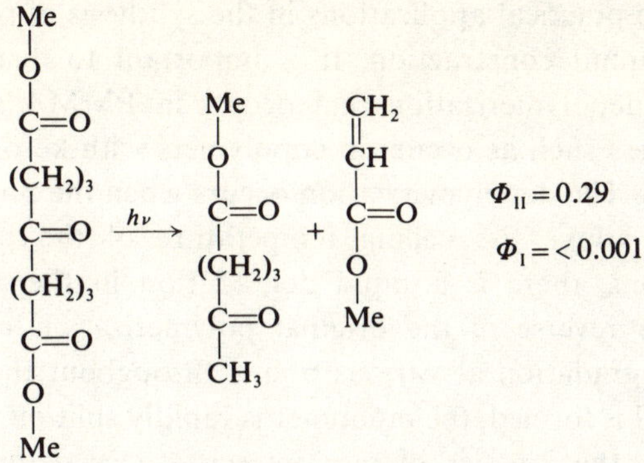

(10.26)

$$\text{P}\!\left[\!-\text{O(CH}_2)_4-\overset{\text{O}}{\overset{\|}{\text{C}}}-\text{CH}_2\text{CH}_2-\overset{\text{O}}{\overset{\|}{\text{C}}}-\text{CH}_2\text{CH}_2-\overset{\text{O}}{\overset{\|}{\text{C}}}\right]\!\text{P} \xrightarrow{h\nu}$$

$$\text{P}\!\left[\!-\text{O(CH}_2)_4\text{O}-\overset{\text{O}}{\overset{\|}{\text{C}}}-\text{CH}_2\text{CH}_2\cdot\right] + \text{P}\!\left[\!-\overset{\text{O}}{\overset{\|}{\text{C}}}\right]\!-\text{CH}_2\text{CH}_2-\overset{\text{O}}{\overset{\|}{\text{C}}}\cdot \xrightarrow{\text{CH}_2=\text{CHX}}$$

$$\text{P}-\overset{\text{O}}{\overset{\|}{\text{C}}}-\text{CH}_2\text{CH}_2\overset{\text{O}}{\overset{\|}{\text{C}}}-(\text{CH}_2\text{CHX})_n-\text{H} \text{ and}$$

$$\text{P}-\text{O(CH}_2)_4\text{O}-\overset{\text{O}}{\overset{\|}{\text{C}}}-\text{CH}_2\text{CH}_2(\text{CH}_2\text{CHX})_n-\text{H} \qquad (10.27)$$

10.2.11 *Photolysis of ester-containing polymers*

Aliphatic esters absorb at shorter wavelengths than the ketone and aldehyde carbonyl. As a result, the absorbance often overlaps those of other structures in the polymer and the chemistry is less easy to determine. The most important work in this area relates to the photodecomposition of polymers such as poly(methyl methacrylate) and other polyacrylates. Much of this work has been done by Grassie and co-workers. When polymers containing ester groups are irradiated with short-wavelength radiation, usually 254 nm, both the Norrish type I and type II processes occur but with lower efficiency than in ketone or aldehyde groups. In the methacrylate series, if the temperature is high enough, the polymer depolymerizes, giving off monomer by the procedure shown below:

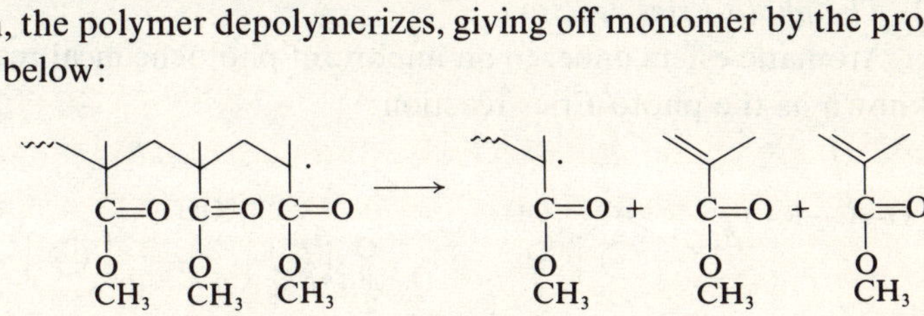

(10.28)

This process has some practical applications in the synthesis of photoresists used for microcircuit construction. It is important to distinguish between the kind of depolymerization that occurs in PMMA and the random scission process such as occurs in copolymers with ketones and other vinyl monomers. The depolymerization occurs when the photodegradation is initiated above the ceiling temperature of the polymer. Under these conditions, there is a rapid degradation in the polymer which is essentially a reverse of the original polymerization reaction. Instead of random degradation at various points throughout the chain, once a growing radical is formed, the monomer is rapidly split off, unit by unit. In this instance, the kinetics of random scission as described in earlier chapters does not apply. The theory has been developed extensively by Burnett *et al.* (1954). Later work in this field has been reviewed by Grassie (1972).

The occurrence of the Norrish type I and type II processes in ester groups has considerable practical importance in relation to the weathering degradation due to solar ultraviolet of important polymers such as the polyethylene terephthalates and styrene-based polyesters. As shown by Wiles and coworkers (Wiles 1973) polyesters degrade by both the Norrish I and II although the Norrish type I α-scission is usually more important in the solid phase. Alexandru and Guillet (1976) showed that when ketone-containing comonomers are included in a polyester chain, the quantum yield of chain breaking increased depending on the ketone structure. They suggested that an important mechanism for the breakdown in molecular weight of polymers in the solid phase was due to the reaction of oxygen with the primary radicals produced by the Norrish type I process. The reaction with oxygen essentially prevented the recombination of these radicals and resulted in a permanent break in the main chain of the polymer. For this reason, polymers degraded in oxygen decreased in molecular weight more rapidly than those where the photolysis took place under nitrogen. Similar factors would be expected to take place during the normal photodegradation of polymers containing ester groups in the backbone of the polymer chain.

10.2.12 *The photo-Fries reaction*

Aromatic esters undergo an important photochemical rearrangement known as the photo-Fries reaction:

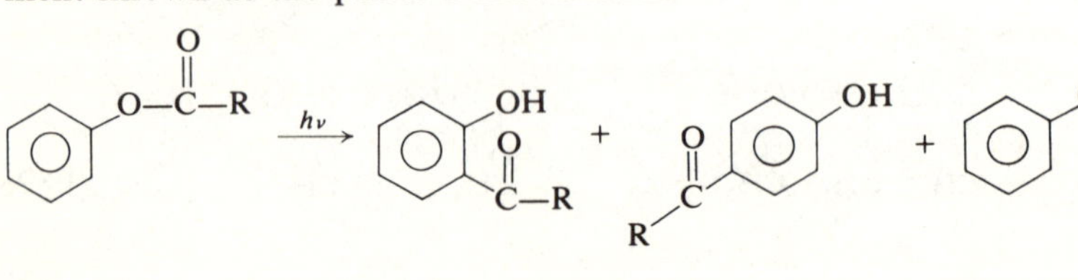

(10.29)

This reaction, which appears to be quite general and is often used in organic synthesis, was first discovered by research to determine the cause of the photochemical yellowing which occurred with polymers prepared by condensing bisphenols with aromatic dicarboxylic acids (Kobsa 1962, 1964).

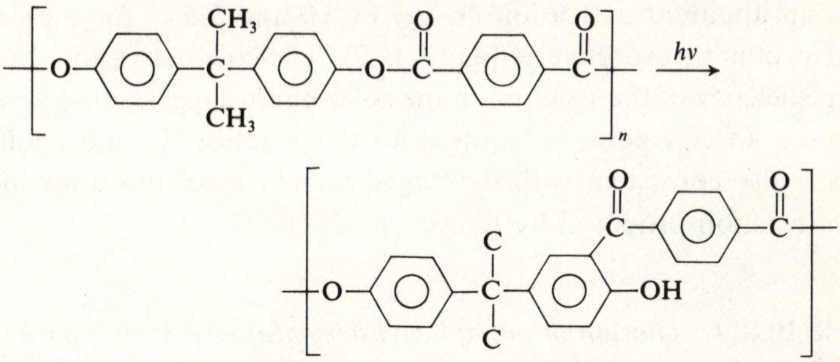

(10.30)

Inspection of thin film uv spectra showed a strong absorbance characteristic of the aromatic hydroxy ketone at 355 nm when the polymer was irradiated in sunlight or uv with wavelengths less than 320 nm. This observation led to the study of model systems which showed the generality of the reaction in organic chemistry (Kobsa 1962, Anderson and Reese 1960, 1963).

The reaction is of some importance in polymer photochemistry because it appears to occur with relatively high efficiency, even in solid polymers at temperatures well below their glass transitions. Li and Guillet (1977) reported a systematic study of the photo-Fries reaction in poly(phenylacrylate) and model compounds related to it, both in the solid phase and in solution:

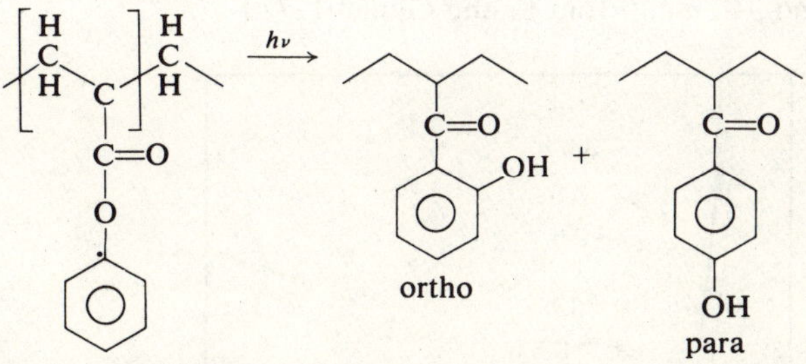

(10.31)

The two main products, the ortho and para substituted phenyl ketone groups, were readily identified and could be estimated quantitatively from their uv absorption spectra (figure 10.6), the ortho compound having a strong absorbance with $\lambda_{max} = 335$ nm and the para, an absorbance at $\lambda_{max} = 265$ nm.

The total quantum yields were quite high (~0.4) and nearly independent of whether the irradiation was carried out in solution or in the solid phase (table 10.6). When cooled below the glass transition ($T_g = 50\,°C$) no change in quantum yield of either product was observed down to the γ transition (§8.7), below which the quantum yields decreased with an apparent activation energy of 5.0 and 7.5 kJ for the ortho and para product, respectively (figure 10.7). The low activation energy and high efficiency of the reaction in the solid phase suggests that a very small amount of free volume is required for this reaction (Li and Guillet 1977). This is also consistent with the 'caged radical' mechanism for the photo-Fries reaction proposed by Kobsa (1962) (10.32).

Table 10.6. *Irradiation of poly(phenyl acrylate) (PPA) and phenyl acetate (PA)*[a]

	Solvent	ϕ_o	ϕ_p	ϕ_p/ϕ_o
PA	Cyclohexane	0.18±0.02	0.20±0.02	1.1
PA	1,2-Dichloroethane	0.18±0.02	0.22±0.02	1.2
PPA	1,2-Dichloroethane	0.12±0.02	0.23±0.02	1.9
PPA	Solid film (air)	0.10±0.02	0.13±0.02	1.3
PPA	Solid film (N_2)	0.10±0.02	0.14±0.02	1.4

[a] Irradiation at 250 nm in air at 23 °C.
Source: Li and Guillet (1977).

Figure 10.6. Absorption spectra of a PPA film after different periods of irradiation at room temperature, using light of wavelengths between 220 and 340 nm. (From Li and Guillet 1977.)

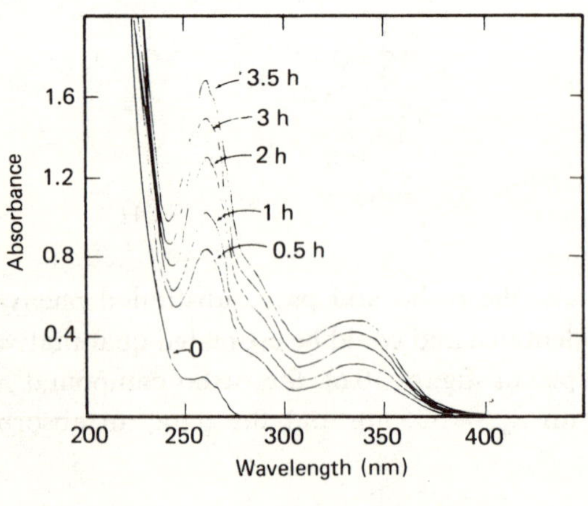

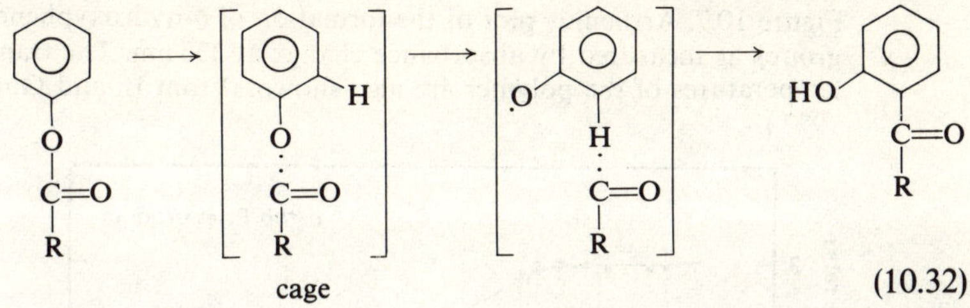

(10.32)

The relatively small rotation of the phenyl or acyl group within the cage should require very little excess free volume. The slightly higher activation energy for the para product, which requires a larger rotation, also fits this hypothesis.

An alternative mechanism involving a concerted hydrogen transfer has been proposed for the photo-Fries by Anderson and Reese (1963), but this would also require little excess free volume.

A commercially important polymer (Lexan polycarbonate) also appears to photodegrade by the photo-Fries reaction (Gupta *et al.* 1978). These authors proposed the following sequence of reactions:

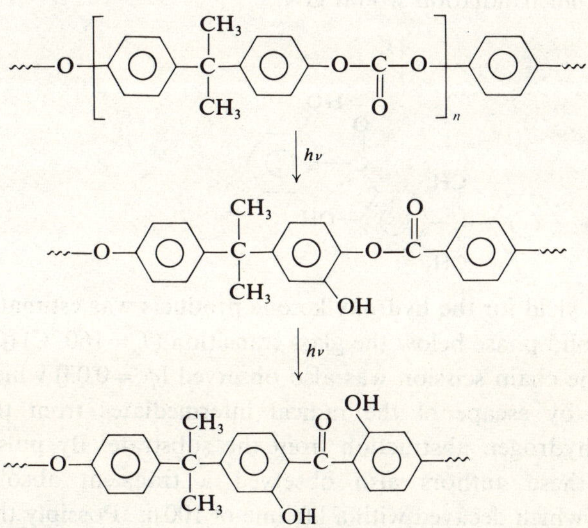

(10.33)

Although it seems more likely that the first product would be

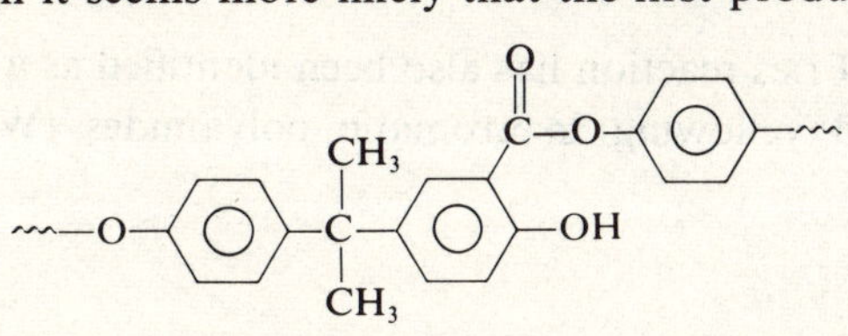

Figure 10.7. Arrhenius plot of the formation of *o*-hydroxyphenone groups as measured by absorbance change at 335 nm. The transition temperatures of the polymer are also shown. (From Li and Guillet 1977.)

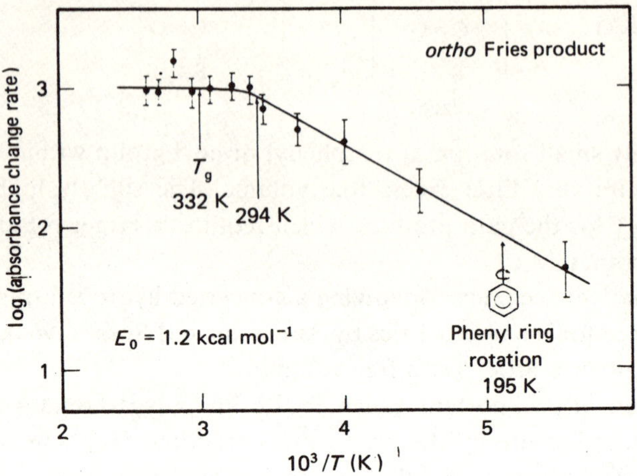

which, on further irradiation would give

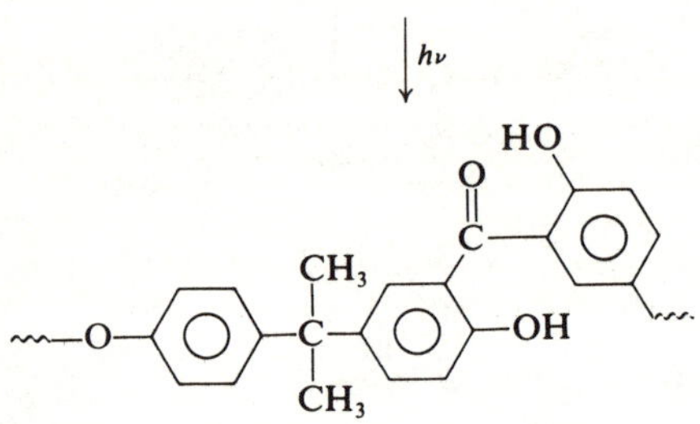

(10.34)

The quantum yield for the hydroxy ketone products was estimated to be ~0.2 in the solid phase below the glass transition ($T_g = 160\,°C$) (Gupta *et al.* 1980). Some chain scission was also observed ($\phi_s = 0.03$) which could be explained by escape of the radical intermediates from the cage, followed by hydrogen abstraction from the substrate. By pulsed laser experiments these authors also observed a transient absorbing at $\lambda_{max} = 450$ nm which decayed with a lifetime of 100 ns. Possibly this could be one of the free radical species responsible for the observed chain scission.

Recently the photo-Fries reaction has also been identified as a cause of photodegradation and yellowing in aromatic polyamides (Wiles and Carlsson 1980).

11

Photopolymerization

11.1 Introduction

The use of light to initiate the polymerization of vinyl monomers is of long-standing interest to polymer chemists. In fact, photoinitiated reactions were used to improve understanding of the basic kinetics of vinyl polymerization, and photochemistry has since underlaid much of our fundamental knowledge about the polymerization process. In conventional photopolymerizations, light is involved only in the initiating step to generate either free radicals or ions, which then propagate by a chain mechanism to give high molecular weight products. This work has been well reviewed, and will not be dealt with further in this monograph (Burnett 1954, Ledwith 1979, Bevington 1961).

Recently, however, new types of photopolymerizations have been described in which photochemically excited states are involved in each step of the polymerization reaction. Prominent among the groups involved in these investigations have been those of Hasegawa in Japan and De Schryver and Smets in Belgium. DeSchryver *et al.* (1970) have defined this type of photopolymerization as a process in which a macromolecule is obtained from a low molecular weight material by a stepwise condensation reaction in which, for each of these chain-extending steps, the absorption of a quantum of light is necessary.

Reaction of this type can be envisaged to occur in at least three ways, involving (i) singlet excited states, (ii) triplet excited states and (iii) reaction of excited intermediates formed subsequent to the absorption of light (for example, biradicals). The following sections show representative examples of each of these types of photopolymerization.

11.2 Photopolymerization of anthracene derivatives

Anthracene is known to dimerize under the action of ultraviolet light to give I.

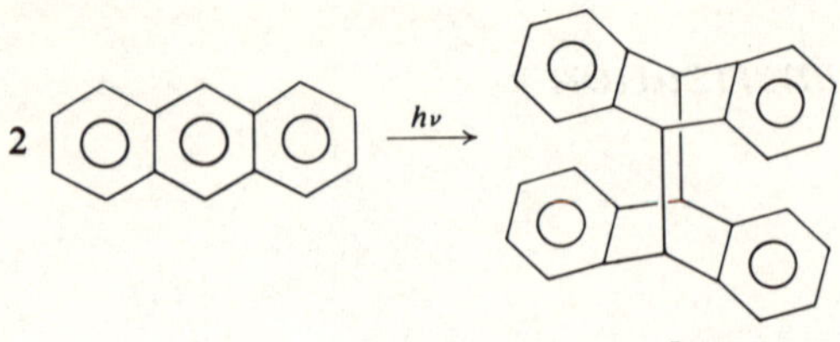

(11.1)

I

De Schryver *et al.* (1971*b*) synthesized a series of bisanthracenes, II, in which *n* varied from 9 to 12 and obtained relatively high molecular weight material (20 000–50 000) by irradiating the compounds in solution using light of wavelengths above 330 nm. Like the anthracene dimer itself, the polymers are thermally and photo reversible under suitable reaction conditions.

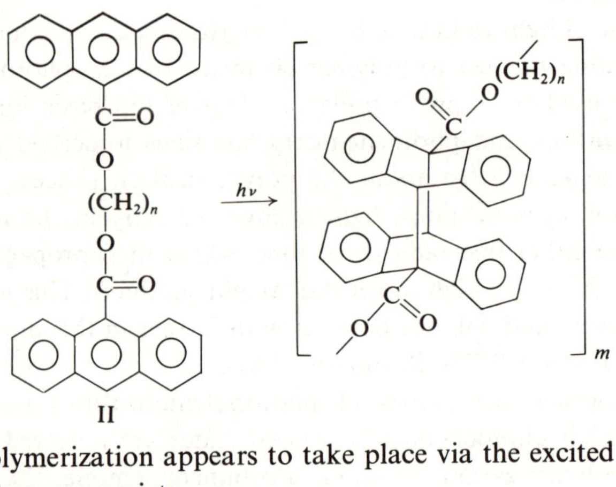

(11.2)

II

This polymerization appears to take place via the excited singlet state of the anthracene moiety.

11.3 Bismaleimides

In a second series of experiments (De Schryver *et al.* 1974*a*,*b*) N,N′-polymethylene bismaleimides were synthesized which could be polymerized both by sensitization with benzophenone and by direct irradiation to form polycycloaddition products.

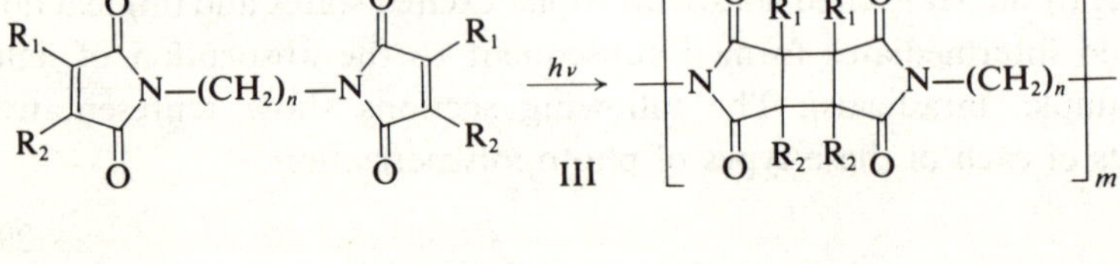

(11.3)

In these structures the R groups can be chlorine, bromine or methyl, and molecular weights of the order of 20 000 are reported. In this case the reaction is believed to occur primarily through the intermediacy of the excited triplet state.

11.4 Biscoumarines

De Schryver *et al.* (1971*b*) have also reported on the triplet photopolymerization of biscoumarines

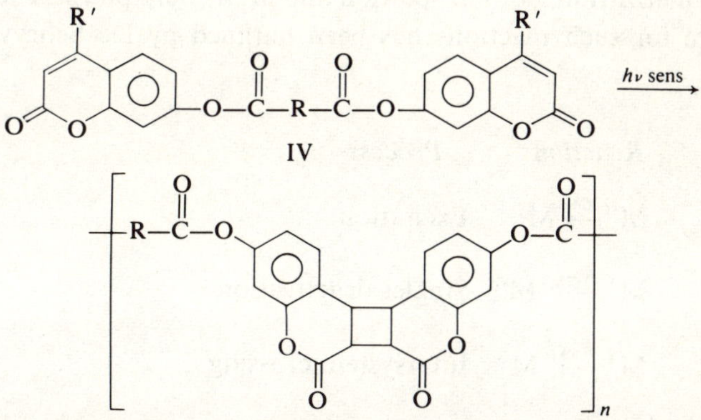

(11.4)

Photodimerization of the coumarine groups takes place after energy transfer from the excited triplet of benzophenone to yield polymers with molecular weights of approximately 20 000.

11.5 Bisbenzophenones

Finally, another class of photopolymers is produced by irradiation of bisbenzophenones in isopropanol solution through the formation of polybenzpinacols (McCombs *et al.* 1971, Higgins *et al.* 1970, DeSchryver *et al.* 1971 *a*).

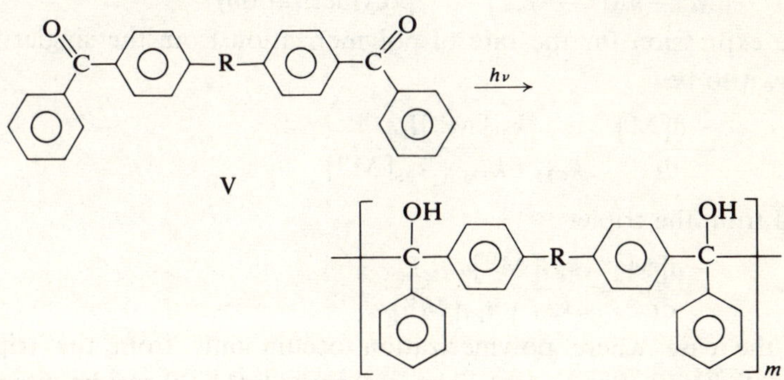

(11.5)

Presumably this reaction goes by the intermediacy of a triplet biradical but because of the nonspecificity of the radical and its high reactivity, a number of side reactions can also occur.

11.6 Kinetics and mechanism of photopolymerization of non-conjugated bichromophoric systems

The above reactions constitute a class of non-conjugated bichromophoric compounds which have been demonstrated to undergo photopolymerization, both in solution and in the bulk phase. The general mechanism for such reactions has been outlined by De Schryver *et al.* (1975*a*).

Reaction	*Process*	
$M^0 \xrightarrow{h\nu} M^1$	Excitation	(11.6)
$M^1 \xrightarrow{k_{GM}} M^0$	Singlet deactivation	(11.7)
$M^1 \xrightarrow{k_{TM}} M^3$	Intersystem crossing	(11.8)
$M^1 + M^0 \xrightarrow{k_{RS}} M_2^0$	Singlet addition	(11.9)
$M^3 \xrightarrow{k_{GT}} M^0$	Triplet deactivation	(11.10)
$M^3 + M^0 \xrightarrow{k_{RT}} M_2^0$	Triplet addition	(11.11)

Equations (11.9) and (11.11) represent the formation of dimer from the singlet and triplet manifolds, respectively. These represent the first step of the polymerization reaction and if continued *n* times will build up a polymer of *n* units in length, i.e.,

$$M^* + nM^0 \rightarrow M_{n+1}^0 \quad \text{(polymerization)} \tag{11.12}$$

The expression for the rate of polymerization from the singlet is easily shown to be

$$\frac{-\mathrm{d}[M]}{\mathrm{d}t} = \frac{k_{RS}[M^0]I_{abs}}{k_{GM} + k_{TM} + k_{RS}[M^0]} \tag{11.13}$$

and from the triplet

$$\frac{-\mathrm{d}[M]}{\mathrm{d}t} = \frac{k_{RT}[M^0]\phi_{TM}I_{abs}}{k_{GT} + k_{RT}[M^0]} \tag{11.14}$$

In the case where polymerization occurs only from the triplet and essentially all the incident light is absorbed, (11.14) can be integrated to give

$$(k_{GT}/k_{RT})\ln[1/(1-P)]+P[M^0]_0=I_0\phi_{TM}t \quad (11.15)$$

where P is the extent of reaction and $[M]_0$ is the original monomer concentration.

The degree of polymerization $\bar{X}_n$ is given by

$$\bar{X}_n=1/(1-P) \quad (11.16)$$

and hence (11.15) can be written

$$(k_{GT}/k_{RT})\ln \bar{X}_n+[(X_n-1)/\bar{X}_n][M^0]_0=I_0\phi_{TM}t \quad (11.17)$$

Three types of kinetic behavior can be identified, depending on the relative values of the rate constants. Since this type of photopolymerization is not a chain reaction, but an example of a special kind of condensation polymerization, the molecular weights will be low until very high conversions (P) are achieved. Typical experimental results on intrinsic viscosities and molecular weights for the photopolymerization of *N,N′*-undecamethylene-bis(bromomaleimide) are shown in figure 11.1. In dilute solution, the second term in (11.17) is small and $\ln \bar{X}_n$ is a linear function of t, as shown in figure 11.2.

11.7 Four-centre photopolymerization

Another very interesting type of photopolymerization involves the use of monomers with two polymerizable double bonds, which are

Figure 11.1. Changes in molecular weight, $\bar{M}_n$, during the photopolymerization of N,N′-undecamethylene-bis(dimethylmaleimide). (From De Schryver 1973.)

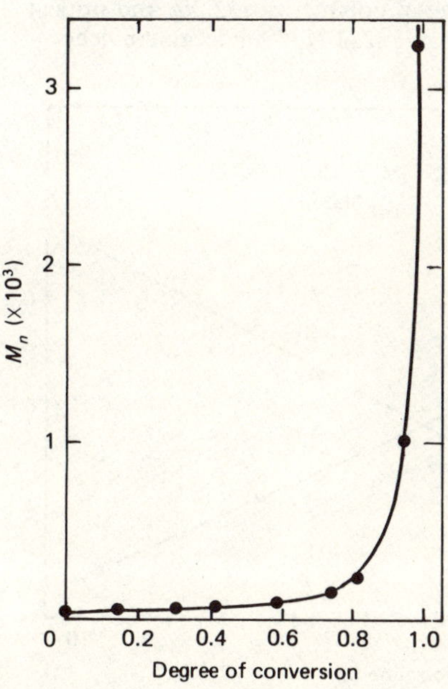

polymerized in the solid state, the so-called 'four-centre photopolymerization'. The earliest examples of this type of photopolymerization appear to be due to Hasegawa and Suzuki (1967) and Nakanishi and Hasegawa (1970) who reported the photopolymerization of 2,5-distyryl pyrazine (DSP, VI) in the crystalline solid phase.

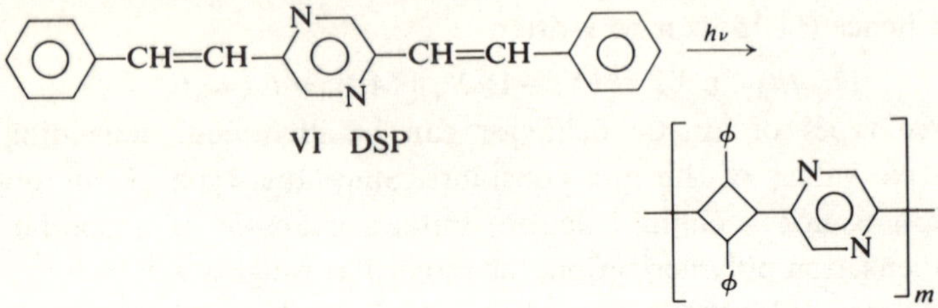

VI DSP

Later they extended this research to photopolymerization of biscinnamates VII and phenylene diacrylic acid derivatives VIII to give polymers of the general structure IX

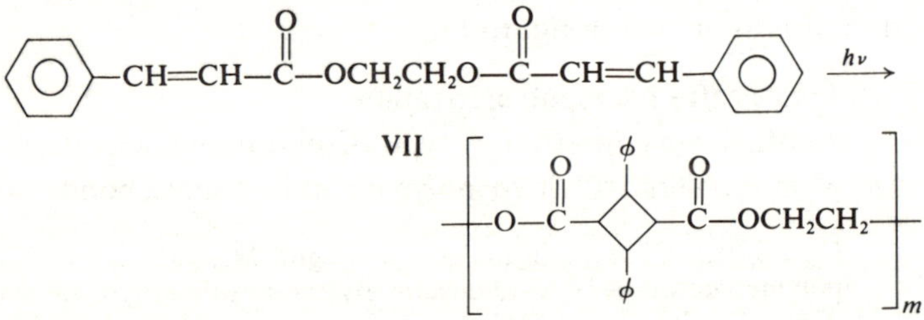

VII

Figure 11.2. Changes in degree of polymerization, $\bar{x}_n$, and optical density during photopolymerization of N,N′-undecamethylene-bis(bromomaleimide).

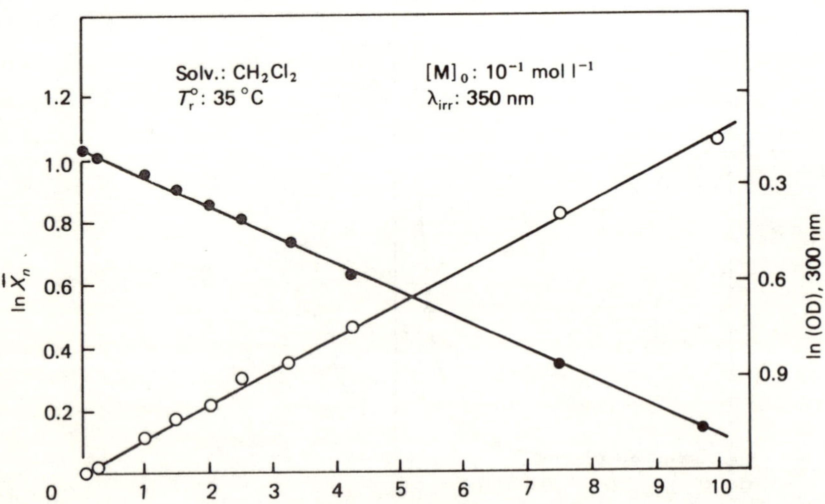

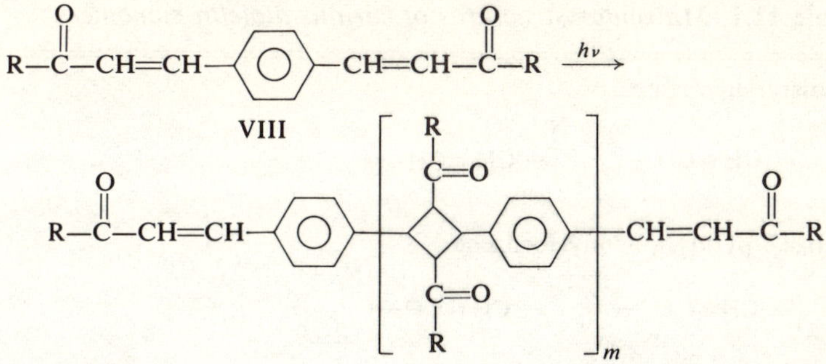

VIII

IX

A variety of other monomers which can be polymerized in this way are listed in table 11.1.

11.8 Photopolymerization of diacetylenes

Gerhard Wegner and his collaborators (Wegner 1969) showed that monomers of the general structure

$$R{-}C{\equiv}C{-}C{\equiv}C{-}R'$$

X

will photopolymerize in the crystalline state to give, in some cases, extended chain crystals of polymers with crystal structures very similar to those of the monomers. Since Wegner's original discovery, a great many diacetylene monomers have been prepared and polymerized, including those listed in table 11.2. These, along with the 'four-center type photo-polymerizations' of Nakanishi and Hasegawa (1970), are examples of a very important new field of topochemical polymerization, where the crystal structure of the polymer is controlled by that of the monomer. The principles of topochemical polymerization of solid diolefins were first elucidated by Hirshfield and Schmidt (1964), but the successful application of these principles did not occur until the more recent work by Wegner (1969) and Nakanishi and Hasegawa (1970).

11.9 Topochemical polymerization

Topochemical polymerization takes place within the solid state of organic crystals, which are made up by packing molecules into a three-dimensional regular lattice. The individual molecules are separated from each other by van der Waals distances which are usually much too large to allow the formation of the transition state of a chemical reaction. There are many examples of photochemical reactions which do not occur because of this restriction on mobility in the solid state. However, reactions can occur if the molecules are mobile enough to approach each

Table 11.1. *Molecular structures of various diolefin monomers*

2,5-distyrylpyrazine

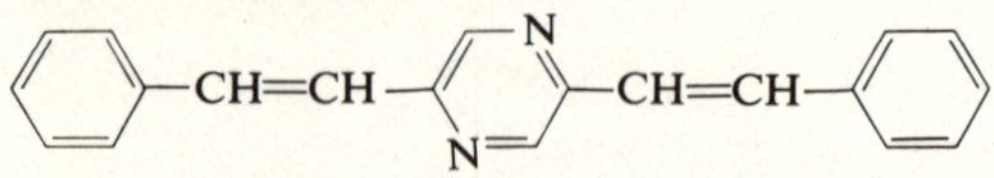

1,4-bis[β-pyridyl-(2)-vinyl]benzene

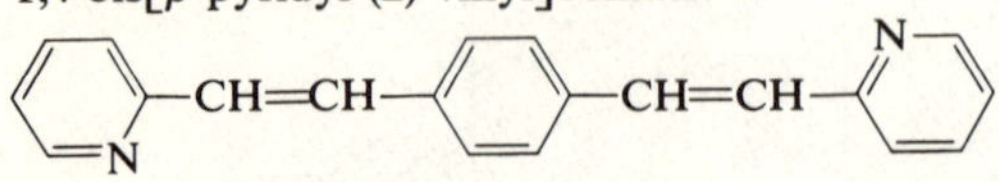

1,4-bis[β-pyridyl-(3)-vinyl]benzene

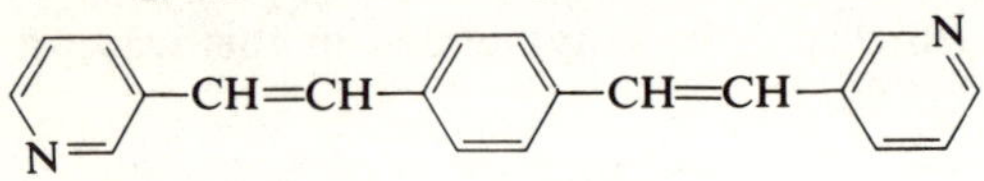

1,4-bis[β-pyridyl-(4)-vinyl]benzene

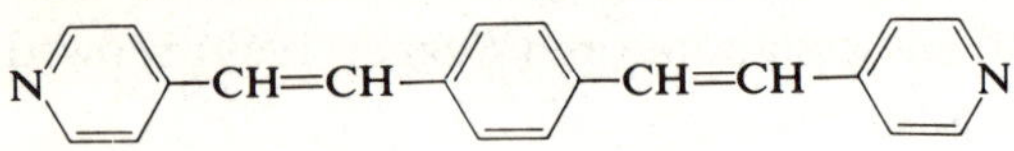

p-phenylenediacrylic acid dimethyl ester

p-phenylenediacrylic acid diethyl ester

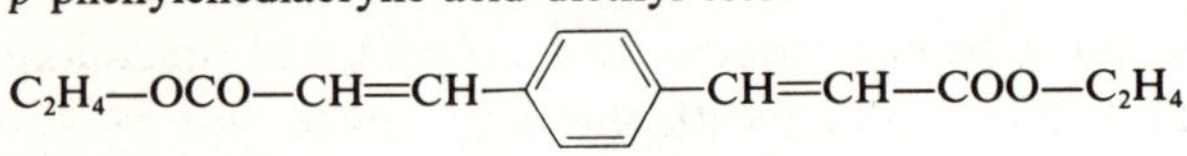

p-phenylenediacrylic acid diphenyl ester

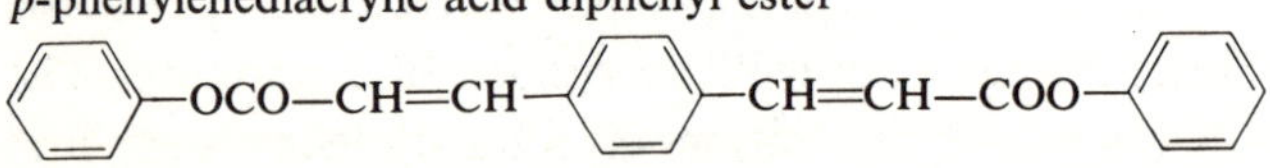

p-phenylene-di(α-cyanoacrylic acid) di-*n*-propyl ester

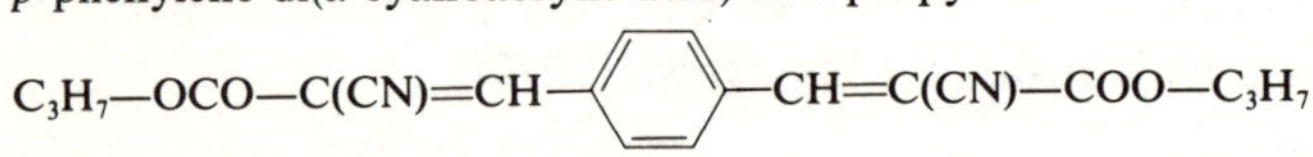

m-phenylenediacrylic acid dimethyl ester

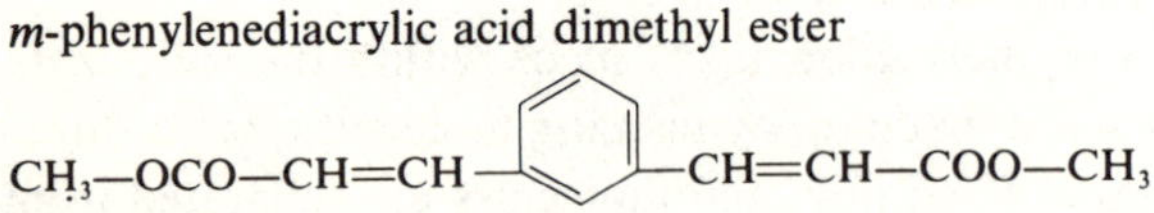

4-(β-carboxyvinyl)-α-cyanocinnamic acid dimethyl ester

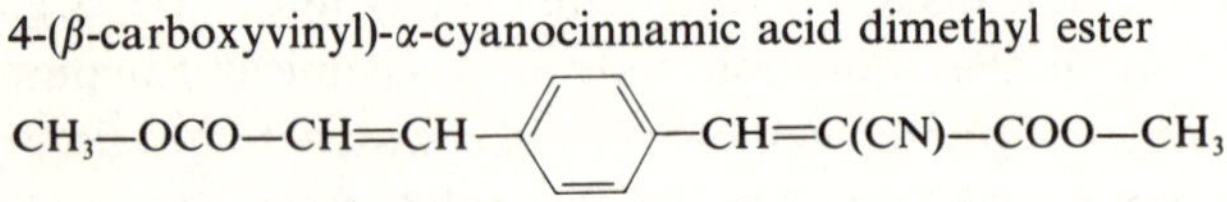

4-(*β*-*sec*-butoxyvinyl)-*α*-cyanocinnamic acid methyl ester

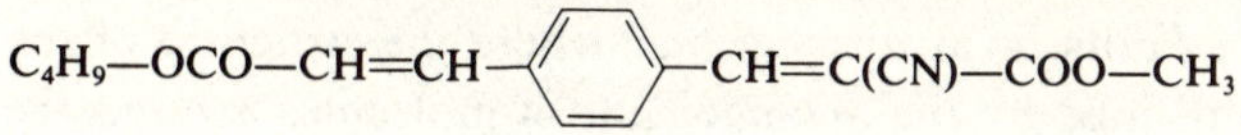

4-(*β*-*sec*-butoxyvinyl)-*α*-cyanocinnamic acid ethyl ester

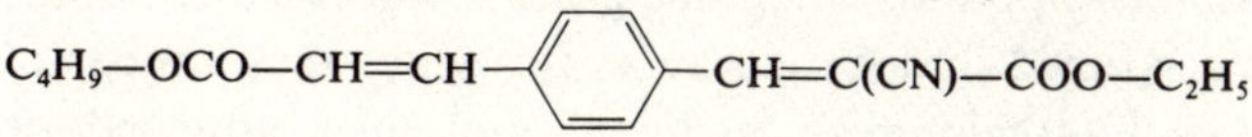

other to a distance of less than 3 Å. In some cases rotational diffusion is sufficient to allow the necessary contact between reactive groups and neighboring molecules, as illustrated in figure 11.3. Ideally, in such cases the center of mass of the molecule does not move from its lattice site, and the reaction proceeds homogeneously inside the parent crystal. Such a reaction is said to be proceeding under topochemical control.

Thomas (1974) has defined topotactic reactions as being those in which the lattice of the solid product shows one or a small number of crystallographically equivalent orientations relative to the lattice of the parent crystal, and in which the reaction has proceeded throughout the bulk of the reactant in a homogeneous manner. Topochemical reactions may lead to topotactic polymerization, but not in all cases. The difference between homogeneous (topotactic) and non-homogeneous initiation is illustrated in figure 11.4.

Table 11.2. *Some examples of diacetylenes which undergo topochemical polymerization. General structure IX and R with R′ as shown below*

Symmetrical (R = R′)	Unsymmetrical	
R	R	R′
$-(CH_2)_n-CH_3$	$-CH_2OH$	$-C{\equiv}C-CH_2OH$
$-(CH_2)_n-OH$	$-CH_2-O-CO-NH-Ph$	$-C{\equiv}C-CH_2-O-CO-NH-Ph$
$-(CH_2)_n-CO_2H$ and salts	$-(CH_2)_n-OH$	$-(CH_2)_m-CH_3$
$-(CH_2)_n-O-CO-NH-Ph$	$-(CH_2)_n-CO_2H$	$-(CH_2)_m-CH_3$
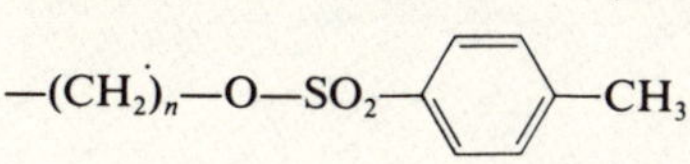		
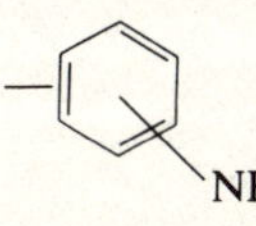 (*o*-, *m*-, not *p*-)		

Topochemical reactions may occur (1) inside a perfect crystal where the structure of the reaction product is defined by the packing of the monomer molecule in the lattice, (2) at well defined sites in crystals, such as packing sequence faults or at dislocations, where the structure of the reaction product is defined by the arrangement of molecules around the defect, (3) inside a composite structure of host and guest molecules such as in channel inclusion compounds where the guest molecules are mobile, but the mobility is restricted in such a way as to reduce the number of possible reactions, and (4) along an array of molecules attached or absorbed in a well defined manner to a rigid molecule in solution. Such a reaction is often called matrix polymerization and some information held by the template molecule, such as molecular weight, stereoregularity, etc., may be transferred to the reaction product.

Wegner (1977*a*) has recently proposed that the solid-state polymerizations of a number of the polyacetylenes are true topochemical polymerizations in which a direct transition occurs from the monomer molecules to polymer chains without destruction of the crystal lattice, and without the formation of non-crystalline intermediates (figure 11.5). On the other hand, the four-centre photopolymerization of DSP, for example, appears to occur by decomposition of the parent crystal into a polycrystalline

Figure 11.3. Illustration of the Hirshfield & Schmidt model of polymerization of compounds such as diacetylene dicarboxylic acid. The solid-state process is diffusionless, and rotation of the monomers enables them to link up with their neighbors without any change in linear separation of the repeat units upon polymerization. (From Nakanishi *et al.* 1980.)

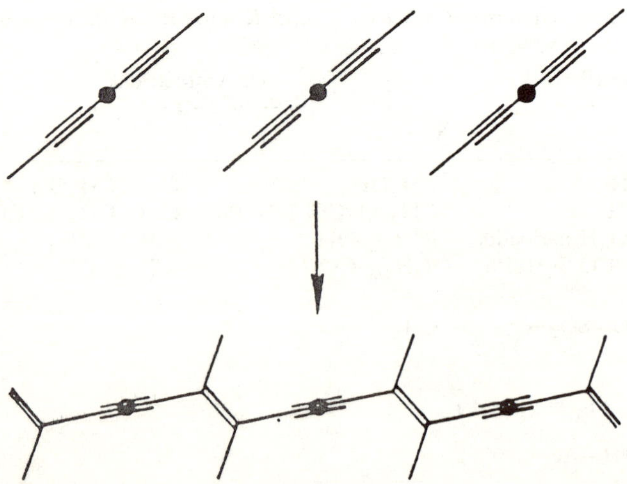

aggregate. In this case, nucleation is heterogeneous, rather than homogeneous, as in the case of the diacetylenes.

Because of the requirement for precise geometry and close approach of the reactive or functional groups, only certain crystal forms of the various monomers will undergo photopolymerization. Many of the monomers synthesized by Hasagawa, Wegner and others will crystallize in a variety of habits and will also undergo crystalline phase transition from one form to another. Only those forms which meet the stringent stereochemical requirements for polymerization will give crystalline polymers.

Detailed studies of the crystalline morphology of both monomers and polymerizable crystals have been made by a number of investigators. Excellent summaries of these data can be found in recent review articles by Wegner (1979) and Nakanishi *et al.* (1980).

Figure 11.4. Two different mechansisms of phase transformation in solid-state polymerization: (*a*) homogeneous reaction in form of solid solution, (*b*) heterogeneous growth by nucleation. (From Wegner 1977 *b*.)

(*a*)

(*b*)

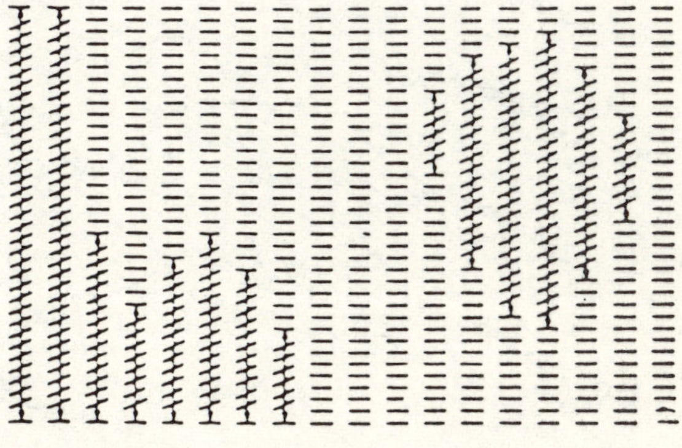

11.10 Kinetics and mechanism of four-centre photopolymerization

The four-centre type photopolymerization of DSP in the crystalline state is shown schematically in figure 11.6. As can be seen, the polymer chains are held together with cyclobutane rings. The relation between the monomer packing and the polymerization is such that the monomer molecule is only slightly tilted from its starting position, so that the adjacent double bonds can lead to cyclobutane rings. The space

Figure 11.5. (*a*) Scheme of a topochemical polymerization: transformation of a single crystal of monomers into a single crystal of polymers. (*b*) Topochemical polymerization of diacetylenes. (From Wegner 1977 *b*.)

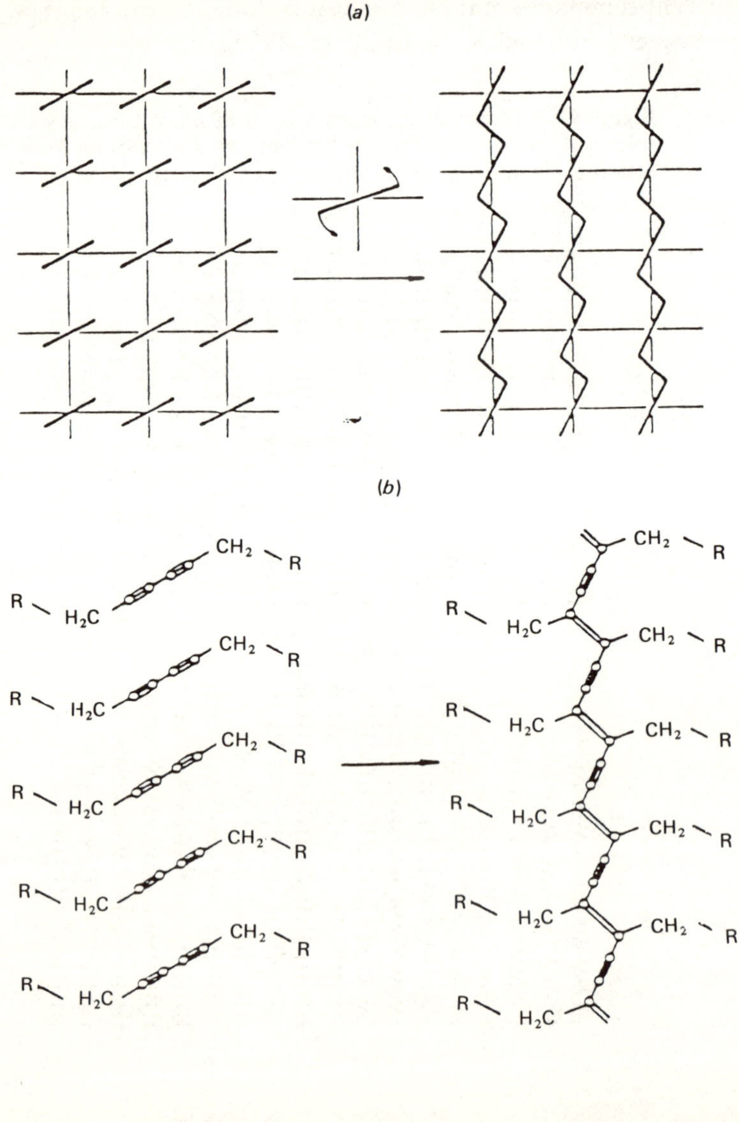

group of the crystal does not change and the total volume change during polymerization is very small. The polymer does not absorb at the same wavelength region as the monomer, and if the monomer crystals are irradiated only in the long wavelength tail of the absorption, only oligomers, such as dimers and trimers are formed. If these oligomer crystals are then irradiated at a shorter wavelength, high polymer is then formed with molecular weights similar to those obtained by direct irradiation of the monomer by short wavelength light. Quantum yields for the polymerization are extremely difficult to measure, but are estimated to be greater than unity for polymerization at $\lambda = 365$ nm. However, it should be pointed out that reaction rates and light absorption are vectors in anisotropic media, which makes it difficult to evaluate the absolute values of both quantum yields and rate constants. Typical conversion time curves for DSP and polydiacetylenes are shown in figure 11.7. On the basis of this evidence, Wegner has proposed that monomers like DSP polymerize in a heterogeneous manner, whereas the diacetylenes polymerize homogeneously. However, this contention has been the subject of some debate, and Nakanishi *et al.* (1980) consider that in certain cases four-center photopolymerization is also a homogeneous topotactic process.

11.11 Mechanism of photopolymerization of diacetylenes

In the parent crystal of a reactive diacetylene monomer, the molecules are packed in a ladder-like fashion such that the ends of one triple-bond system approach the adjacent triple-bond system within a distance of about 4 Å or less. Furthermore, the long axis of the triple-

Figure 11.6. Four-centre polymerization of DSP. (From Wegner 1977*a*.)

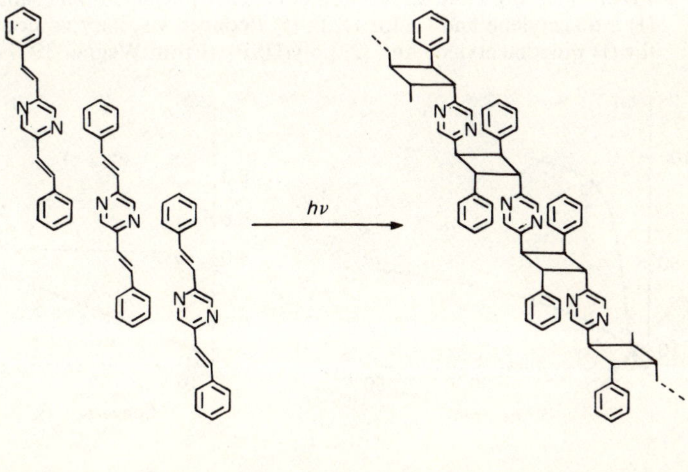

bond system is inclined with regard to the translational axis at an angle γ of about 45°. Polymerization occurs by successive tilting of each molecule along the ladder with the minimum of movement of the atoms from their lattice sites (figure 11.5). For this reason the mode of packing and the lattice symmetry is retained throughout the reaction. Polymerization can occur simply by annealing the colorless monomer crystals below their melting point or by irradiation with high-energy γ-ray photons, or uv light. Because the polymerization involves the formation of a conjugated backbone, the crystals turn deeply colored after polymerization, and finally turn dark red, blue or green with a typical metallic lustre. In many cases, quantitative conversion is reached within a few hours. Polymerization of the diacetylene usually occurs by a homogeneous solid state reaction. That is, the individual chains grow independently from each other starting from points randomly distributed throughout the lattice, as shown in figure 11.4*a*. The polymerizing crystal is essentially a solid solution of extended macromolecules in a matrix of unpolymerized monomer. A particular substituted diacetylene may have as many as five different crystal modifications, only one of which may show photochemical reactivity. It is now believed that the reactive intermediates in the polymerization of the diacetylenes are carbenes. In the earliest work, Wegner (1979) proposed a butatriene structure shown below (XI). Later

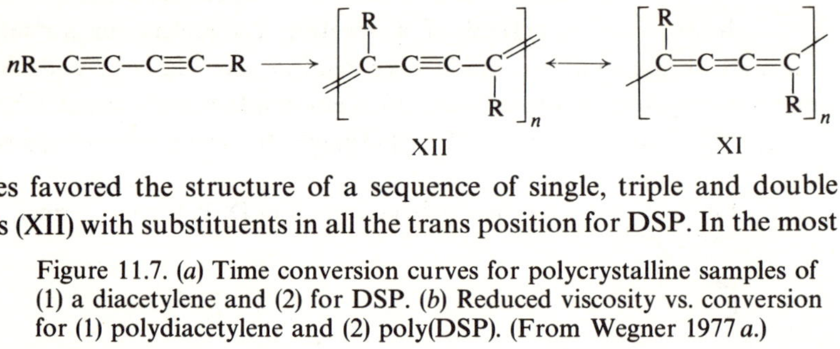

XII XI

studies favored the structure of a sequence of single, triple and double bonds (XII) with substituents in all the trans position for DSP. In the most

Figure 11.7. (*a*) Time conversion curves for polycrystalline samples of (1) a diacetylene and (2) for DSP. (*b*) Reduced viscosity vs. conversion for (1) polydiacetylene and (2) poly(DSP). (From Wegner 1977*a*.)

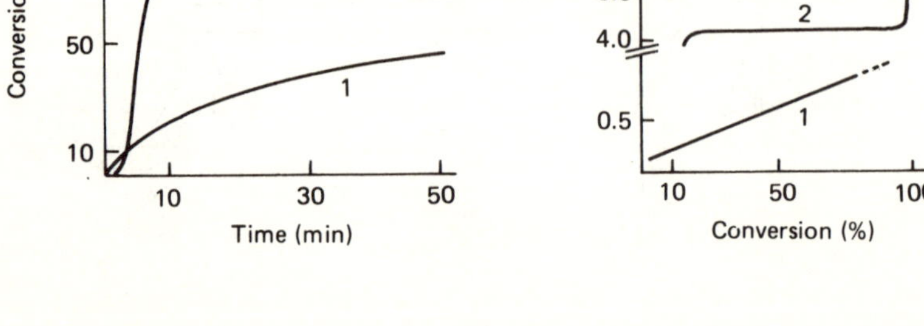

recent work, the structure is usually considered to be mesomeric, with the triene and ene-yne formulations as limiting cases, and this seems to be favored by recent studies by resonance Raman spectroscopy.

11.12 Experimental studies of diacetylene polymerization

The polymer which has been most extensively studied is from the para-toluene sulphonate of 2,4-hexadiene-1,6-di-ol (PTS, XIII) which can

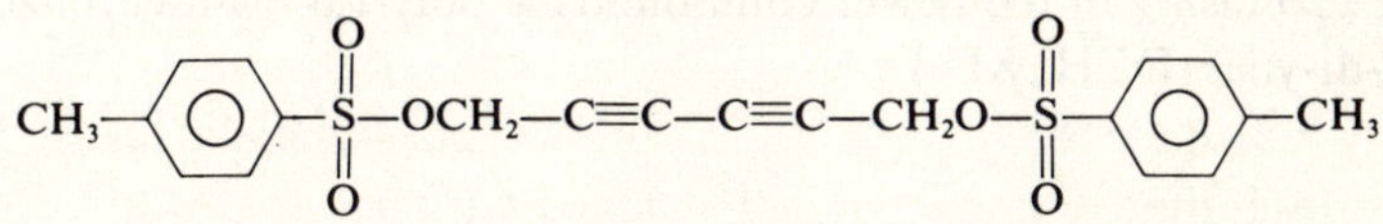

XIII PTS

be readily synthesized in high purity, and large single crystals can be grown by slow evaporation of a concentrated solution in acetone. In general, the polymerization occurs by photoirradiation over the whole absorption spectrum of the monomer. It proceeds homogeneously and nucleation effects are not observed, although sometimes phase changes occur when partial conversion is reached. In almost all cases, high molecular weight polymers are formed and these do not act as sensitizers or quenchers during photopolymerization. It is now known that triplet carbenes are detected and are assumed to be the active species in the polymerization. The molecular weight M and degree of polymerization P increase with the extent of reaction, as shown in figure 11.8.

Figure 11.8. Dependence of the average degree of polymerization on conversion for PTS. (From Wegner 1979.)

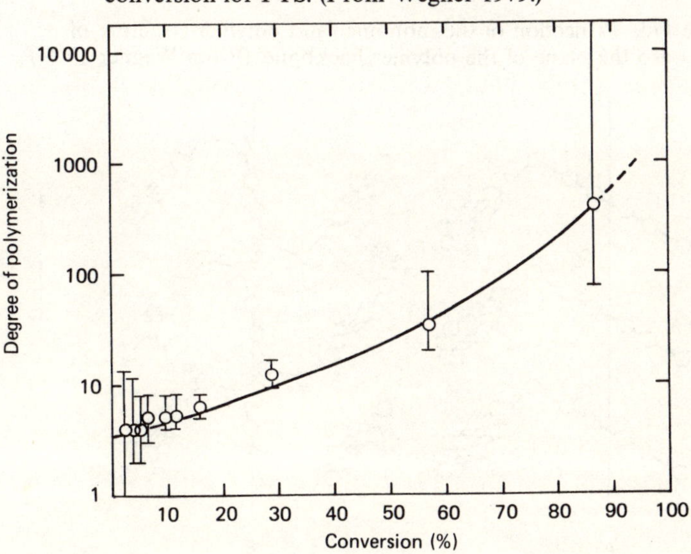

11.13 Other polydiacetylenes

The solid-state polymerization of diacetylenes has now been extended to more than 200 different compounds, most of which polymerize to some extent; however, only a few compounds have been investigated in any detail, and many diacetylenes which are reactive in principle do not polymerize quantitatively, or without phase transitions which destroy the single crystal character of the polymer. Among the more promising of the newer compounds is poly(1,6-bis-n-carbazolyl 2,4-hexa-di-yne) (DCH, XIV)

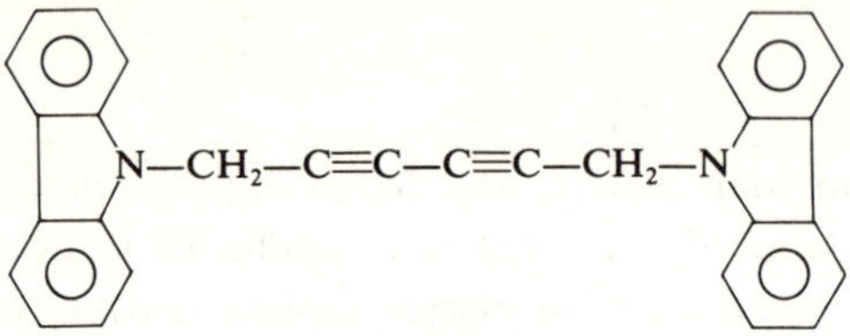

XIV DCH

The synthesis and polymerization were described independently by Yee and Chance (1978) and Enkelman *et al.* (1977). Large single crystals of the monomer can be obtained by slow evaporation of a concentrated solution of dimethyl formamide. The monomer and polymer structure of this compound is shown in figure 11.9. Unlike most of the other reactive diacetylenes, DCH crystals are quite stable to ordinary laboratory light and the polymerization induced in the X-ray beam of a diffractometer is low enough to allow its structural determination at room temperature.

Figure 11.9. Projection of the monomer and polymer structure of DCH onto the plane of the polymer backbone. (From Wegner 1979.)

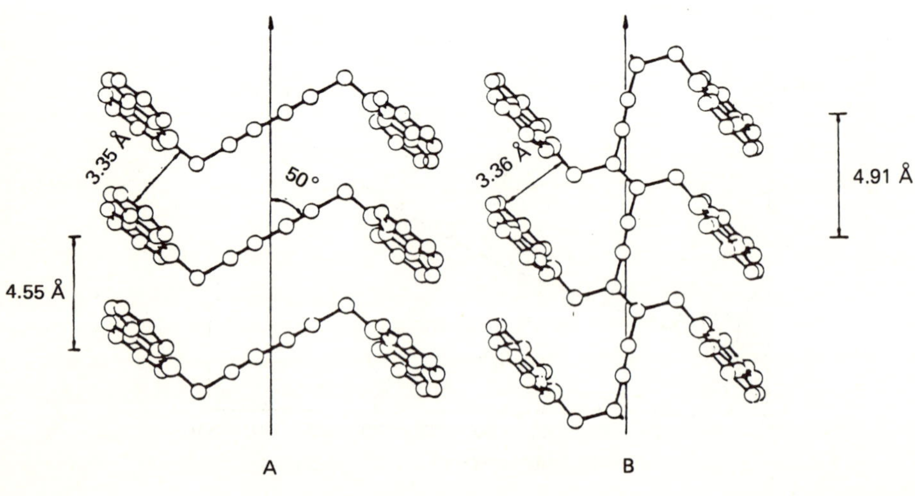

Polymerization occurs by annealing the crystals at temperatures between 120 and 240 °C or by irradiation by cobalt-60 γ-rays. A dose of about 50 megarad is sufficient to reach quantitative conversion. The polymer contains an infinite array of substituted carbazole units stacked along the *b*-axis of the crystal with a distance of 3.35 Å. This is about 0.6 Å greater than the carbazole distance in the excimer-forming sites in polycarbazole and therefore it is not surprising that both fluorescence and phosphorescence disappear in the polymer crystals. It is not clear whether this is caused by self-absorption, self-quenching or by interaction with defects in the crystal structure.

11.14 Copolymerization of diacetylene monomers

Certain of the diacetylene monomers can be copolymerized if their crystal habits are sufficiently similar. Wegner (1979) has reported that DCH can form a continuous series of mixed crystals with a similar monomer, 1,(*n*-carbazolyl)-4-(9 anthracenyl)2,4,hexa-di-yne (ACH, XV).

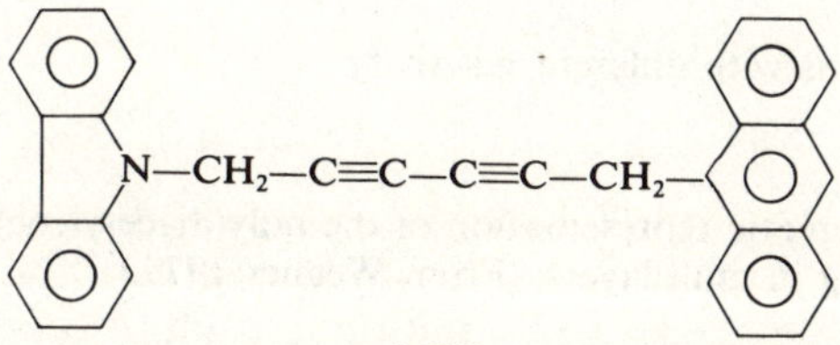

ACH
XV

ACH does not itself polymerize easily, but does copolymerize in mixed crystals, with DCH. Chemical and spectroscopic evidence of copolymer single crystals showed that the copolymers have the identical composition to the parent phase.

11.15 Photopolymerization in monolayers and multilayers

A further and important extension of the polymerization of diacetylene involves the synthesis of molecularly thin films prepared by solid-state polymerization of suitable monomers in mono- or multilayers. Diacetylene monomers are synthesized with long hydrocarbon chains involving 20 or more carbon atoms, and having melting points greater than 45 °C. These are spread on a Langmuir balance and photopolymerized either in single or multilayers. By consecutive dipping and withdrawing of the solid substrate from the balance it is possible to prepare multilayers whose thickness is exactly given by the number of dippings multiplied by the thickness of the individual layer. A number of diacetylene monomers capable of undergoing monolayer polymerization are listed in

Table 11.3. *Spreading and polymerization behavior of long-chain diacetylenes with ability to form monomolecular films on water or on* 10^{-3} *M* $CdCl_2$

Compound R–C≡C–C≡C–R′				Photopolymerization $\lambda < 300$ nm	
R	R′	mp (°C)	Multilayer formation	Crystal	Multilayer[a]
n-C_9H_{19}–	–$(CH_2)_8$–COOH	48		+ +	
n-$C_{10}H_{21}$–	–$(CH_2)_8$–COOH	57	yes	+ +	+
n-$C_{12}H_{25}$–	–$(CH_2)_8$–COOH	61	yes	+ +	+
n-$C_{14}H_{29}$–	–$(CH_2)_8$–COOH	46/59[b]	yes	+ +	+
n-$C_{12}H_{25}$–	–$(CH_2)_3$–COOH	57	no	+ +, +[c]	–
n-$C_{16}H_{33}$–	–$(CH_2)_2$–COOH	93	yes	+ +	–
n-$C_{16}H_{33}$–	–COOH	75	yes	+ + +	+
n-$C_{12}H_{25}$–	–COOH	58	no	+ + +	–
n-$C_{10}H_{21}$–	–$(CH_2\cdot)_9$–OH	49	no	+	–

[a] + + + dark blue < 10 s, + + dark blue 10–30 s, + dark red after exposure > 2 min, – no reaction.
[b] Phase transition at 46 °C.
[c] There are two modifications with different reactivity.

Source: Wegner (1979).

Figure 11.10. Schematic representation of the poly(diacetylene) structure occurring in multilayers. (From Wegner 1979.)

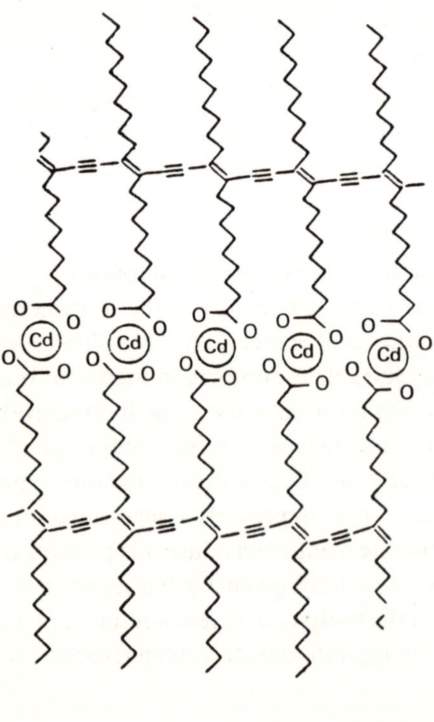

table 11.3. The structure of multilayers prepared from these is shown in figure 11.10. These new polymeric films are now being investigated for a variety of applications ranging from solar energy conversion (Grätzel 1981) to the synthesis of synthetic liposomes for cancer drug delivery (Hub *et al.* 1981).

12

Photocyclization

One of the most common photoinduced reactions in organic chemistry is cyclization. Indeed, in many systems cyclization between two reactive species, one of which is electronically excited, is predicted by the Woodward–Hoffman rules, in contrast to the thermally excited ground-state molecules, which will react by other pathways. In recent years such reactions have been studied in polymer systems as well.

12.1 Photodimerization

12.1.1 *Cinnamic acid derivatives*

Historically the first of these photoreactions, and perhaps the one with the most important practical application, is the photodimerization of cinnamic acid derivatives which leads to crosslinking in polymer systems and is the basis of commercial photoresists. Applications of this chemistry to photolithography have been reviewed by Williams (1969) and Delzenne (1971, 1974).

Cinnamic acid undergoes photodimerization, even in the solid state, forming cyclobutane derivatives. The fact that the reaction can take place within the matrix of a solid crystal is evidence that product formation requires a minimum of molecular motion. *Trans*-cinnamic acid crystallizes into two crystal forms, a stable α and a metastable β form, and photodimerization occurs to give the corresponding α-truxillic (I) and β-truxinic (II) acids:

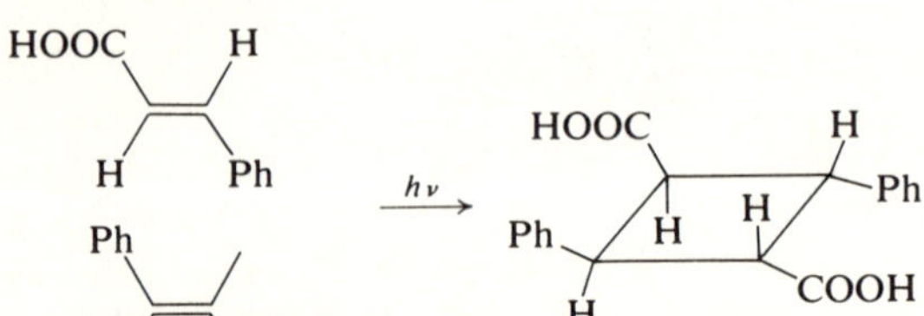

(12.1)

I

α-crystalline form

α-truxillic acid

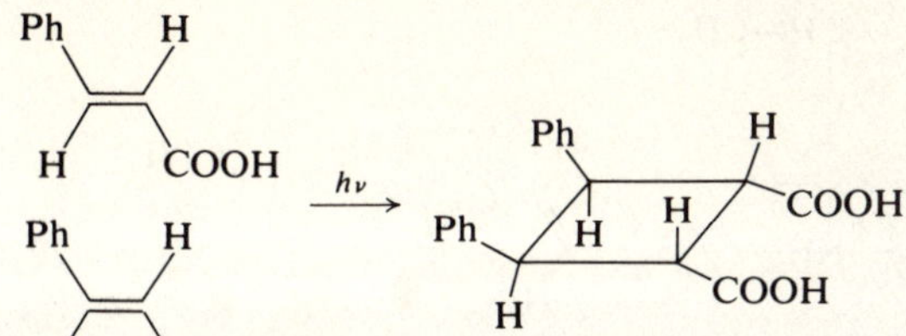

β-crystalline form β-truxinic acid II (12.2)

Similar reactions can take place in polymers containing cinnamyl groups. For example, poly(vinyl alcohol) esterified with cinnamic acid will undergo crosslinking when cinnamate groups on adjacent polymer chains form a cyclodimer:

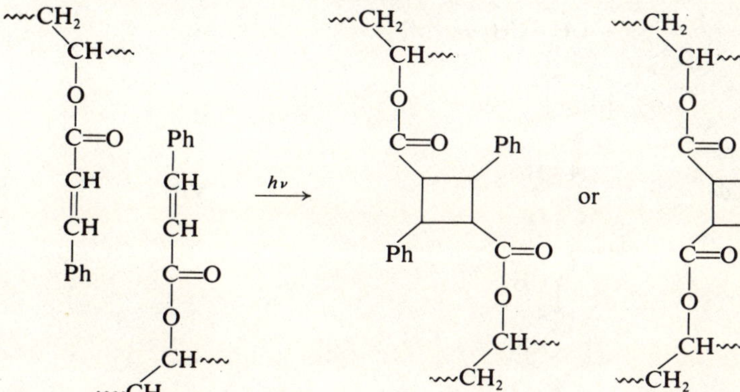

Vinyl cinnamate groups III Truxillic IV Truxinic Crosslink (12.3)

Similar crosslinks are formed when cinnamic acid is bonded to cellulose derivatives or epoxy resins. Williams (1969) has reviewed other possible structures incorporating cinnamate groups, for example:

Acetal

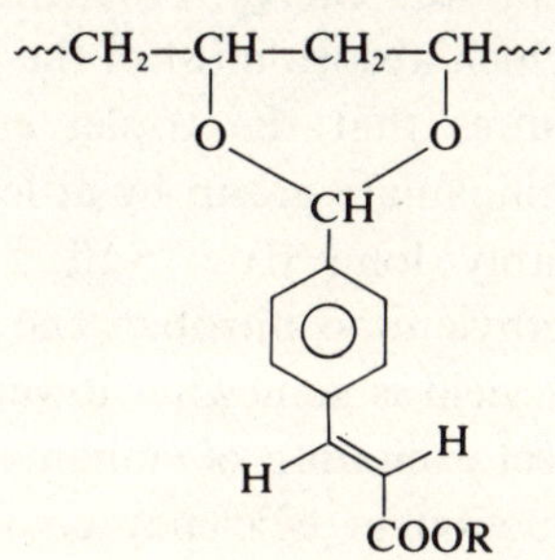

V

Urethane 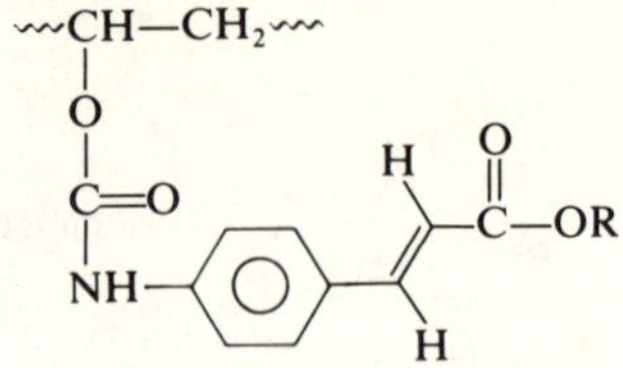 VI

Ester 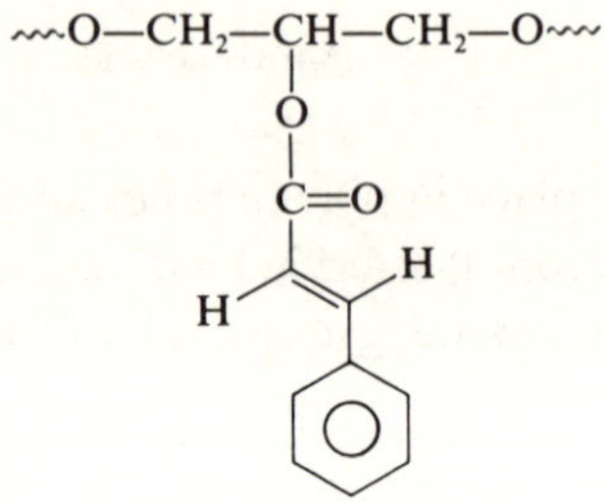 VII

Amide 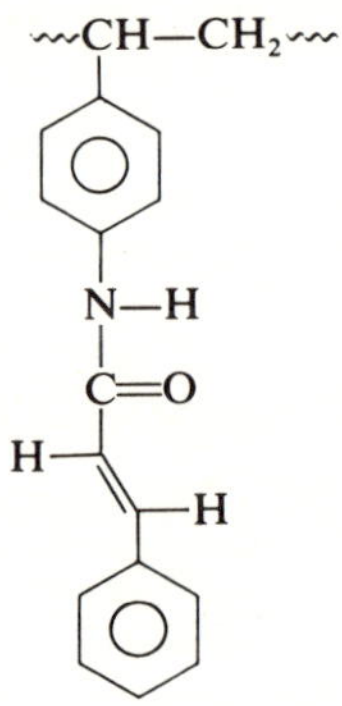 VIII

The photodimerization of cinnamic acid derivatives occurs both by direct absorption of a photon, and by energy transfer from a suitable sensitizer. In both cases the major reacting species is the n-π^* triplet of the cinnamyl group. However, for practical purposes, sensitization is extremely important because it extends the sensitivity of the photoresist into the near uv and blue visible where the mercury arcs used for exposure are most efficient in the production of photon energy. Furthermore, it permits the use of glass optical systems which absorb most of the light of $\lambda < 300$ nm.

Efficient sensitization requires that the triplet energy level of the sensitizer exceed that of the cinnamate group by at least 8 kJ mol^{-1} and that its lifetime be reasonably long (i.e., $>10^{-2}$ s). Benzophenone, Michler's ketone and *p*-dimethylamino nitrobenzene are effective sensitizers, although the quantum yield is somewhat lower, typically $\sim$ 0.35, as compared to $\sim$0.5 for direct excitation of cinnamyl groups. Delzenne (1971) suggests that the reduction in efficiency could be due to some contribution from the singlet state in the case of direct reaction, or an inefficiency in triplet energy transfer in the case of the sensitized systems. In any case, it is not possible to equate the disappearance of cinnamyl groups with the number of crosslinks, since some reaction may take place intra- rather than intermolecularly, giving rise to truxillic or truxinic groups between adjacent groups, or even across loops in the same polymer chain.

12.1.2 *Chalcone derivatives*

Other analogous groups such as chalcones (IX) can be used to make photoactive polymers where R can be such groups as *p*-Br, *p*-$(CH_3)_2N$, *m*-($-NO_2$), etc. In these cases, however, crosslinking apparently

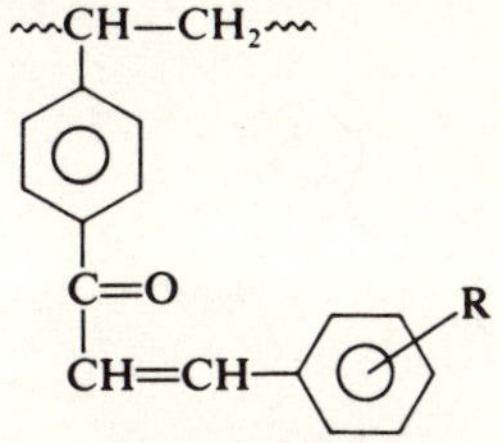

IX

Poly(vinyl chalcone)

occurs via the formation of a biradical derived from the double bond of the cinnamoyl group and abstraction of a proton from neighboring CH or CH_2 groups to form units of the following structure (Lyalikov *et al.* 1970):

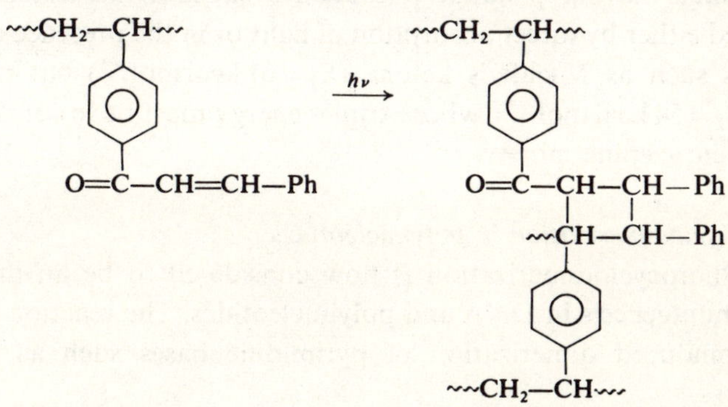

(12.4)

12.1.3 *Coumarins*

Other structures which form crosslinks by photocyclization are polymers including the coumarin group (X), which can be considered an internal ester of cinnamic acid. Dimers are formed via a cyclobutane ring (Anet 1962):

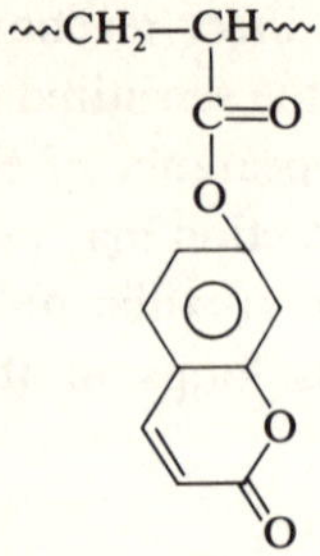

X

12.1.4 *Dibenzazepines*

Hyde *et al.* (1973) have demonstrated the photodimerization of N-acyl derivatives of the dibenz[b,f]azepine nucleus (XI) to give cyclobutane dimers(XII):

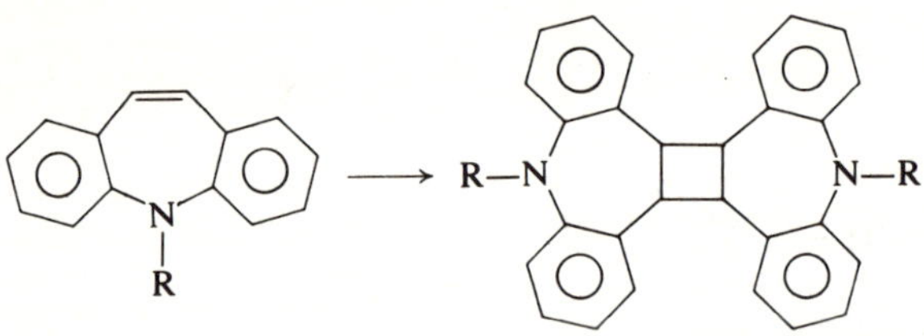

(12.5)

XI XII

When R is an acryloyl group, for example, the resulting monomer can be copolymerized with styrene, methyl methacrylate or N-vinyl carbazole to give polymers having pendant photoactive substituents which can be crosslinked either by direct absorption of light or in the presence of triplet sensitizers such as Michler's ketone ($E_T = 61\,\mathrm{kcal\,mol^{-1}}$) but not fluorenone ($E_T = 54\,\mathrm{kcal\,mol^{-1}}$), whose triplet energy must be lower than that of the dibenzazepine moiety.

12.1.5 *Cyclodimerization in polynucleotides*

Photocyclodimerization is now considered to be an important cause of mutagenesis in DNA and polynucleotides. The reaction involves the photoinduced dimerization of pyrimidine bases such as thymine (XIII):

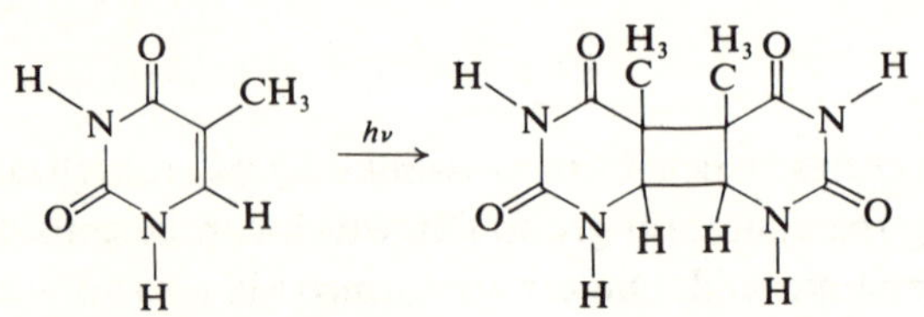

(12.6)

XIII XIV

The resulting crosslinking of DNA represents a defect in the coding sequence and hence can cause mutations as well as other cell damage.

12.2 Cycloadditions

12.2.1 *Paterno–Buchi reaction*

The cyclodimerization reactions discussed in the previous section represent only a very small selection of the possible cyclizations which can occur via photochemically excited intermediates. In particular they refer to reactions between an excited molecule or group and a ground-state molecule or group otherwise identical chemical structure. A much greater range of possibilities exists if the two groups are chemically different. One example of this type of reaction is known as the Paterno–Buchi reaction. This is a typical example of what is classified as an 'ene' reaction:

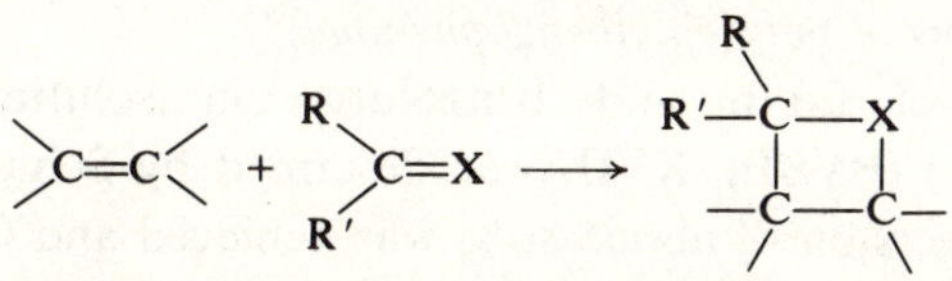

(12.7)

In polymers we can distinguish two types of structures produced by this reaction, depending on whether the ring is produced in the backbone of the polymer chain or in a side chain, thus if one of the unsaturated groups is in the backbone, a ring is formed in the chain (structure XV):

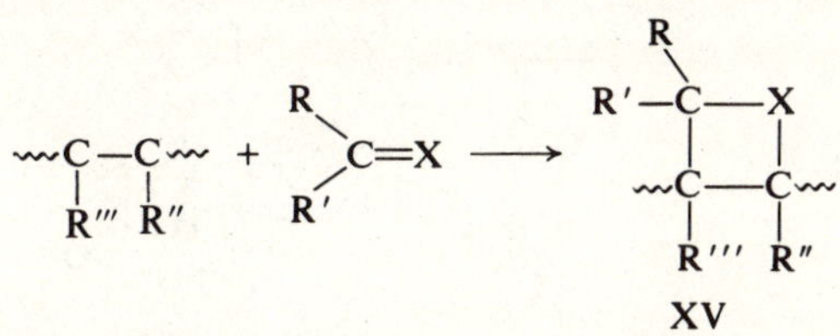

XV

(12.8)

On the other hand, if the unsaturated group is in a side chain, cyclic groups are formed pendant to the polymer main chain (XVI):

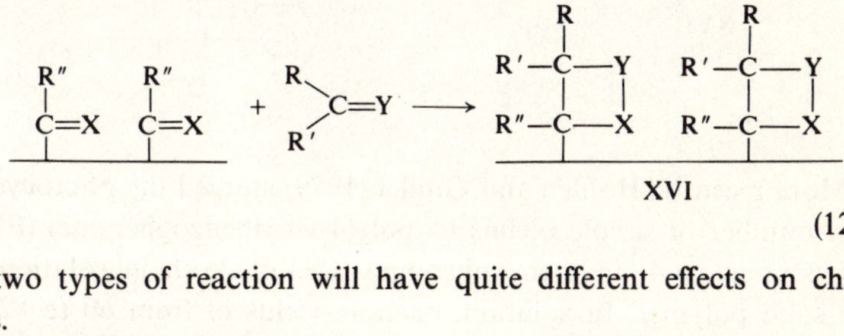

XVI

(12.9)

These two types of reaction will have quite different effects on chain stiffness.

The Paterno–Buchi reaction involves the photocycloaddition of carbonyl compounds to olefins and has been studied extensively in small molecules (Arnold 1968). The general mechanism of the reaction involves the attack of the electrophilic oxygen of the n-π^* triplet carbonyl on an electron-rich olefin to generate preferentially the most stable biradical intermediate, followed by ring closure to yield an oxetane:

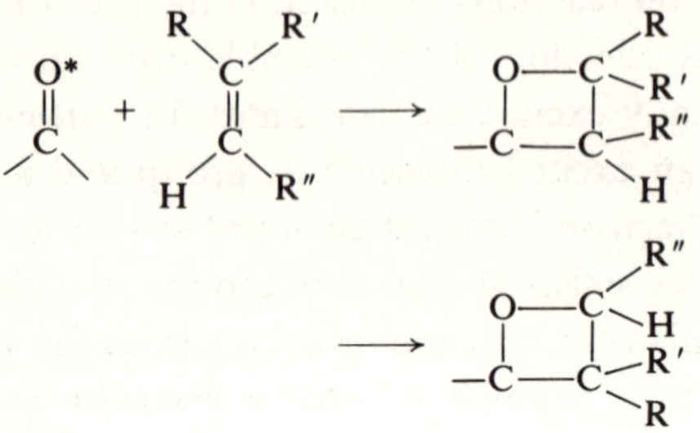

(12.10)

12.2.2 *Photocyclizations of poly(vinylbenzophenone)*

The reaction of furan and benzofuran in solution with poly(vinylbenzophenone) (PVBP), XVII, was described by Schulz *et al.* (1969). In this case conversion of about 80% was achieved and furan or benzofuran groups were attached to the side chains of PVBP via an oxetane ring:

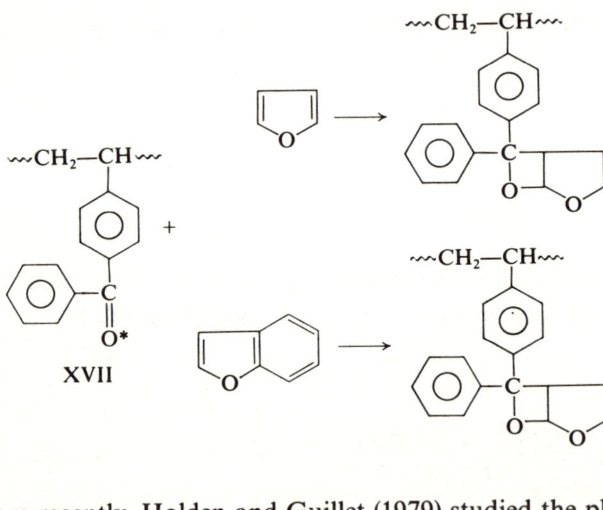

XVII XVIII XIX

(12.11)

More recently, Holden and Guillet (1979) studied the photocyclization of a number of simple olefins to poly(4-vinylbenzophenone) (PVB) and poly(styrene-co-4-vinylbenzophenone) (PSVB), both in solution and in the solid polymer. In solution, oxetane yields of from 60 to 82% were obtained with a variety of di-, tri- and tetrasubstituted olefins. Isobutene, for example, adds to both PVB and PSVB in solid films with conversions greater than 90%:

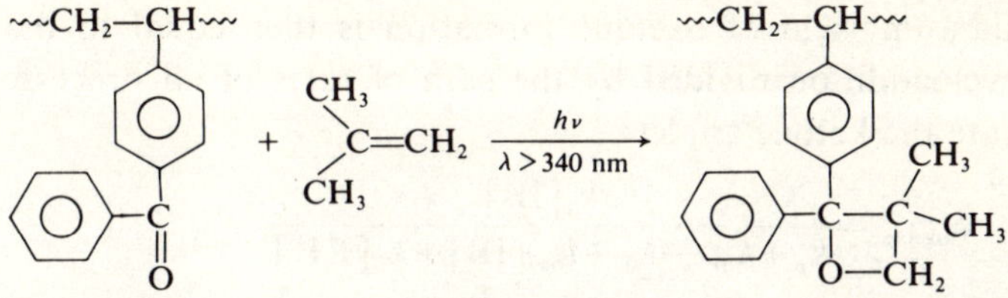

XVII XX (12.12)

While the quantum yield for reaction of vinyl benzophenone with 2-methyl-2-hexene in benzene solution was 0.13±0.02 which is nearly identical to that of a small molecule model compound (4-methyl benzophenone 0.12±0.02) the quantum yields of oxetane formation in the solid film were reduced by a factor of 10 to about 0.013 at 366 nm. In the solid, the quantum yields were independent of film thickness between 2 and 10 micrometers, and of temperature from 23 to 65 °C, which encompasses the glass transition temperature of the polymer. The lack of a pronounced effect at the glass transition temperature suggests that the diffusion constant of the small molecule isobutylene through the polymer does not change drastically at T_g. A rationalization of the lower quantum yields in the solid phase can be made on the basis of the lower equilibrium solubility of isobutene in the solid polymer matrix as compared to solution.

The kinetic scheme proposed by Holden and Guillet (1980) is given below:

$$\phi_2C{=}O \xrightarrow[ISC]{h\nu} \phi_2C{=}O^* \ (T_1) \qquad \text{Excitation} \qquad (12.13)$$

$$T_1 \xrightarrow{k_p} \phi_2C{=}O + h\nu' \qquad \text{Phosphorescence} \qquad (12.14)$$

$$T_1 \xrightarrow{k_{nr}} \phi_2C{=}O \qquad \text{Non-radiative decay} \qquad (12.15)$$

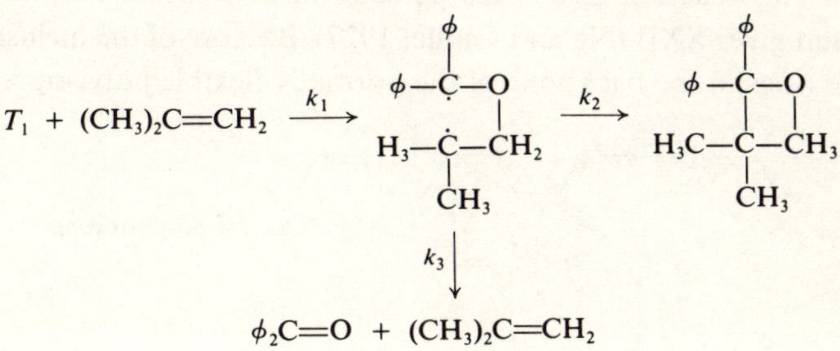

(12.16)

$$T_1 + (CH_3)_2C{=}CH_2 \xrightarrow{k_4} \phi_2\dot{C}{-}OH + \dot{C}H_2{-}\overset{\overset{\displaystyle CH_3}{|}}{C}{=}CH_2 \qquad (12.17)$$

$$T_1 + \mathrm{RH} \xrightarrow{k_5} \phi_2\dot{\mathrm{C}}\text{—OH} + \mathrm{R}\cdot \qquad (12.18)$$

H-atom abstraction
from polymer

The quantum yield of oxetane formation is then equal to the rate of photocycloaddition divided by the sum of rates of all processes which deactivate the ketone triplet:

$$\phi_{\text{oxetane}} = \frac{\alpha k_1[\mathrm{IB}]}{k_p + k_{nr} + (k_1 + k_4)[\mathrm{IB}] + k_5[\mathrm{RH}]} \qquad (12.7)$$

where $\alpha = k_2/(k_2 + k_3)$, and [IB] and [RH] are the effective concentrations of isobutene and of polymer hydrogen, respectively. It is clear from (12.7) that ϕ_{oxetane} is a function not only of the rate of photocycloaddition, but of all other triplet deactivating processes. However, if we make the first-order approximation that the sum of $k_p + k_{nr}$ is larger than the other terms in the denominator of (12.7) a ten-fold decrease in isobutene concentration would lead to a similar decrease in quantum yield.

This reaction can also be used to crosslink polymer systems if a polyfunctional olefin is used. For example, it was shown that films of PVB containing squalene (the isoprene hexamer) crosslink relatively rapidly under the action of ultraviolet light. The quantum yield of oxetane formation was estimated to be approximately 0.03 as compared with a value of 0.34 for poly(vinylcinnamate), the most widely studied photo-crosslinkable polymer. While the squalene-PVB system is less sensitive than some other photo-crosslinkable polymers, it has the advantage that the Paterno–Buchi reaction leads inherently to inter-rather than intramolecular crosslinking. Furthermore, the oxetane cross-links can be reversed by hydrolysis with acid, which may be of interest in certain applications.

12.2.3 *Photocyclization of polyisoprene*

This reaction also takes place both in solution and in the solid phase and gives XXII (Ng and Guillet 1977). Because of the inclusion of the oxetane rings in the backbone of the normally flexible polyisoprene chain,

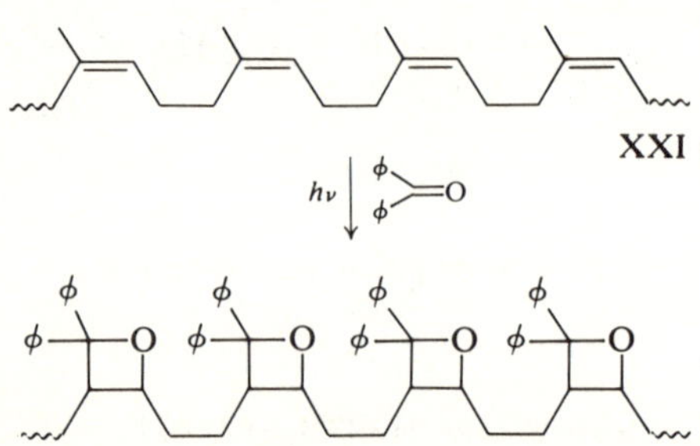

XXI *cis*-polyisoprene

(12.19)

XXII

the glass transition temperature of the polymer is raised from about −65 °C to nearly 46 °C when only half of the double bonds have been reacted with benzophenone. The variation of the glass transition temperature with per cent conversion to oxetane rings is shown in table 12.1. Further substitution is possible but difficult in the solid phase because of the reduced solubility of the benzophenone in the solid polymer below its glass transition temperature. It is interesting to note that since the molecular weight of the benzophenone is five times that of the isoprene unit, complete conversion to oxetane rings would give a polymer which is 80% benzophenone and 20% isoprene by weight.

Since each step in the reaction involves the addition of a photoexcited molecule, this is somewhat analogous to 'photopolymerization', as redefined by De Schryver (see chapter 11), and Ng and Guillet (1977) suggested the name 'photoreplica polymerization' for this type of process. One can think of this as a polymerization of benzophenone on a preformed template or matrix of polyisoprene. Unfortunately, this process is not without side reactions since there is some abstraction of hydrogen by the photoexcited benzophenone leading to benzpinacol, and crosslinking through recombination or addition of the resulting radicals to the double bond of the polyisoprene.

12.2.4 *Reactions of maleic copolyesters*

Another example of photoaddition was reported by Schulz (1973) in which a number of ring ethers and cycloacetals were added to the double bonds in a copolymer of maleic anhydride and ethyleneglycol. The reagents used were tetrahydrofuran, tetrahydropyran, 1,3-dioxolane, 1,4-dioxan or trioxan and these compounds were used as solvents for the reaction. Some crosslinking was observed in the case of tetrahydropyran,

Table 12.1. *Glass transition temperatures of polyisoprene and its photochemical adducts with benzophenone*

Irradiation time (min)	Residual double bonds (mol%)	T_g (°C)
0	100	−64.9
15	70.4	31
30	61.8	37.8
45	58.0	37.6
60	51.5	41.2
120	42.6	45.8

but all of the others give soluble products and were reported to have considerably higher softening ranges than the starting copolymer.

12.2.5 *Addition of maleic derivatives to polystyrene*

A further example of photocyloaddition in polymers is the classic reaction of the addition of maleic anhydride to benzene rings, first described by Angus and Bryce-Smith (1959):

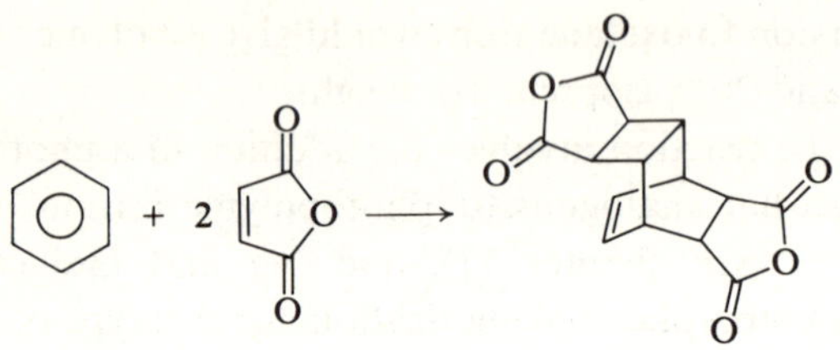

(12.20)

XXIII

Two molecules of maleic anhydride react with one molecule of benzene by means of a charge-transfer complex to form the cyclo adduct (XXIII). Substitution of the benzene ring tends to reduce the efficiency and quantum yield of the reaction so it is not surprising that both Schulz (1973) and Hrdlovic and Lukac (1974) found that the maximum conversion was only 8 to 10% when maleic anhydride was reacted with polystyrene in solution to give XXIV. It is not known which carbon atom of the ring system is attached to the polymer backbone but it is likely to be the one attached to the vinylidine group. The maleic anhydride groups can be saponified with alcoholic potassium hydroxide (XXV) and titrated:

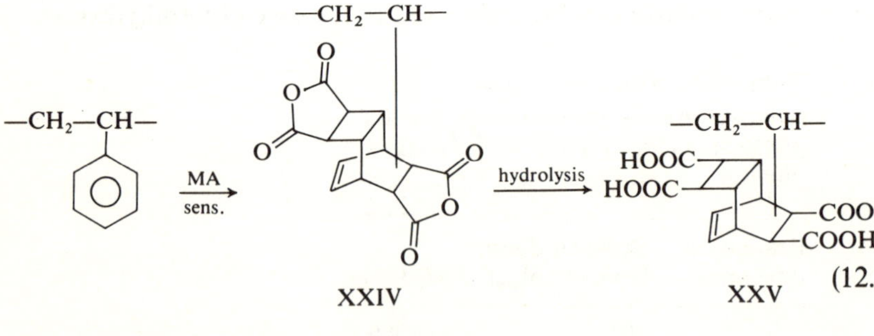

XXIV XXV (12.21)

Schulz (1973) has also reported the photoaddition of maleamide to polystyrene with a conversion of only 4% under similar conditions.

The maleic anhydride addition to polystyrene was further studied by Hrdlovic and Lukac (1974). The reaction requires a triplet sensitizer such as benzophenone or acetophenone and it was shown that the efficiency of the sensitizer was up to three times as great if the sensitizer was bonded to the polymer chain. They suggested that the sensitized photoaddition proceeded by an attack of the excited maleic anhydride in its triplet state on the aromatic nucleus and not by the excited state of the charge-transfer complex between maleic anhydride and benzene.

12.2.6 *Diketones with tetramethyl allene*

A further example of the Paterno–Buchi reaction in polymers is the reaction of aromatic diketones with tetramethyl allene reported by Andrews and Feast (1976) to give, among other structures:

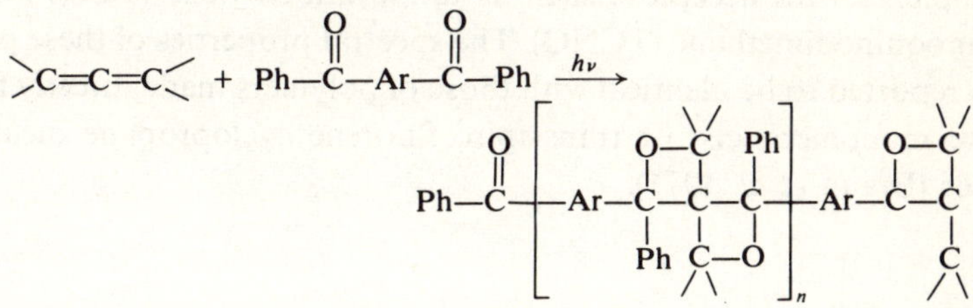

(12.22)

and these authors have also reported similar reactions using furan as the diolefin.

12.2.7 *Reactions of carbenes with unsaturated polymers*

Another type of photocyclization reaction involves the reaction of a carbene with an unsaturated system. For example, Schulz (1973) has reported the reaction of diazofluorene with an unsaturated polyester prepared from the copolymerization of maleic anhydride and ethylene glycol:

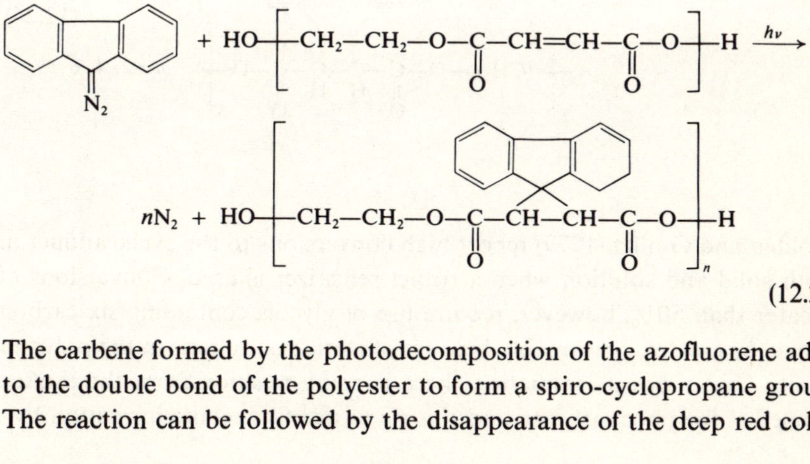

(12.23)

The carbene formed by the photodecomposition of the azofluorene adds to the double bond of the polyester to form a spiro-cyclopropane group. The reaction can be followed by the disappearance of the deep red color of the diazofluorene and the evolution of nitrogen, and conversions of 90% of the double bonds were reported. Because of the spirane structure, the plane of the fluorene group in the polymer is perpendicular to the main chain which should reduce the flexibility of the polymer considerably. As a result, the melting or softening range of the polymer is raised from about 25 °C to 140–180 °C, depending upon the degree of substitution. Because photoisomerization of the double bonds in the polyester takes place during the photoaddition, polymeric photoadducts contain both cis- and trans-cyclopropane units. Since the fluorene group is a strong π-donor, these polymers can form strong polymer charge-transfer complexes with acceptors such as tetracyanoethylene (TCNE) or tetracyanoquinodimethane (TCNQ). The spectral properties of these polymers are reported to be identical with those of polymers made directly from the pure monomers, cis- or trans-spiro fluorene cyclopropane dicarboxylic acids (Braun *et al.* 1971).

12.2.8 *Photoadditions of phenanthrene*

Another photoreaction which proceeds with high efficiency in maleic or fumaric polyesters is the photocycloaddition of phenanthrene.

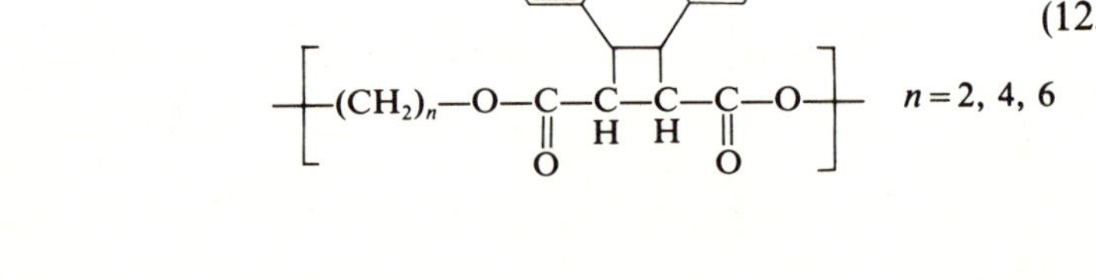

(12.24)

Holden and Guillet (1979) report high conversions to the cyclic adduct in both solid and solution when a triplet sensitizer is used. Conversions of greater than 50%, however, require use of glycols containing six carbon atoms or greater, presumably because of steric restrictions in the polymer chain. Under direct irradiation there is some reaction from the excited singlet of the phenanthrene and this gives rise to oxetane formation by

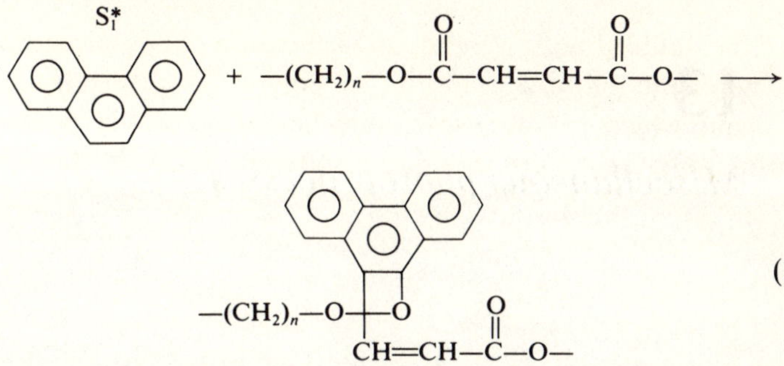

(12.25)

When a triplet photosensitizer is used, this reaction appears to be one of the cleanest so far studied in polymer systems and, in fact, it appears to be more effective in the polymer than in small molecule analogs, possibly because side reactions, such as hydrogen abstraction, are disfavored by the high concentration of groups in the polymer chain.

Estimates of the amount of oxetane adduct in the polymer products were obtained from changes in the uv spectrum, glass transition temperature, and intrinsic viscosity of the products following treatment with methanolic HCl. Since even prolonged treatment with methanolic HCl failed to produce a marked decrease in T_g or $[\eta]$, it appears that the oxetane is only a minor product. This result is in contrast with the phenanthrene-dimethyl fumarate photocycloaddition, where oxetane and cyclobutanes are formed in comparable yields in the unsensitized reaction.

The precursor to the oxetane adduct appears to be a weakly emitting singlet exciplex between phenanthrene and dimethyl fumarate. Quenching of phenanthrene fluorescence and growth of exciplex emission were observed with both dimethyl and diethyl fumarate in several solvents.

12.2.9 *Summary*

As these examples show, photocyclization reactions in polymers can be made to proceed with high efficiency. Structures can be produced which would be difficult or impossible to make by other synthetic methods and the possibility exists of making marked changes in polymer properties in localized regions by exposure to patterns or other optical devices. It seems likely that the potential of this type of photochemistry is considerable for a wide variety of practical applications and one might expect considerable research activity in this field in the future.

13

Miscellaneous photoprocesses

13.1 Photoconductivity in polymers

One of the most interesting recent applications of polymeric materials relates to their use in xerography. This resulted from the discovery that polymers, when exposed to light, develop charged species which can migrate under an electric field and thus conduct electricity. In the absence of light most polymers, with the exception of those containing long sequences of double bonds, are fairly good insulators and conduct charge only with difficulty. However, a surprising number of common vinyl and condensation polymers show a substantial increase in conductivity when exposed to light. These include poly(vinylfluoride), poly(vinylacetate), poly(vinylalcohol) and poly(N-vinylcarbazole) (PVCz). By doping PVCz with dyes, particularly with compounds which form charge-transfer complexes, this photosensitivity can be increased and extended into the visible region of the spectrum.

13.1.1 *Photoconductivity of poly(vinylcarbazole)*

In this chapter we will deal primarily with studies of PVCz. A more detailed discussion of PVCz and other polymers is given by Patsis and Seanor (1976).

The conductivity σ is equal to the current density J divided by the applied field strength ε where J is in ampere/unit electrode area. This will be related to n, the number of negative-charge carriers (usually electrons) per unit volume, and p, the number of positive-charge carriers (or positive holes) per unit by

$$\sigma = J/\varepsilon = ne\mu_n + pe\mu_p \tag{13.1}$$

where e is the charge on the electron, and μ_n and μ_p are the mobilities of the negative and positive carriers, respectively. A typical experimental set-up is shown in figure 13.1. A thin film of PVCz is coated on an aluminium substrate and a thin, conducting, but transparent film of aluminum is evaporated on the upper surface. A potential gradient is established across the two conducting surfaces of the film and when the film is exposed to a flash of light the passage of current can be displayed on an amplifier. From such measurements one can determine both photoconductivity and the mobility of the charge-carrying species using the relation

$$\mu = d^2/Vt \tag{13.2}$$

where d is the thickness of the film, V is the applied voltage and t is the carrier drift time. This simple theory assumes that the density of injected carriers is low and the field in the insulator is essentially unaffected by the injected charge. At higher light intensities the field at the injection electrode may fall to zero due to space-charge limitations.

Theory of photoconductivity. The photoeffect is usually evaluated in terms of the effective gain G, which is defined as the number of generated carriers reaching the external circuit per unit time compared to the number of photons absorbed during the same time:

$$G = J_p/eI_0(1-T)A \tag{13.3}$$

where J_p is the photocurrent, e the electronic charge, I_0 the number of incident photons per $cm^2\,s^{-1}$, T the optical transmittance of the film and A the area of sample illuminated.

13.1.2 *Mechanism of carrier generation in PVCz*

In PVCz the primary mechanism for charge-carrier generation appears to be due to the excitation of the carbazole rings to their first

Figure 13.1. Typical experimental arrangement used for transient photoconductivity studies. Evaporated film of aluminum serves as the illuminated, semitransparent electrode.

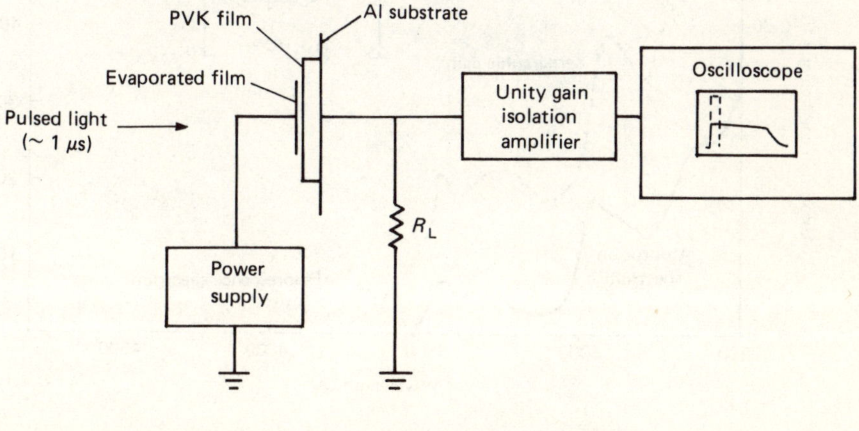

excited singlet states. The singlet excited state itself is not a conductive species but can be considered as an associated electron–positive hole pair which can migrate through the solid. For conduction to take place this exciton pair must dissociate to produce a separated electron and a positive hole in such a way that the electron ends up in the conducting state. This involves the acquisition of more energy and can be accomplished, for example, by exciton-surface interactions (Goodwin and Lyons 1967). These have been extensively studied in single crystals and it is reported that other processes which can also induce an electron to enter the conducting state are singlet-singlet interactions (Braun 1968), singlet–triplet interactions (Fourny *et al.* 1968), singlet–photon interactions (Kepler 1967), triplet–photon interactions (Holtzman *et al.* 1967), and two-photon interactions (Strome 1968).

Because of the relationship to the singlet absorption in the first mechanism of photoconductivity it is to be expected that the response of the photoconductor to the wavelength of the light flash will bear a very simple relationship to the absorption spectrum of the polymer. Figure 13.2 shows that the wavelength response of the xerographic gain is almost

Figure 13.2. Absorption spectrum, fluorescence spectrum and photocurrent spectrum (expressed as xerographic gain) for a 7.6 μ poly-N-vinylcarbazole film. Applied voltage 630 V; illuminated surface, positive. (After Regensburger 1968.)

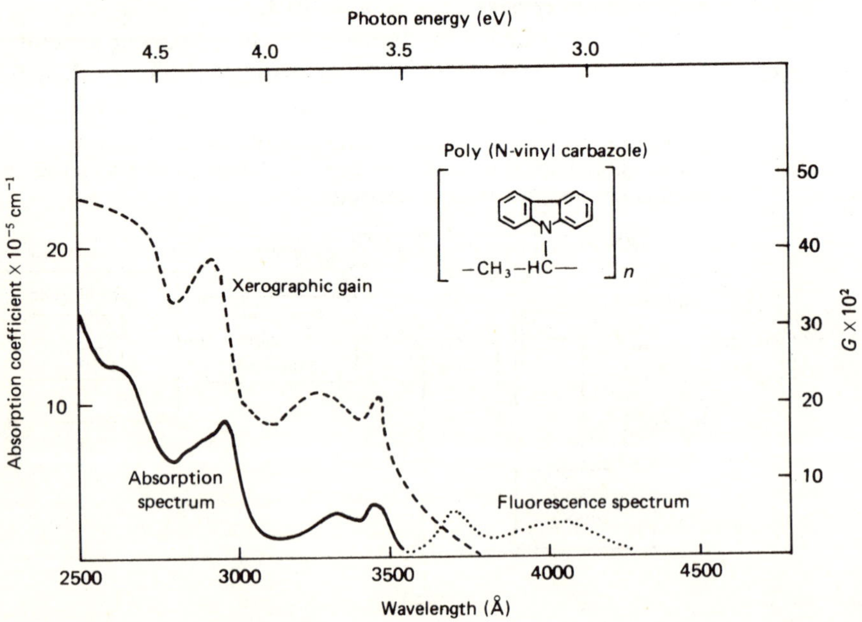

identical in shape to the absorption spectrum. Similar effects have been observed with pyrone polymers and polyacetylenes as well as with poly(acenaphthalene). Based on fluorescence quenching studies, Klöpffer (1969) concluded that exciton migration is the most probable energy transfer mechanism for PVCz, and that the exciton visits about one thousand monomer units during its lifetime, corresponding to an average random walk distance of about 200 Å. The photocurrent action spectrum shown in figure 13.2 was explained by Bauser and Klöpffer (1970) as arising from interactions of the singlet excitons with trapped holes.

13.1.3 *Sensitization of photoconductivity*

The practical application of PVCz to xerography resulted from the discovery that the photocurrent sensitivity could be extended into the visible range by the addition of dyes or other compounds capable of inducing charge-transfer complexes with PVCz. These represent two separate cases, the first, often called 'optical sensitization', reproduces the visible absorption spectrum of dye (figure 13.3*a*) and the process presumably occurs by energy transfer from the excited state of the dye molecule to that of the carbazole unit of the polymer by mechanisms described in chapter 9.

The second type of sensitization is illustrated in figure 13.3*b* where neither the donor nor acceptor absorbs in the visible and the response is due to the absorption of the charge-transfer complex between the two.

Figure 13.3. Effect of sensitizers on photoconduction action spectrum in polymers. (From Reucroft 1976.)

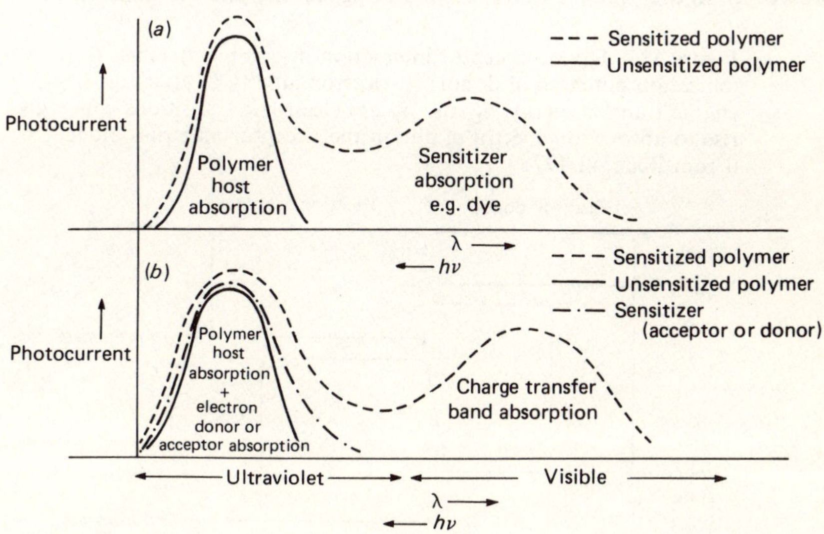

The energy levels are illustrated in figure 13.4. PVCz itself is a good electron donor and hence it should be doped with an electron acceptor in order to form a charge-transfer complex. A good electron acceptor is a material with a high electron affinity, in other words, a large amount of energy is released when the material acquires an electron.

In addition to the usual electronic excitation of the individual donor and acceptor, the complex provides other modes of electronic excitation in which electrons can be excited from states in the donor molecule to states in the acceptor molecule. This usually requires lower photon energies and hence can occur at longer wavelengths. This type of sensitization is sometimes referred to as chemical sensitization.

Detailed studies on PVCz film doped with 1.6 mol-% 2,4,7-trinitro-9-fluorenone (TNF, I) were first described in 1967 (Lardon *et al.* 1967).

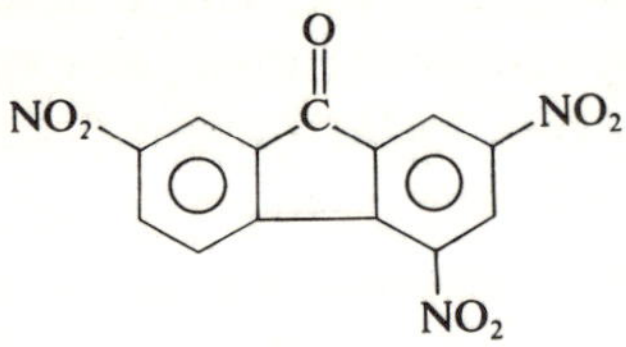

TNF I

Later it was reported (Shattuck and Vahtra 1969, Schaffert 1971) that increasing the TNF content increases the sensitization and that PVCz–TNF films containing 1:1 molar ratio PVCz to TNF monomer units display the maximum sensitization effect. The PVCz–TNF system is of particular interest because of its high photosensitivity to visible light, as shown in figure 13.5. Figure 13.6 compares its photosensitivity with

Figure 13.4. Donor–acceptor interaction in organic systems: I, ionization potential of donor; A, electron affinity of acceptor; $h\nu_{CT}$, charge transfer energy. S_D and S_A are electronic transitions which give rise to absorption spectra of donor and acceptor materials alone. (From Reucroft 1976.)

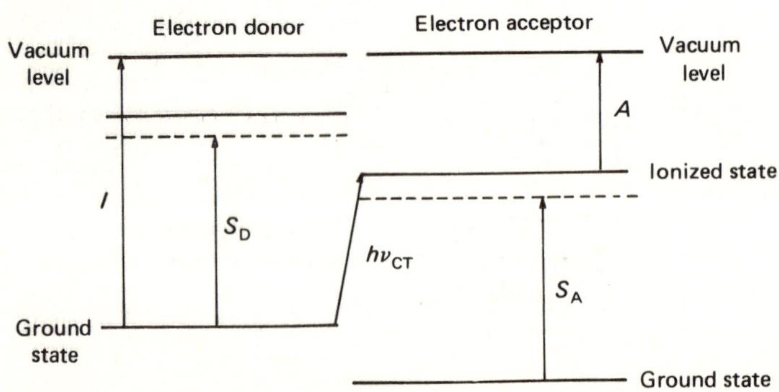

commonly used inorganic photoconductors such as amorphous selenium and ZnS–CdS. It is interesting to note that charge-transfer complex crystals of N-isopropylcarbazole with TNF do not give photoconduction response in the charge-transfer band which implies that traps unique to the polymer are responsible for the effect in the polymer film.

Based on studies of the photocarrier generation efficiency in PVCz–TNF Melz (1972*a*, *b*) found that the carrier generation efficiency was the same for electrons and holes and depended strongly on the electric field.

Figure 13.5. Photoconductive gain *G* and absorbance for poly(N-vinylcarbazole) and poly(N-vinylcarbazole) + 1.6% 2, 4, 7-trinitrofluorenone. (From Lardon *et al.* 1967.)

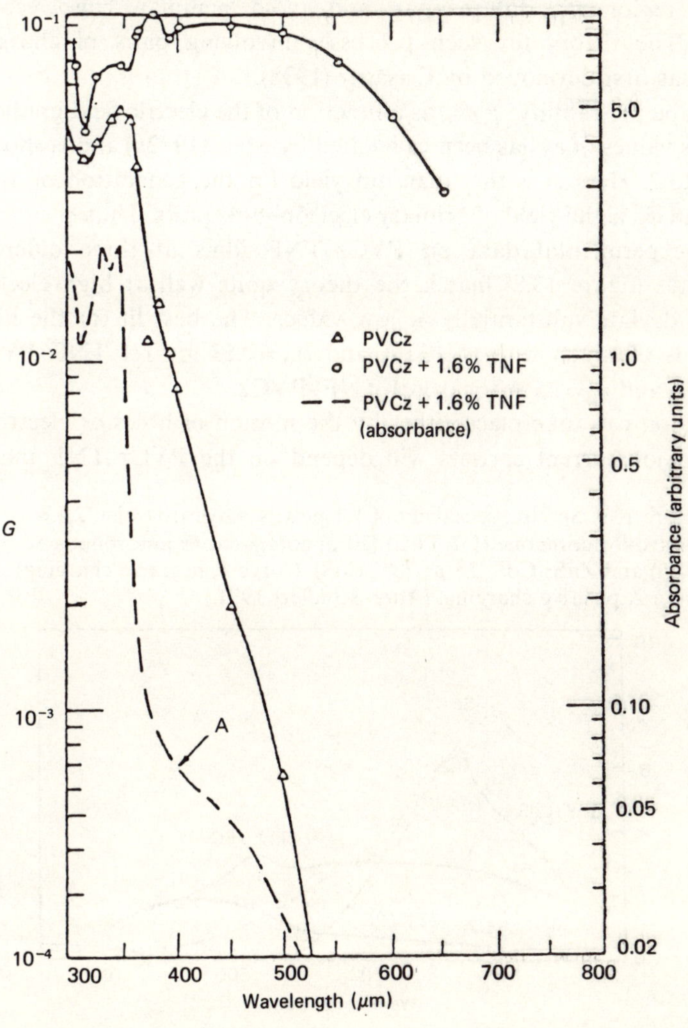

The generation process was interpreted as taking place in two steps, the first being the promotion of an electron from its ground state to an excited bound state, presumably the singlet. From this it undergoes autoionization into a continuum or conducting state, or may decay back to the ground state. If ionization occurs, an electron–hole pair is created in a region of space defined by a radius r_0, the distance at which the coulombic attractive force between the electron and the positive hole is equal to the thermal energy kT, thus,

$$r_0 \approx 2e^2/3\kappa kT \tag{13.4}$$

where κ is the dielectric constant. In the presence of a strong electric field, there is a finite probability that an electron–hole pair originally generated in such a region will diffuse apart and avoid 'geminate cage recombination'. The theory for such processes involving pairs of charged particles was first developed by Onsager (1938).

The escape probability, ϕ/ϕ_0, as a function of the electric field gradient for various values of r_0 has been calculated by Melz (1972*b*) and is shown in figure 13.7. Here ϕ is the quantum yield for the generation of 'free' carriers and ϕ_0 is the yield of primary electron–hole pairs. Thus $\phi/\phi_0 = G$.

Actual experimental data on PVCz–TNF films at three different temperatures (figure 13.8) match the theory quite well at high electric fields, but deviate substantially at low values. The best fit for the high field data is obtained with $r_0 = 35$ Å and $\phi_0 = 0.23$ for 1:1 TNF–PVCz (figure 13.7) and $r_0 = 25$ Å for 0.06:1 TNF–PVCz.

Conduction can take place either by the motion of holes or electrons and the major current carriers will depend on the PVCz–TNF molar

Figure 13.6. Spectral sensitivity of 1:1 poly(N-vinylcarbazole): 2,4,7-trinitro-9-fluorenone (1:1 TPC) (20 μ) compared to amorphous Se (50 μ) and ZnS: CdS (25 μ; 70% CdS). Curve 1, negative charging; curve 2, positive charging. (After Schaffert 1971.)

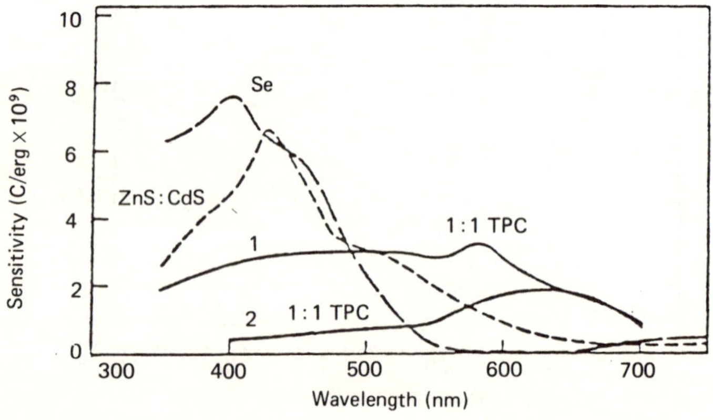

ratio. At low values of this ratio, below about 0.2, the conductivity is primarily by the movement of positive holes, whereas at high values most of the conductivity is due to electrons. This is shown in figure 13.9.

The behavior shown by the PVCz–TNF system can be duplicated with a number of other polymers and although the major application of this phenomenon at present is primarily in the field of xerography it seems likely that other applications will develop in the field of switching and microcircuitry, and possibly for photographic purposes.

13.2 Mechanophotochemistry

The chemical changes induced in organic molecules by the absorption of light can also induce changes in the mechanical properties of these materials. For example, in chapter 12 we have described the changes in the glass transition of certain polymers as a result of photocyclization reactions which stiffen the chain of a polymer molecule and alter its state from that of an amorphous rubber to a glassy solid.

Changes in molecular structure of groups in polymers can induce contractions or expansions both on a microscopic or molecular scale and

Figure 13.7. Ratio of ϕ to ϕ_0 vs. electric field for various initial separations r_0 and experimental data for TNF–PV (0.06:1). (From Melz 1972*b*.)

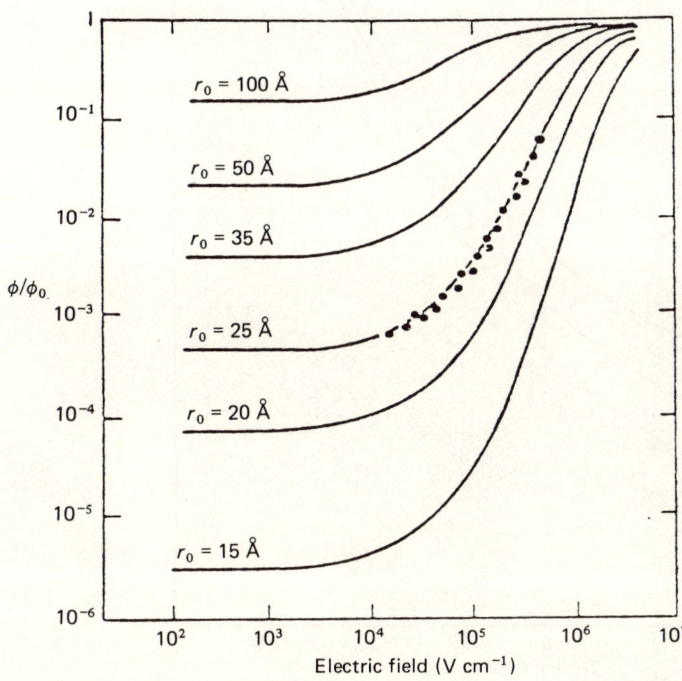

a macroscopic scale. For example, Agolini and Gay (1970) prepared polymers containing azo groups which can undergo cis–trans isomerization when exposed to light. The equilibrium distribution between cis and trans isomers in the dark is displaced by exposure to light, which results in a contraction of the material by about 0.5% of its original length:

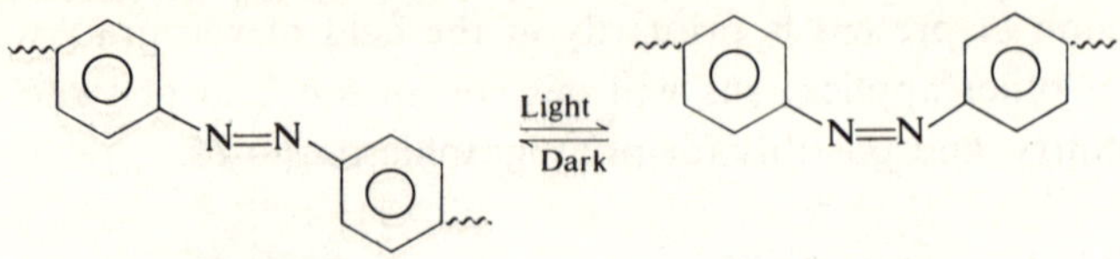

trans II cis III

Figure 13.8. Experimental results (points) and theoretical predictions (solid curve) for the electric field dependence of the photocarrier generation efficiency in TNF doped PVCz films (0.06:1 mole ratio of TNF to PVCz monomer unit) at three temperatures. Theoretical computations use $r_0 = 25$ Å and $\phi_0 = 0.23$. (After Melz 1972*a*.)

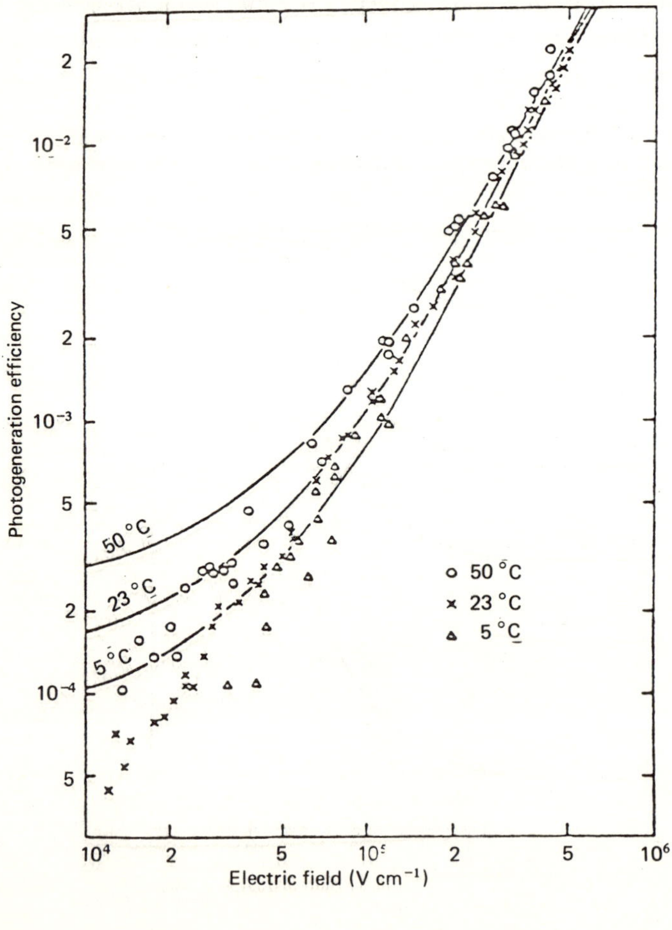

The material was shown to resume its original length on standing in the dark.

In azobenzene itself, the free energy difference between the higher energy cis form and the trans is about 40 kJ mol^{-1} and the activation energy for thermal cis-trans interconversion is 92 kJ mol^{-1}. Similar transitions are believed to be an essential element in the process of mammalian vision (Wald 1968, Heller 1972). Bieth *et al.* (1969) attached azobenzene groups to the enzyme acetyl cholinesterase using the compound *p*-phenyl-azophenyltrimethyl-ammonium chloride (PTA, IV) and

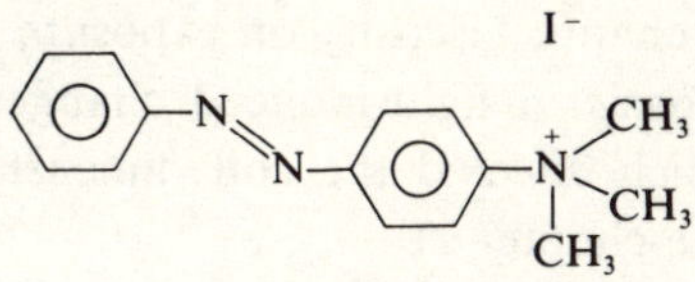

PTA IV

showed that the cis isomer is a less effective inhibitor of the enzyme than the trans, and that the two forms could be interconverted in a reversible manner with light of different wavelengths. In principle this provides a means of controlling enzyme activity with light.

Van der Veen and Prins (1971) incorporated the azo dye chrysophenin (CHP, V) into a polymeric membrane which, after 10 minutes

Figure 13.9. Dependence of hole and electron mobility on TNF–PVCz mole ratio. (From Gill 1972.)

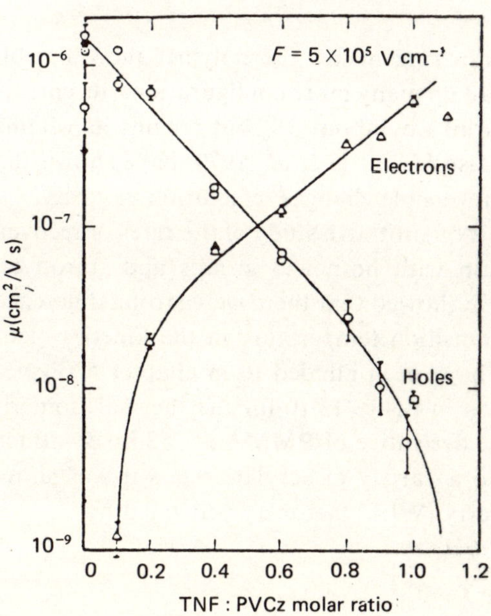

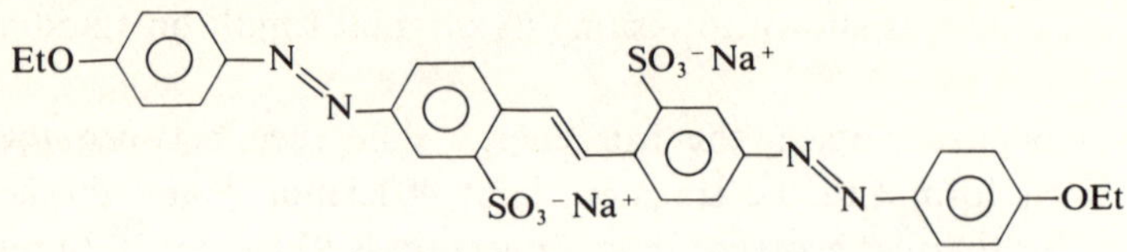

CHP V

irradiation in light, contracted by 1 to 2%. Again, the material recovered its original length in the dark. However, these processes are relatively slow. Lovrien (1967) had reported that V forms complexes with both poly(vinylpyrrolidone) (PVP) and poly(methacrylic acid) (PMAA) in aqueous solution which reversibly change viscosity on exposure to light. The effect was explained on the basis of light-induced changes in the conformation of the polymers which affected the coil dimensions and hence the intrinsic viscosity $[\eta]$ (see chapter 2).

Smets (1972) synthesized a series of methacrylate esters with spiropyran groups VI in the backbone and side chains. The rearrangement of the dye

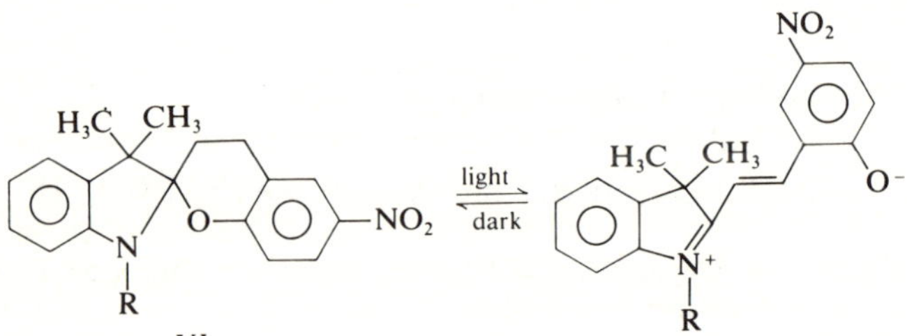

VI

which opens up the spiropyran ring makes the polymer more flexible. Since the polymer can now take up many more configurations, its entropy increases and the polymer shrinks by about 3% but regains its original dimensions in the dark. (See also §5.7.) Irie *et al.* (1979) have shown that such polymers also undergo reversible changes in solution viscosity.

Eisenbach (1980) has made a quantitative study of the rates of recovery of the equilibrium distribution with both azo groups and spiropyran polymers in the solid phase. He showed that there were strong differences above and below the glass transition temperature in the kinetics of the photoreaction. These results have been alluded to in chapter 5. Typical experimental results are shown in figure 13.10 for the thermal isomerization of a 4-aminoazobenzene derivative of PMMA at 283 K. By attaching the same chromophore to a variety of acrylate esters it was shown that the Williams–Landell–Ferry (WLF) factor a_T as defined by

$$\ln a_T = \frac{-B}{f_g}\left(\frac{T-T_g}{(f_g/\Delta a)+T-T_g}\right) \tag{13.5}$$

gives a smooth composite curve for the entire series of esters, as shown in figure 13.11. Here f_g is the fractional free volume at the glass transition temperature and Δa is the difference in thermal expansion coefficients of the liquid and glassy polymer. The change of slope at the glass transition temperature is characteristic of such plots and is taken as evidence for the important role of free volume in controlling the rate of recovery to the equilibrium distribution of conformations.

Aviram (1978) has reported a gel that expands by 35% in linear dimensions when exposed to light. Poly[*p*-(N,N-dimethylamino)-N,γ,D-glutamanalide) (VII) was prepared and crosslinked after coating in a thin film. When this film is exposed to a non-reactive liquid containing a certain amount of carbon tetrabromide the film swells to form a gel. The material expands when irradiated with uv light of 360 nm due to photoionization of the charge-transfer complex formed between the functional groups on the polymer and carbon tetrabromide. An electron leaves the functional group on the polymer chain to become attached to one of the bromine atoms in the carbon tetrabromide, leaving the

Figure 13.10. First-order plot for the thermal cis-trans isomerization of 4-amino-azobenzene derivative as pendent group (0.1 mol %) in poly(methylmethacrylate) film at 283 K. (From Eisenbach 1980.)

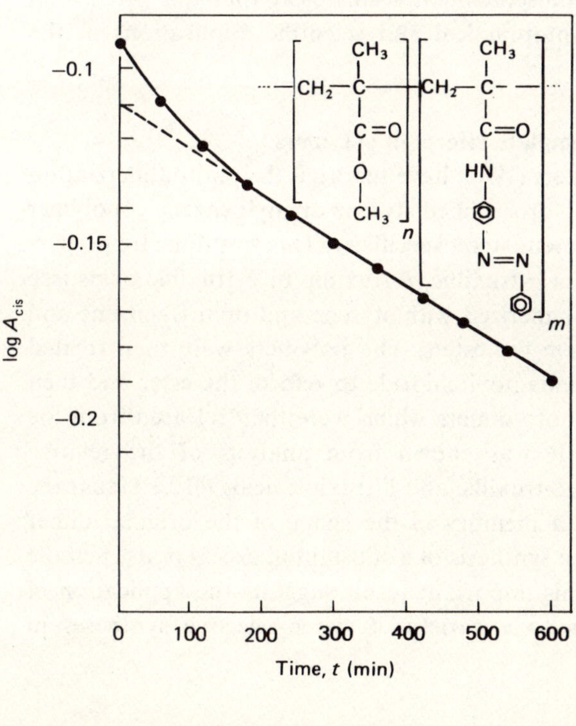

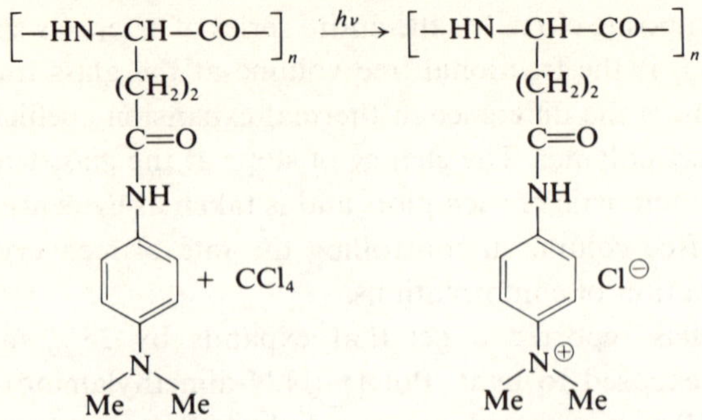

VII VIII

polymer side chains positively charged. The repulsion of the positive charge caused the chains to expand so that more liquid molecules penetrate the gel to minimize the electrostatic repulsion between the chains. This caused a swelling of the gel which gives rise to up to 35% changes in the linear dimension (figure 13.12).

Aviram suggests that these new photomechanical gels could be used for making relief images in 3 D photography or for light-activated actuators, triggers and relays. It should also be possible to control the permeability of a membrane with light. The mechanism for expansion and contraction of the gel caused by charge transport is very similar to that postulated to occur in mammalian muscle and it seems clear that this interesting effect may have important practical and scientific applications in the future.

13.3 Photochemical template effects in polymers

Damen and Neckers (1980) have reported the photodimerization of trans-cinnamic acid in a crosslinked styrene divinylbenzene copolymer containing holes with different stereospecificity. This was done by preparing polymerizable esters of α-truxillic, β-truxinic or δ-truxinic acids (see §12.1.1) which were copolymerized with styrene and divinylbenzene and then hydrolyzed to remove the esters. The polymers were then treated with an excess of trans-cinnamoyl chloride to reform the ester and then photolyzed to produce photo-dimers which were then released from the polymer by hydrolysis. It was shown from analysis of the relative proportion of the dimers, α-truxillic and β-truxinic acids (§12.1.1) that the polymer matrix retained a memory of the shape of the original dimer present and resulted in the synthesis of a substantial excess of it when the reaction was repeated. This important result suggests the application of polymeric photocatalysts to a variety of stereo-selective syntheses in organic chemistry.

13.4 Photochemistry of polystyrene

The photochemical reactions induced in solid films of polystyrene upon irradiation are complex. In the earliest studies (Grassie and Weir 1965) and in subsequent work (Fox 1967, Rabek and Ranby 1974) it has been shown that the major processes observed are the formation of crosslinks, chain scission, development of unsaturation in the chain, and photoisomerization to substituted fulvenes. More recently, Weir and Milkie (1979) extended these studies to those of the *p*-halogenated styrenes, *p*-fluoro, *p*-chloro and *p*-bromo styrene. Other studies have been made of poly(*p*-methyl styrene) (Weir 1973) and *p*-2-tert-butylstyrene (Weir *et al.* 1979). Because of the complexity of determining the exact structure and concentration of the complex structures produced in polystyrene the most

Figure 13.11. WLF-plot for the thermal cis-trans isomerization of 4-aminoazobenzene derivatives in polyacrylates and -methacrylates. (From Eisenbach 1980.)

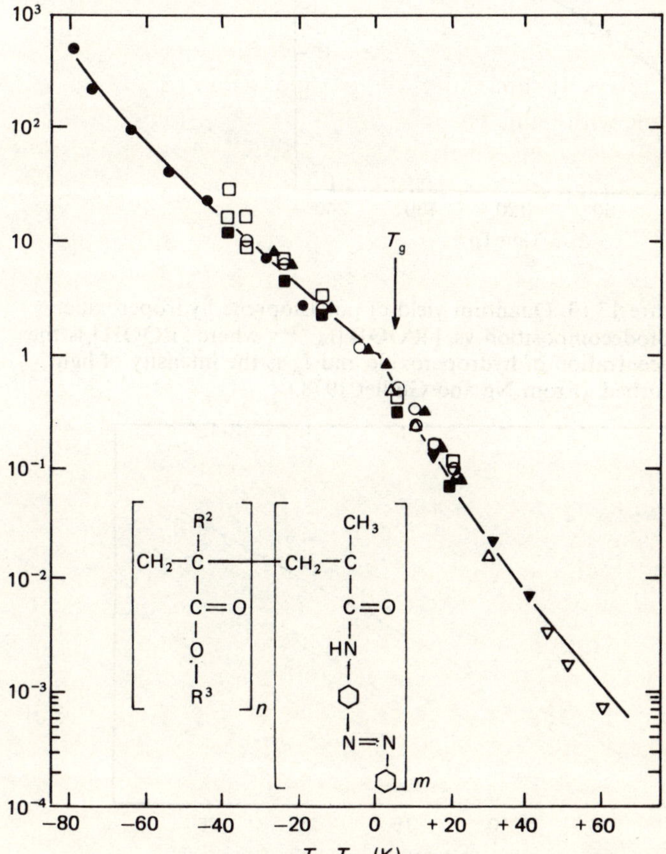

quantitative results are those for gas evolution, either hydrogen or HX, and crosslinking and scission.

The relative rates of chain scission and crosslinking observed by Weir and Milkie are shown in table 13.1. It can be seen that scission rates are comparable in polystyrene and the three *p*-halogenated styrenes but crosslinking efficiency is increased dramatically in the chloro- and bromo-substituted polymers. A similar effect is shown in the yields of hydrogen and HX, indicated in table 13.2. The high yields of crosslinking

Figure 13.12. Plot of observed deformations of gels from VII versus exposure time. l_0 represents initial length and l represent final length. (From Aviram 1978.)

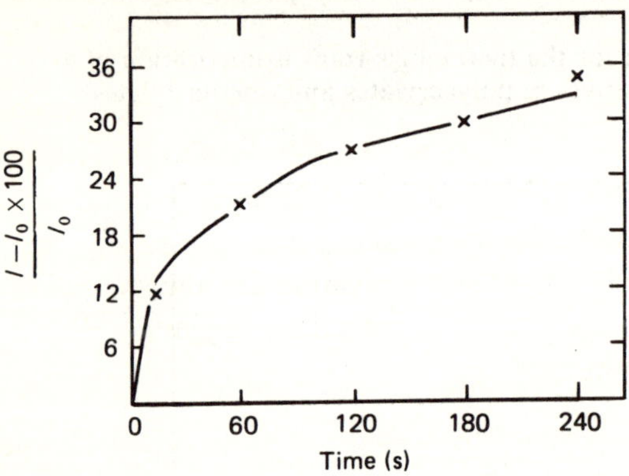

Figure 13.13. Quantum yield of poly(isoprene hydroperoxide) photodecomposition vs. $[R'OOH]I_A^{-1/2}$, where [ROOH] is the concentration of hydroperoxide and I_A is the intensity of light absorbed. (From Ng and Guillet 1978.)

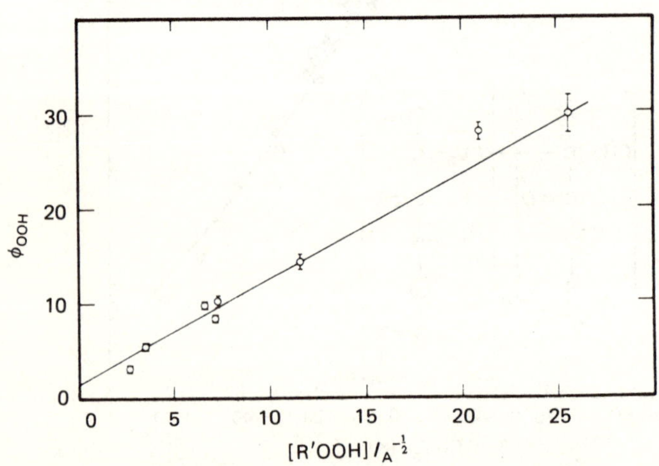

suggest that at least some of the crosslinks that are formed by dissociation of the halogen atoms to give a phenyl radical. The relatively low values of the quantum yields for hydrogen formation is a demonstration of the fact that polystyrene is relatively stable to direct photolysis and that most of the photon energy absorbed is dissipated by photophysical processes. For this reason, the photophysics of polystyrene and other vinyl aromatic polymers retains more scientific interest than their photochemistry (see chapters 6, 7 and 8).

13.5 Photochemistry of hydroperoxides

The hydroperoxide group is a central intermediate in the photooxidation of macromolecules. In spite of its importance in this respect, very few quantitative studies have been made of its photochemistry in polymer systems.

In early work, Martin and Norrish (1953) showed that tertiary butyl hydroperoxide, which can be considered a model for the hydroperoxide groups introduced on the tertiary carbons of polypropylene and low density polyethylene, decomposes rapidly when exposed in solution to uv light. The quantum yields for peroxide decomposition are always greater than unity and increase with peroxide concentration, suggesting that a chain mechanism is involved:

Table 13.1. *Rates of chain scission and crosslinking of poly(*p*-halegeno styrene)s*

Polymer	Scissions per monomer unit ($h^{-1} \times 10^4$)	Crosslinks per monomer unit ($h^{-1} \times 10^4$)
Poly(*p*-fluorostyrene)	0.98	1.2
Poly(*p*-chlorostyrene)	1.1	4.5
Poly(*p*-bromostyrene)	1.3	13.0
Polystyrene	1.44	1.8

Table 13.2. *Quantum yields for product formation during the photolysis of poly(*p*-halogeno styrene)s*

Polymer	$\phi_{H_2} \times 10^4$	$\phi_H \times 10^4$
Poly(*p*-fluorostyrene)	6.0	0.6 (HF)
Poly(*p*-chlorostyrene)	24.0	30.0 (HCl)
Poly(*p*-bromostyrene)	60.0	20.6 (HBr)
Polystyrene	6.0	0.0

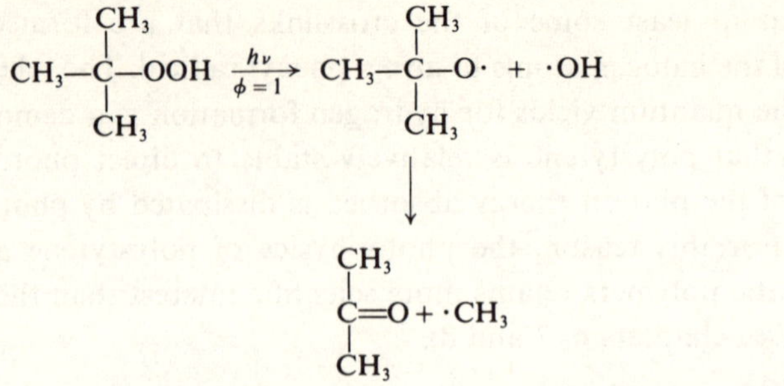

(13.6)

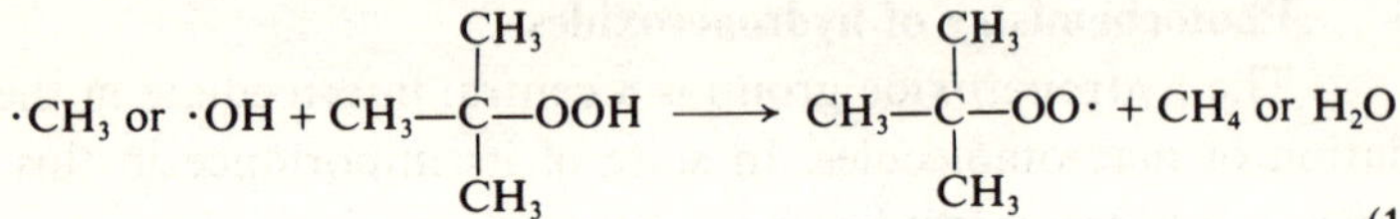

(13.7)

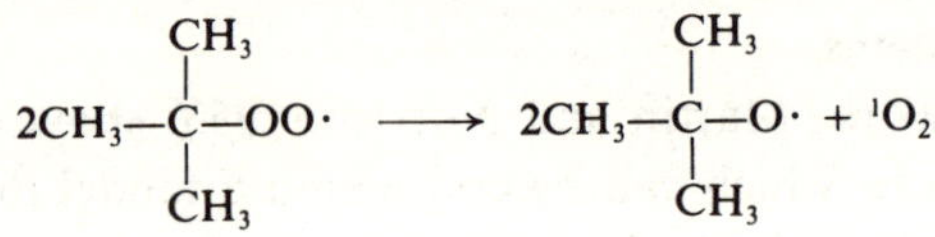

(13.8)

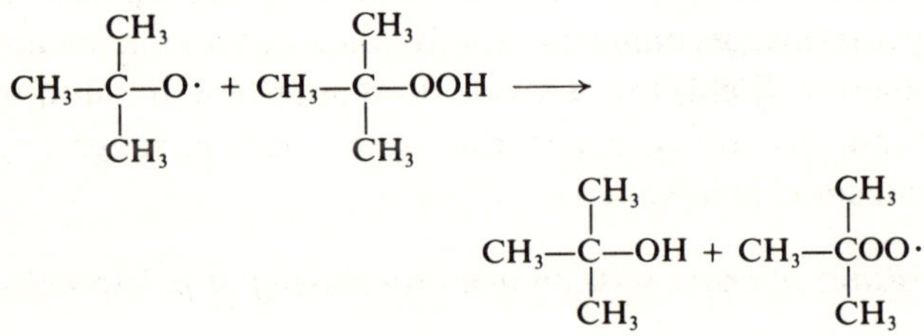

(13.9)

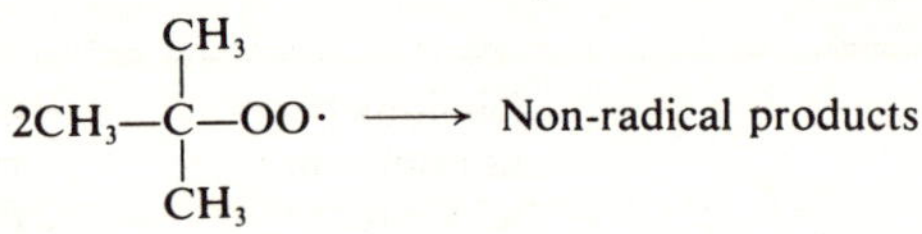

(13.10)

The major products of the photodecomposition are acetone, methane and t-butyl alcohol, if the photolysis is carried out in a non-interacting solvent.

The variation in quantum yield with concentration (Guillet 1978) is shown in table 13.3. Ng and Guillet (1978) have reported quantitative studies of the photolysis of hydroperoxide groups introduced into polyisoprene by reaction of singlet oxygen

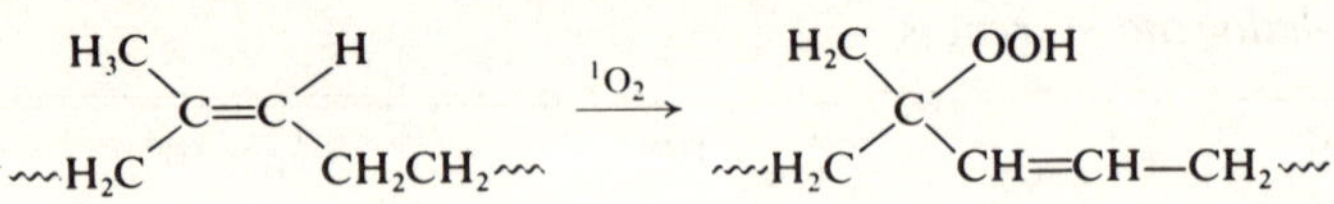

(13.11)

In this case the hydroperoxide is in a position α to a double bond, which makes it somewhat different from that in saturated hydrocarbons. The hydrogen atoms in the allylic position will be readily extractable by any radical species. Nevertheless, the polymer shows qualitatively similar effects. The quantum yield for hydroperoxide decomposition is greater than unity, and depends both on concentration and light intensity (see figures 13.13 and 13.14). The decrease in quantum yield as the light intensity increases is a characteristic expected of a radical chain decomposition.

Chain scission in this polymer occurs with much lower quantum yield, presumably by β scission of the alkoxy radical formed in the primary decomposition of the hydroperoxide, thus

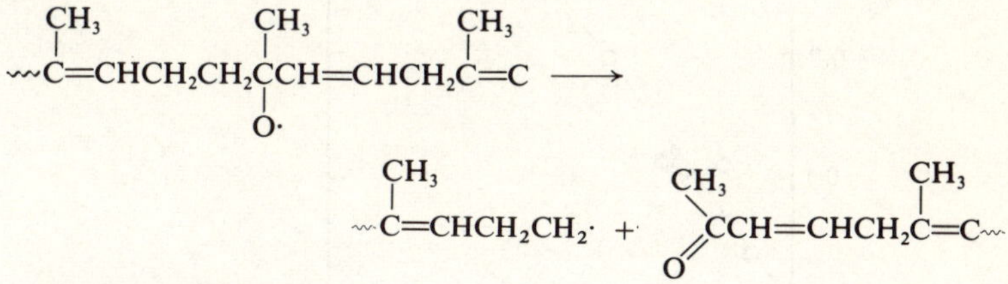

Table 13.3. *Quantum yields for photolysis of* tert-*butyl hydroperoxide in hexane, 36 °C,* $\lambda = 313$ *nm*

C(%)	ϕ
1	3.5 ± 0.2
3	5.2 ± 0.3
5	9.9 ± 0.5

Figure 13.14. Dependence of poly(isoprene hydroperoxide) photodecomposition quantum yield on light intensity: $[OOH] = 2.5 \times 10^{-3}$ mol/l; 30°C at 313 nm; in 1,2-dichloroethane. (From Ng and Guillet 1978.)

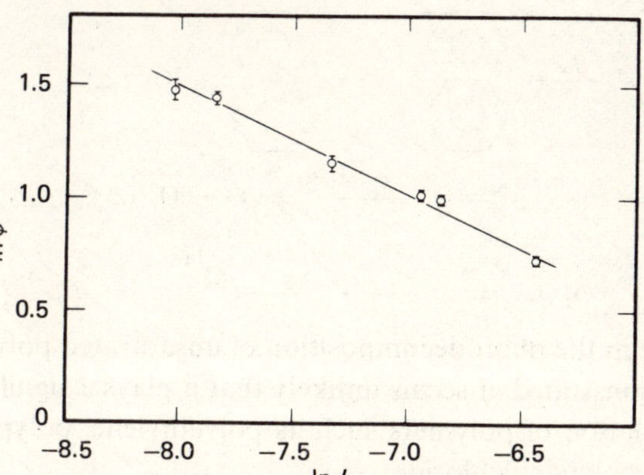

Figure 13.15. Relation between ϕ_{OOH} and ϕ_S for cis-poly(isoprene hydroperoxide). (From Ng and Guillet 1978.)

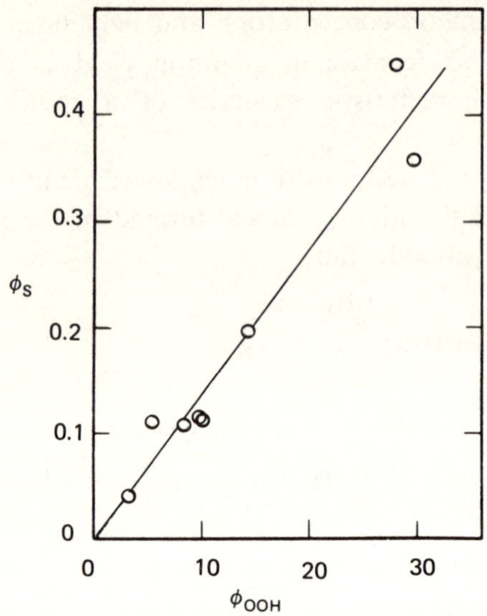

The inefficiency of this process in causing chain scission is illustrated in figure 13.15 which shows that only ~1% of the processes causing peroxide decomposition result in scission of the backbone chain of the polymer.

The ease with which singlet oxygen adds to double bonds has led several authors (Trozzolo and Winslow 1968, Ranby and Rabek 1975) to suggest that it may be a key intermediate in initiating the photodegradation of many polymers. It can be readily generated photochemically by the quenching of photochemically induced triplet states in ketones or aromatic compounds by molecular oxygen:

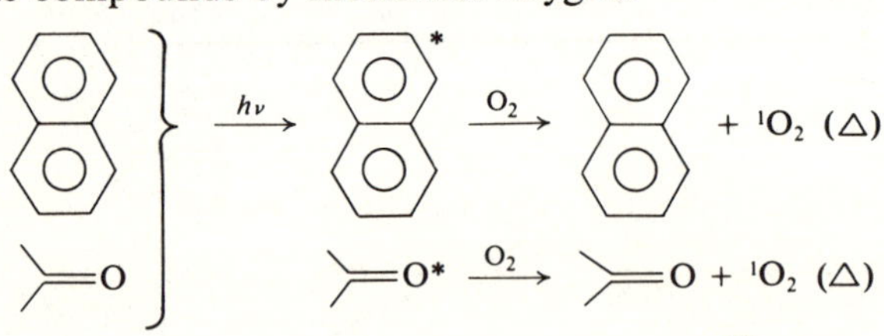

(13.12)

(13.13)

Although its role in the photodecomposition of unsaturated polymers has been clearly demonstrated, it seems unlikely that it plays a significant role in the photooxidation of polymers such as polyethylene, polypropylene, polystyrene and poly(vinylchloride).

14

Photo- and radiation chemistry of polymers

14.1 Introduction

Conventional photochemistry as it has been broadly outlined earlier is concerned with valence or outer shell electronic excitation associated with radiation lying in the approximate wavelength range 200–750 nm. Below 200 nm, higher energy processes become progressively more important; photoionization, for example, becomes predominant at 150 nm and below. In effect, photochemistry ceases at around 90–100 nm and the region below this belongs to the province of radiation chemistry, i.e., that chemistry induced by absorption of α, β, γ or X-rays.

The recent availability of high intensity ir lasers allows experiments in which multiple absorptions in the region above 750 nm can excite organic molecules to 'electronic' energy levels. But this area of study will not be dealt with here.

The field of radiation chemistry of polymers has been investigated extensively over the past three decades, largely as a result of the interest in the ionizing radiation obtained from nuclear reactors and because of the extensive work done on nuclear weaponry and power systems. It is perhaps unfortunate that the early work in this field was done almost exclusively without benefit of information about the photochemistry of the same molecules and in fact this remains a major problem in the interpretation of the results of radiation chemistry, even at the present time.

The field of radiation chemistry of polymers has been reviewed by Chapiro (1962), Ausloos (1968) and Dole (1972, 1973). In this chapter we will not attempt to summarize the field at its present stage but will simply review briefly a number of accounts of experimental investigations which attempt to draw correlations between radiation chemistry and conventional photochemistry of macromolecules. We will restrict our attention to chemical and physical changes caused by electromagnetic radiation, although similar effects can also be observed by the use of high energy particles such as electrons, α-particles and other accelerated nuclei as well as neutrons.

14.2 Absorption of high energy radiation

As the wavelength of radiation is reduced below that of the far ultraviolet the energy of the incident photon increases drastically. Table 14.1 shows the wavelength ranges for various types of high energy radiation and the associated energy per quantum. As a result of the high energy of the X- or γ-ray photons, it is unlikely that the first step of interaction with matter will be with a valence or outer shell electron. In fact, the first step in the absorption of high energy radiation by matter is usually the formation of a high energy ion by collision with one of the inner electrons surrounding an atom and the recombination of this ion gives rise to a stream of Compton electrons. These electrons cascade by collision with other atoms in the system until they reach lower energy levels and can be captured by the outer shell electrons. For example, a single γ-ray photon can give rise to many excitations resulting from the collision of Compton electrons with the matter surrounding the original absorption site. For this reason chemical events induced by high energy irradiation often follow tracks or spurs as the particle moves through the condensed phase. Figure 14.1 shows schematically a typical sequence of events for absorption of a γ-ray in polyethylene.

The interaction of a high energy radiation with a molecule AB may take one of two paths, either (14.1), the primary ionization, or (14.2), some form of excitation.

$$AB \rightsquigarrow AB^{+} + e^{-} \quad \text{(ionization)} \qquad (14.1)$$

$$AB \rightsquigarrow AB^{*} \quad \text{(excitation)} \qquad (14.2)$$

Table 14.1. *Relation between wavelength and photon energy*

Wavelength λ (nm)	Type of radiation	Photon energy, ε_p (eV)	(kcal)	(kJ)
1250	Infrared	1	23	96
125	Ultraviolet	10	230	960
12.5		100	2300	9600
1.25	Soft X-rays	1 000	2.3×10^4	9.6×10^4
0.125		10 000	2.3×10^5	9.6×10^5
0.0125	X-rays	10^5	2.3×10^6	9.6×10^6
0.001	γ-rays (^{60}Co)	1.2×10^6	2.9×10^7	12.1×10^7

The electrons ejected during the ionization will lose their excess kinetic energy by initiation of further ionization and excitation until the electrons finally reach thermal energy. At this point, neutralization of a cation or attachment can occur

$$e^- + AB^+ \rightarrow AB^* \quad (14.3)$$

or attachment to a neutral molecule

$$e^- + AB \rightarrow AB^- \quad (14.4)$$

Cationic species may also have sufficient excess energy to decompose further by one of two mechanisms

$$AB^+ \rightarrow A^+ + B \quad (14.5)$$

$$AB^+ \rightarrow A^+ + B\cdot + e^- \quad (14.6)$$

14.3 Chemical processes occurring in polymers

The radiation chemistry of polymers will depend on the reactions of the ions or radicals created by (14.5) and (14.6). In polymers, not only does radiation induce chemical changes but extensive changes of physical properties also occur as a result of crosslinking or chain scission via

Figure 14.1. Schematic showing events following the absorption of a high energy photon or particle.

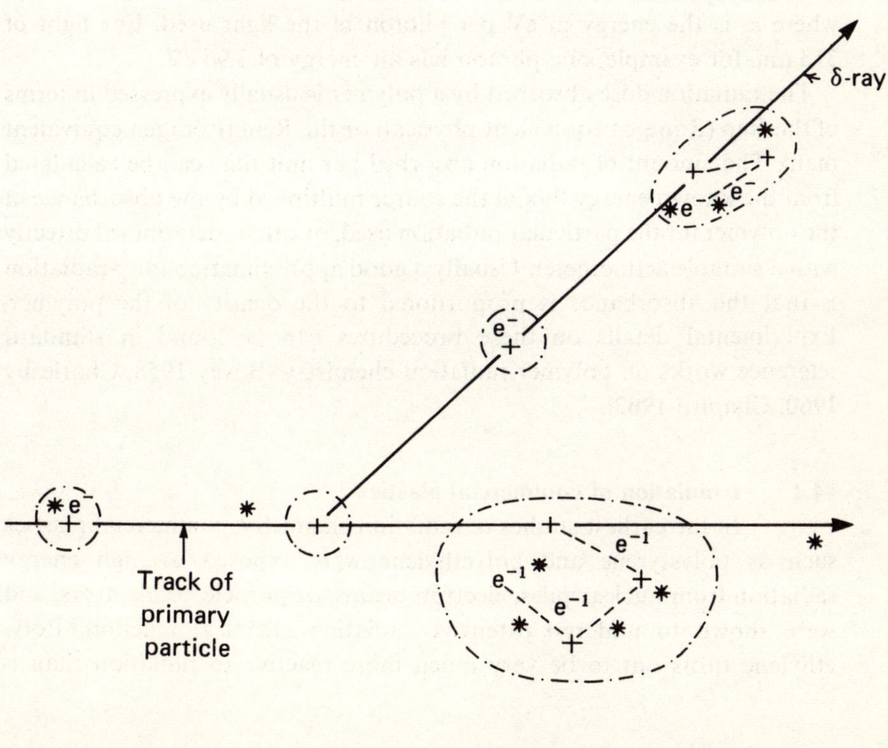

mechanisms similar to those described in earlier chapters (see chapter 4). Concern with these two reactions of crosslinking and scission dominates the earliest studies of the radiation chemistry of polymeric materials. In general, the chemical structure of a polymer determines which of the two processes will predominate.

The efficiency of a radiochemical process is usually described in terms of the G value, which is the number of chemical or physical events per hundred electron volts of energy absorbed (9600 kJ). For a polymer with a normal molecular weight distribution, scission will predominate over crosslinking if the G value for scission, G_s, exceeds the G value for crosslinking, G_x, by a factor greater than four. When crosslinking predominates, the molecular weight rises until a critical dose, called the gel dose, r_g, is reached, above which an insoluble gel network of crosslinked polymer is formed in increasing amounts. As a general rule, vinyl polymers in which each atom of the main chain carries at least one hydrogen atom will tend to crosslink, while, if the polymer contains a tetrasubstituted carbon atom, the polymer will predominantly degrade. There are, however, many notable exceptions to this general rule.

From this definition of the G value, it will be obvious that the relationship of this to a photochemical quantum efficiency is given by:

$$G=\phi(100/\varepsilon_p) \qquad (14.7)$$

where ε_p is the energy in eV per photon of the light used. For light of 313 nm, for example, one photon has an energy of 3.96 eV.

The radiation dose absorbed by a polymer is usually expressed in terms of the Rep (Röntgen equivalent physical) or the Rem (Röntgen equivalent man). The amount of radiation absorbed per unit time can be calculated from the known energy flux of the source multiplied by the absorbance of the polymer for the particular radiation used, or can be determined directly with a suitable actinometer. Usually a good approximation for γ-radiation is that the absorbance is proportional to the density of the polymer. Experimental details on these procedures can be found in standard reference works on polymer radiation chemistry (Bovey 1958, Charlesby 1960, Chapiro 1962).

14.4 Irradiation of commercial plastics

In the earliest studies of radiation chemistry, commercial plastics such as polystyrene and polyethylene were exposed to high energy radiation from nuclear piles, electron beams, or particle accelerators, and were shown to undergo extensive radiation chemical reaction. Polyethylene turns out to be very much more reactive to radiation than is polystyrene, and so these two polymers can be used to represent two general types of polymer reactivity toward high energy radiation.

The major chemical processes in both polymers are crosslinking and the production of molecular hydrogen, with the concomitant formation of unsaturation. It is generally accepted that the primary event is the removal of a hydrogen atom to form a carbon free radical:

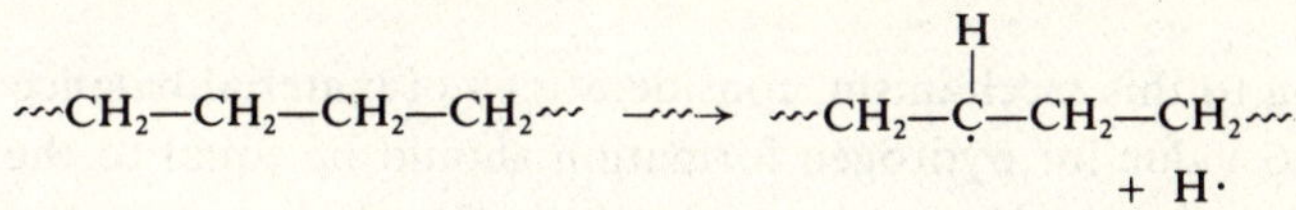

(14.8)

The hydrogen atoms recombine to form molecular hydrogen (14.9) and crosslinks are formed by recombination of the carbon radicals formed on the polyethylene chain (14.10)

$$2H\cdot \longrightarrow H_2 \qquad (14.9)$$

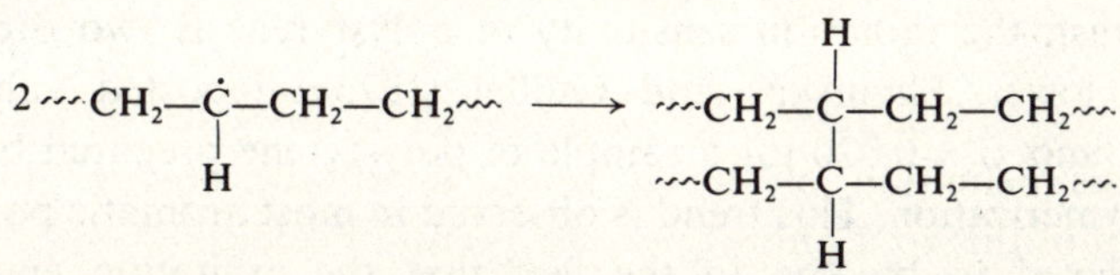

(14.10)

Internal unsaturation (vinylene groups) are formed by disproportionation of two polymeric carbon radicals:

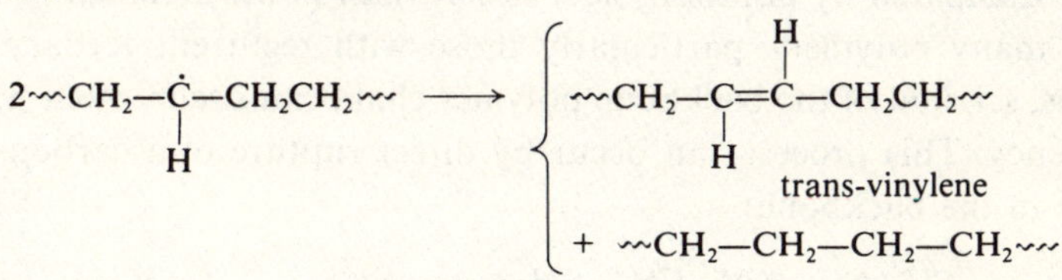

(14.11)

and conjugated double bonds (diene) by removal of a hydrogen adjacent to a preformed double bond to form the very stable allyl radical

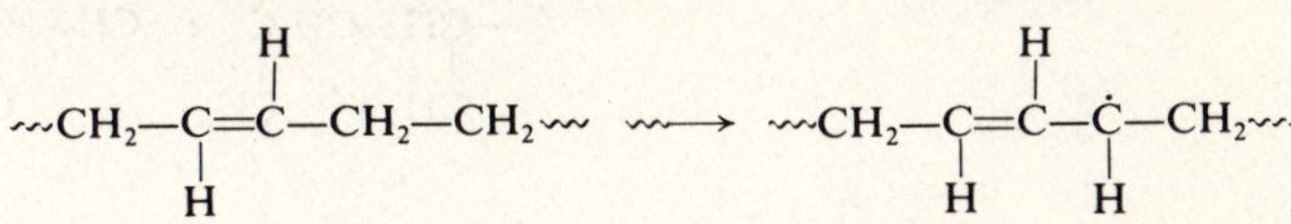

(14.12)

This can be done either by direct absorption of radiation (14.12) or by a radical transfer (14.13) from a primary radical:

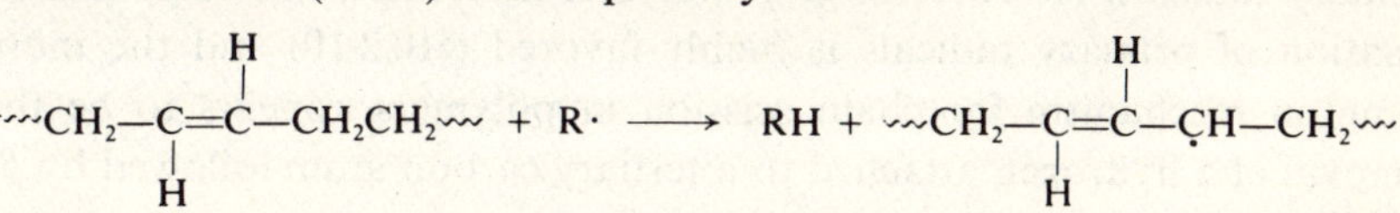

(14.13)

Disproportionation of this structure by reaction with another secondary or allylic radical (14.14) would produce the diene structures observed experimentally:

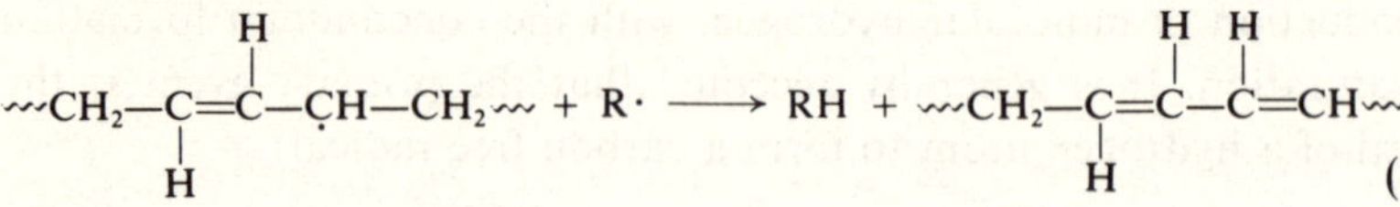

(14.14)

According to this mechanism, considerations of material balance suggest that the G value for hydrogen formation should be equal to the sum of the G values for (14.10), (14.11) and (14.14). Thus

$$G_{H_2} = G_{(vinylene)} + G_{(diene)} + G_x \tag{14.15}$$

Experimental studies of γ-radiolysis of low density polyethylene show that this relation is obeyed within the precision of the analytical determinations. For example Slivinskas and Guillet (1974) found values of $G_{H_2}=3.3$, $G_x=1.7$, $G_{(trans\text{-}vinylene)}=0.8$ and estimated $G_{(diene)}=0.5$.

By contrast, the radiation sensitivity of polystyrene is two orders of magnitude lower. Slivinskas and Guillet (1973*a*) reported values of $G_{H_2}=0.027$ and $G_x=0.020$ for a sample of polystyrene prepared by free-radical polymerization. This trend is observed in most aromatic polymers and is believed to be due to the fact that the excitation energy is delocalized in the excited states of the aromatic chromophore, and tends to be dissipated by photophysical rather than photochemical processes.

In many polymers, particularly those with recurrent tertiary carbon atoms, scission of the backbone polymer chain can occur, often with high efficiency. This process can occur by direct rupture of a carbon–carbon bond in the backbone:

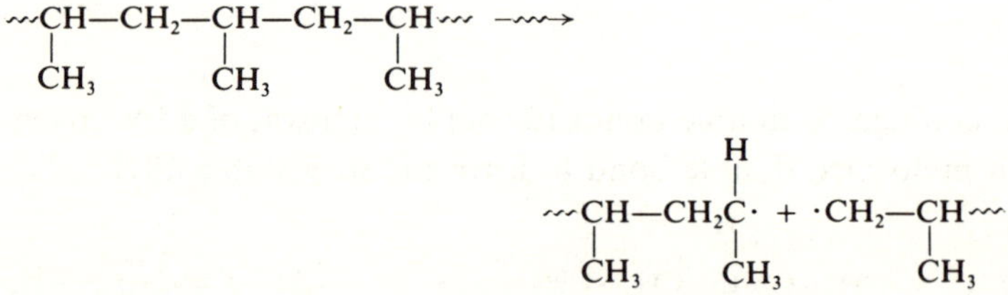

(14.16)

followed by hydrogen abstraction to prevent recombination of the primary radicals. However in large polymer molecules, the 'cage' recombination of primary radicals is highly favored (§10.2.10) and the more common mechanism for chain scission in polymers appears to be the removal of a hydrogen attached to a tertiary carbon atom followed by β-scission. Thus,

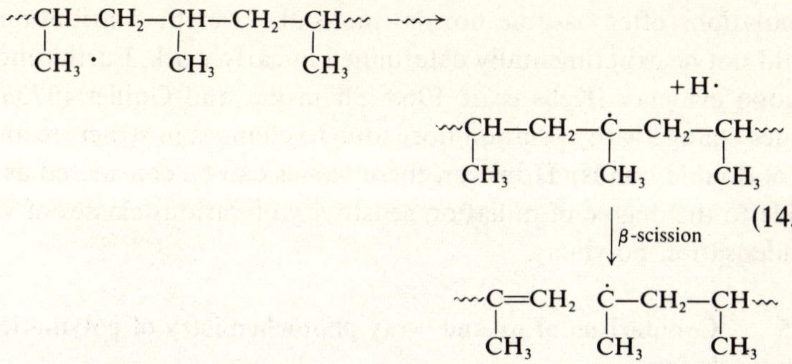

(14.17)

(14.18)

This process can also occur efficiently by hydrogen abstraction by any of the radical species present.

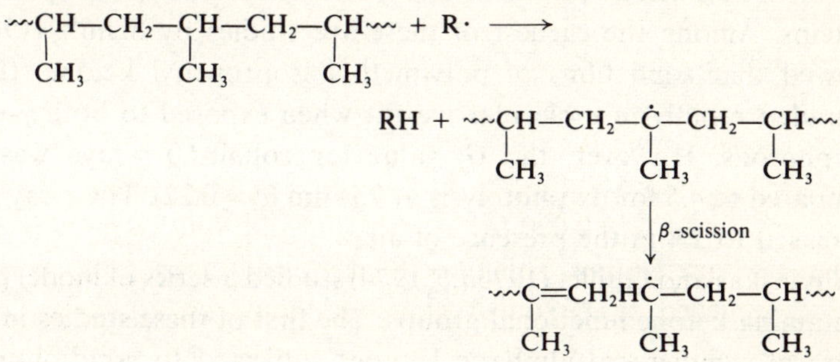

(14.19)

Using the procedures outlined in chapter 4, the *G* values for crosslinking and scission have been determined for a wide variety of commercial and experimental polymers. Representative values are given in table 14.2. It should be pointed out that these values are only approximate. The

Table 14.2. *Representative values of* G_s *and* G_x *for various polymers*

Polymer	G_s	G_x
Polyethylene (low density)	0.2	1.7
Polyethylene (high density)	~0	1.8
Polypropylene	0.6	0.5
Polyisoprene	1.2	~0
Polyisobutylene	5	~0
Polystyrene	0.04	0.02
Poly(dimethyl siloxane)	0	2.5
Poly(methyl acrylate)	0.5	0.76
Poly(methyl methacrylate)	2.5	~0
Cellulose	10	~0
Poly(methyl isopropenyl ketone)	1.1	~0

calculations often assume normal molecular weight distributions which could not be experimentally determined in early work. Furthermore there is good evidence (Kells *et al.* 1968, Slivinskas and Guillet 1973*a*) that G values change with radiation dose (due to changes in structure and build-up of double bonds). However, these values can be considered as a useful guide to the degree of radiation sensitivity of various classes of vinyl and condensation polymers.

14.5 Comparison of uv and γ-ray photochemistry of polymeric ketones

14.5.1 *Studies of model compounds*

Only a few studies have been made of the correlations which can be drawn between uv photochemistry and radiation chemistry in polymer systems. Among the earliest of these are studies by Schulz (1960) who showed that solid films of poly(methyl isopropenyl ketone) (PMIPK) degraded rapidly in molecular weight when exposed to both γ-rays and uv photons. However, the G_s value for cobalt-60 γ-rays was 1.1, as compared to 4.5 for uv photolysis at 254 nm ($\phi = 0.22$). The γ-ray G_s value increased to 2.4 in the presence of air.

Slivinskas and Guillet (1973*a*,*b*, 1974) studied a series of model polymers containing ketone functional groups. The first of these studies involved a series of symmetrical aliphatic ketones subjected to γ-radiolysis (^{60}Co) under vacuum, the products being analyzed by GLC. The results were then compared with the uv photochemistry of the same model molecules.

The absorption of γ-radiation does not follow the selection rules and spin conservation laws for the formation of electronically excited states that pertain in uv-absorption. With high energy γ-ray photons, the probability of absorption is, in fact, directly proportional to the mass of the stopping atoms and consequently a rather random absorption of energy should take place in the first instance. Surprisingly, in spite of this, it appears from the experimental results, that a significant portion of the energy is localized in the same n-π^* excited states as are induced by the absorption of a uv photon. Pitts and Osborn (1961) have observed that the major products arising from γ-irradiated low molecular weight methyl ketones were those of the classical Norrish type I and type II processes. However, as the chain length of the ketone is increased, one should expect to obtain products characterizing paths common to aliphatic hydrocarbons and involving the cleavage of carbon–carbon and carbon–hydrogen bonds as initial steps.

The primary events after interaction of γ-rays with an aliphatic ketone are usually considered to be

$$\mathrm{R{-}\overset{\overset{O}{\|}}{C}{-}R \rightsquigarrow R{-}\overset{\overset{O_{+}}{\|}}{C}{-}R + e^{-}} \tag{14.20}$$

$$\mathrm{R{-}\overset{\overset{O}{\|}}{C}{-}R + R{-}\overset{\overset{O_{+}}{\|}}{C}{-}R \longrightarrow R{-}\overset{\overset{O}{\|}}{C}{-}\dot{R}H^{+} + \cdot R'{-}\overset{\overset{O}{\|}}{C}{-}R} \tag{14.21}$$

$$\mathrm{R{-}\overset{\overset{O}{\|}}{C}{-}\dot{R}H^{+} \longrightarrow R{-}\underset{\bullet}{\overset{\overset{OH_{+}}{|}}{C}}{-}R} \tag{14.22}$$

The free electrons released by the ionization will undergo additional collisions and after moderation will probably react by (14.23) and (14.24):

$$\mathrm{R{-}\overset{\overset{O}{\|}}{C}{-}R + e^{-} \longrightarrow R{-}\overset{\overset{O_{-}}{\|}}{C}{-}R} \tag{14.23}$$

$$\mathrm{R{-}\overset{\overset{O}{\|}}{C}{-}R^{+} + e^{-} \longrightarrow R{-}\overset{\overset{O_{*}}{\|}}{C}{-}R} \tag{14.24}$$

Additional excited molecules will be formed by reactions between cationic and anionic species present. Thus,

$$\mathrm{R{-}\overset{\overset{O_{-}}{\|}}{C}{-}R + R{-}\overset{\overset{O}{\|}}{C}{-}\dot{R}H^{+} \longrightarrow R{-}\overset{\overset{O}{\|}}{C}{-}R + R{-}\underset{\bullet}{\overset{\overset{OH}{|}}{C}}{-}R} \tag{14.25}$$

$$\mathrm{R{-}\overset{\overset{O_{-}}{\|}}{C}{-}R + R{-}\overset{\overset{O}{\|}}{C}{-}R^{+} \longrightarrow 2R{-}\overset{\overset{O_{*}}{\|}}{C}{-}R} \tag{14.26}$$

The multiplicity of excitation processes and the number of possible excited species present in γ-ray photolysis make it difficult to establish a detailed mechanism for the subsequent reactions. When contrasted with the simplicity of direct excitation of states of known energy and geometry, as is usually observed in uv photochemistry, elucidation of the chemistry of high energy radiation effects in polymers represents a formidable challenge.

Although the primary excited states formed by these processes are seldom identical with those created by selective absorption of uv photons, the path of the reaction appears to be nearly the same and it is tempting to assume that only the lower level excited states such as those accessible by uv absorption are populated by a series of collisional processes. In any case, the products from the excited molecules are predominantly those expected from the Norrish type I and type II processes.

In this case, reaction of the carbonyl group competes with reaction of the groups in the hydrocarbon chain. These can be expected, after excitation, to undergo reactions according to (14.8) through (14.14). Analysis of the products of the reaction in simple model aliphatic ketones can give an accurate estimate of the relative proportions of the reactions of the carbonyl and methylene groups. Table 14.3 shows that the carbonyl group is about four times as reactive towards γ-rays as would be predicted from its mass absorption coefficient. For example, although in 4-heptanone the carbonyl would be expected to absorb only 25% of the radiation, the Norrish type I and II processes accounted for 84% of the total reaction products observed. Similarly, the ratio of Norrish type I and II products to total reaction products remains constant at about 4 for the series from 4-heptanone to 7-tridecanone.

It was also noted, both in photochemical and radiochemical investigations, that crystalline ketones, such as 8-pentadecanone and 12-tricosanone, gave no measurable amounts of either type I, or type II or of normal disproportionation products. This is presumably because the molecular motion possible in a crystalline lattice is insufficient to allow escape of the radical products, or the formation of the cyclic six-membered ring intermediate required for the Norrish type II process. When the ketones were dissolved in hexane or other hydrocarbon solvents it was found that the yield of type I and II products was independent of concentration up to about 50%, after which there was some decrease in the products (figure 14.2). This shows that there was an effective mechanism for the transport of the excitation energy from the solvent to the ketone in this instance. The mechanism by which energy originally deposited in the methylene groups of the solvent is transported to the ketone carbonyl is not yet completely understood.

The most important conclusions can be drawn when the results for

Figure 14.2. *G* values for 2-octanone (type II) product from radiolysis of solutions of 7-tridecanone in octane as a function of ketone concentration. (From Slivinskas and Guillet 1973 *b*.)

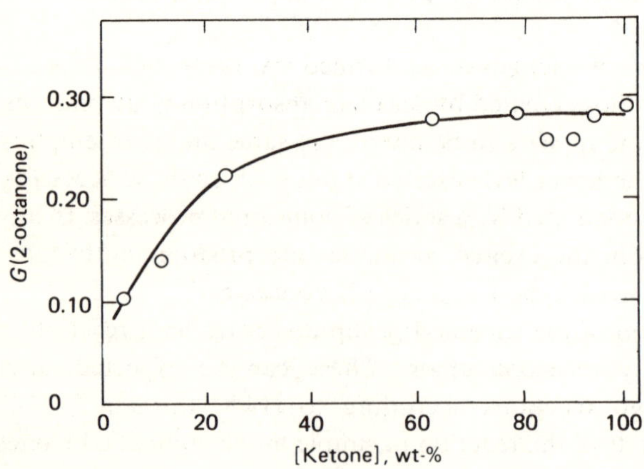

Table 14.3. *Radiolysis yields of symmetrical aliphatic ketones*[a]

			Products from ketone									
			Type I	Type II								
R	Ketone	Phase	G(alkane)	G (methyl ketone)	G(olefin)	$G(A)$[b]	$G(B)$[c]	G_{H_2}	G (total products)	$\frac{G_I+G_{II}}{G_{total}}$	Fraction radiation absorbed by CO	$\left(\frac{G_I+G_{II}}{G_{total}}\right)\left(\frac{1}{f}\right)$
3	4–Heptanone	Liquid	1.42	0.49	0.49	0.15	0.32	0.47	2.87	0.84	0.25	3.4
4	5–Nonanone	Liquid	0.86	0.51	0.30	0.10	0.27	0.37	2.04	0.82	0.20	4.1
5	6–Undecanone	Liquid	0.56	0.45	0.24	0.17	0.40	0.57	1.82	0.69	0.16	4.2
6	7–Tridecanone	Liquid	0.36	0.28	0.17	0.07	0.58	0.65	1.44	0.55	0.14	3.9
7	8–Pentadecanone	Solid	$<1\times10^{-4}$	$<1\times10^{-4}$	$<1\times10^{-4}$							
11	12–Tricosanone	Solid	$<1\times10^{-4}$	$<1\times10^{-4}$	$<1\times10^{-4}$							

[a] Neat, under vacuum, at 35 °C with ^{60}Co γ-rays. [b] A = sum of all low molecular weight products. [c] B = sum of all high molecular weight products.

γ-ray radiolysis are compared with those of uv photolysis of the same ketones. Table 14.4 for example shows comparisons of the G value for photolysis, calculated from (14.7), as compared to that from γ-radiolysis. The ratio of the G value for the type I radical process to the type II intramolecular rearrangement is very much higher in the case of radiolysis than it is with uv photolysis. This was explained by the authors as being due to the fact that, in radiolysis, higher vibrational levels of the electronically excited states of the ketone are achieved, causing a significant increase in the quantum yield or G values for radical formation. It was suggested that this excess energy of excitation allowed more facile escape of the radical pairs from the solvent cage. As expected from earlier uv studies (Hartley and Guillet 1968) the quantum yields for both processes decrease with increasing length of the hydrocarbon chain.

14.5.2 *Studies of ethylene–carbon monoxide copolymers*

In a subsequent paper, Slivinskas and Guillet (1974) studied the radiation chemistry of ethylene–carbon monoxide copolymers (PE–CO) which are large polymeric analogs of the small model compounds studied previously. In this case, the large difference between the G values for radical processes made a substantial difference in the total effect observed in the two types of radiation. Whereas under uv light the PE–CO undergoes a substantial and significant decrease in molecular weight as a result of main-chain scission via the Norrish type II process, even in the solid phase, in contrast, when irradiated with γ-rays, the polymer crosslinks. The degree of crosslinking is proportional to the amount of carbon monoxide present in the polymer. For example, figure 14.3 shows the decrease in molecular weight as the polymer is irradiated with uv light

Table 14.4. *Comparison of* G *values for photolysis (313 nm) and radiolysis of symmetrical aliphatic ketones at 35°C*

	G (photolysis)			G (radiolysis)		
Ketone	$CO_{(I)}$	Methyl $\text{ketone}_{(II)}$	Ratio G_I/G_{II}	$\left(\frac{\text{Alkane}}{2}\right)_{(II)}$	Methyl $\text{ketone}_{(II)}$	Ratio G_I/G_{II}
4-Heptanone	0.44	3.8	0.12	0.71	0.49	1.45
5-Nonanone	0.025	2.8	0.0090	0.43	0.51	0.88
6-Undecanone	0.020	2.4	0.0083	0.28	0.45	0.62
7-Tridecanone	0.017	2.0	0.0085	0.18	0.26	0.69
8-Pentadecanone	0.015	1.7	0.0088	0.065	0.092	0.71
12-Tricosanone	0.012	1.5	0.0080	0.040	0.081	0.69

and figure 14.4 shows the crosslinking effect of γ-radiation. The increase in the molecular weight of the solid polymer film in the latter case is presumably due to the substantial number of free radicals formed by the type I process when γ-radiation is used. Crosslinking can occur by processes (14.8) and (14.10) at an efficiency much higher than scission via the type II process from the carbonyl, which must occur at the same time.

Slivinskas and Guillet also observed that the yields of hydrogen formed via (14.8), were substantially reduced in polyethylene containing carbonyl groups, as compared with pure polyethylene. This is shown in table 14.5. It was shown that the quenching of hydrogen production was significantly greater when the carbonyl group was attached to the polymer chain than when the carbonyl was simply dissolved in the polymer before irradiation. On the basis of these results Slivinskas and Guillet postulated that quenching by carbonyl groups in the polymer chain took place as a result of the trapping of excitons which travelled along the backbone of the polymer via a σ-exciton band mechanism. This was in agreement with theoretical predictions made by Partridge (1970*a*, *b*, *c*) on the theory of excitation energy transfer in alkanes under high energy irradiation. According to the molecular excitation theory of σ bonds it was concluded that very efficient excitation energy transfer would occur along σ C–C

Figure 14.3. Number of bonds broken as a function of Einsteins absorbed per gram of polymer (I_A) during photolysis of 1% CO–PE in solid phase under N_2 at 25°C: $M_n(0) = 25\,800$. (From Slivinskaś and Guillet 1974.)

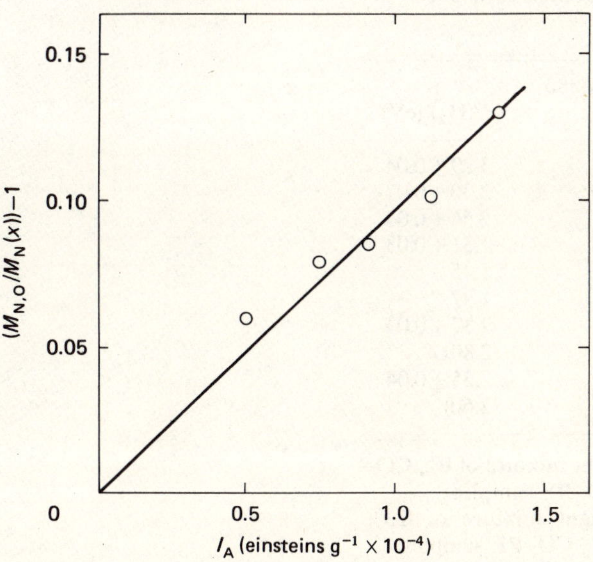

Figure 14.4. Changes in M_V with radiation dose r for CO–PE samples irradiated in solid phase at 35°C under vacuum: (hollow circle) 0% CO–PE; (square) 0.3% CO–PE; (triangle) 0.55% CO–PE; (solid circle) 1.0% CO–PE. (From Slivinskas and Guillet 1974.)

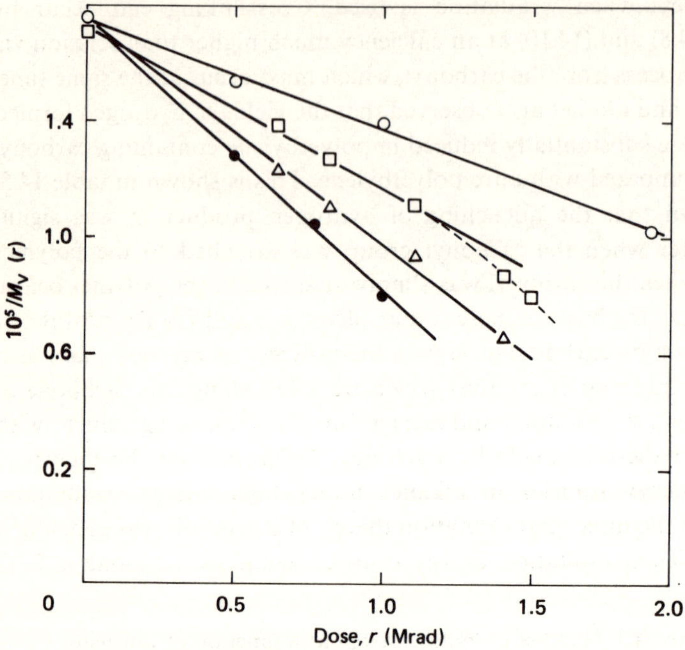

Table 14.5. G(H_2) *values for polymers irradiated in film at 35 °C under vacuum to a 2–3 Mrad dose*

CO in polyethylene (wt-%)	$G(H_2)$ (eV)
0.00	3.29 ± 0.04
0.30	2.90 ± 0.07
0.55	2.56 ± 0.04
1.00	2.31 ± 0.03
3.50	1.45
9.10	1.37
0.50[a]	2.37 ± 0.03
(mixture rule)	(2.80)
0.65[b]	2.35 ± 0.04
(mixture rule)	(2.60)

[a] 50:50 by weight mixture of 0% CO–PE and 1% CO–PE samples.
[b] 50:50 by weight mixture of 0.3% CO–PE and 1% CO–PE samples.

bonds of the alkane polymer chain, and that this excitation could be localized from time to time in C–H bonds, thus causing dissociation to form free radicals and hydrogen atoms.

The PE–CO copolymers also showed an increased value for G_x which was attributed by the authors to the localization of absorbed energy by preferential trapping of thermalized secondary electrons in the carbonyl group because of their lower ionization potential as compared to the methylene groups in the polymer. The efficiency of high energy radiation in promoting the crosslinking of high polymers has been utilized successfully for the crosslinking of polyethylene and many other polymers for coatings and insulation for electrical wiring. The inclusion of small numbers of carbonyl groups in the polymer is potentially a simple and inexpensive way of increasing the efficiency of the crosslinking process (Guillet 1974).

14.5.3 *Styrene–vinyl ketone copolymers*

Polystyrene (PS) is one of the most resistant polymers to radiochemical damage and for this reason has often been used as a base for scintillators and other plastic devices exposed to high energy radiation.

Early studies on the crosslinking of polystyrene have been reviewed by Chapiro (1962). In subsequent work, Slivinskas and Guillet (1973*b*) studied a series of copolymers of styrene with methyl vinyl ketone. As in the case of polyethylene, substantial differences were observed between uv photolysis and γ-radiation, particularly in that the polymers are very much less sensitive to γ-radiation than to uv photolysis. PS itself is known to undergo crosslinking very slowly when exposed to γ-radiation, but when progressively copolymerized with vinyl ketones, the amount of crosslinking decreases such that, when the amount of ketone exceeds about 2%, the polymer degrades rather than crosslinks. This is shown in figure 14.5. In contrast to the results with PE–CO, no change in the yield of hydrogen as a function of carbonyl content occurred, an observation explained on the basis that the excitation of polystyrene is localized in the phenyl rings and thus no large-scale migration of the energy is possible in this polymer by the exciton band mechanism which applies to polyethylene. It is interesting to note that the quantum yield for chain scission in PS – 0.5% MVK copolymer, calculated from the G values, is 0.002 whereas, for uv photolysis, the quantum yield would be 0.2, two orders of magnitude higher. It is obvious that, in the polystyrene copolymer, the phenyl rings are protecting the carbonyl groups from excitation by the high energy radiation whereas no such protective effect is observed when the carbonyl group is directly excited by uv light.

14.6 Radiolysis of polyacrylates and methacrylates

The γ-radiolysis of polymers of esters of acrylic and methacrylic acid have also been studied extensively by David, Geuskens and co-workers (David 1975, David *et al.* 1970, 1972*a,b,c*, Geuskens and David 1973, David *et al.* 1975). In the solid state at 25 °C the mechanism of radiolysis proposed for poly(methyl methacrylate) (PMMA), the most commonly studied polymer, is the initial formation of a radical ion (II) which subsequently splits to form a polymeric cation and a small radical.

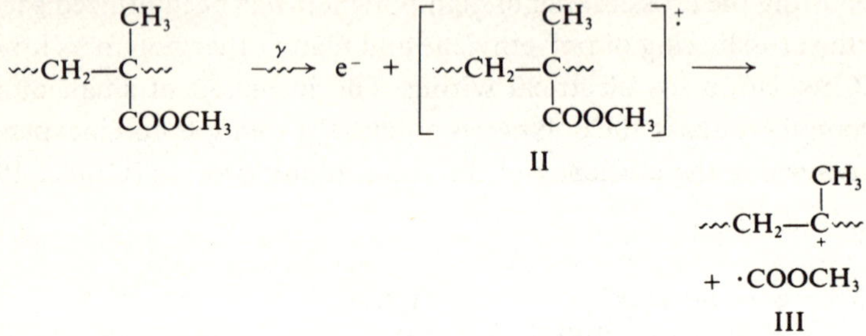

(14.27)

The volatile products can be accounted for by the subsequent reactions of the carbomethoxy radical.

$$\cdot COOCH_3 \overset{RH}{\rightarrow} HCOOCH_3 + R\cdot \qquad (14.28)$$

$$\cdot COOCH_3 \rightarrow CO + \cdot OCH_3 \overset{RH}{\rightarrow} HOCH_3 + R\cdot \qquad (14.29)$$

$$\cdot COOCH_3 \rightarrow CO_2 + \cdot CH_3 \overset{RH}{\rightarrow} CH_4 + R\cdot \qquad (14.30)$$

Figure 14.5. Changes in $\bar{M}_V$ vs. radiation dose r for MVK–PS copolymers irradiated under vacuum in solid phase at 35°C.

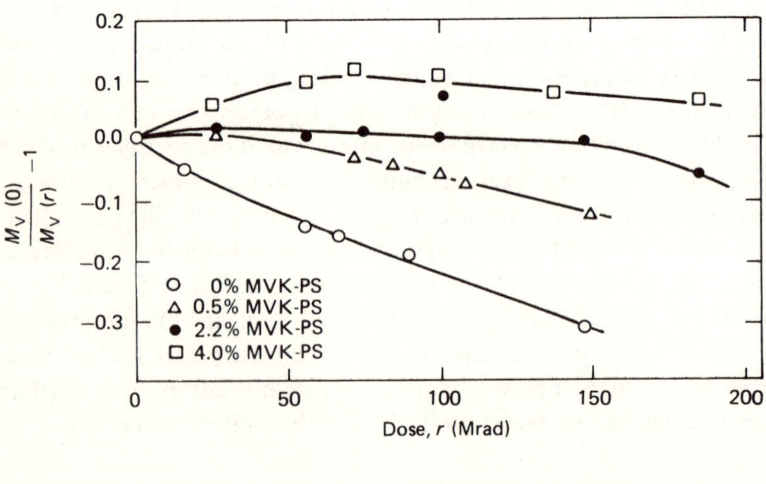

Considerable evidence for the radical and ionic species proposed was adduced from ESR and spectroscopic studies after irradiation at 77 K in the glassy state. There is also evidence for some energy migration, possibly by electron transfer processes.

Chain scission, which is efficient in PMMA is proposed to occur via electron capture by the polymer cation III to give an excited free radical IV which subsequently splits via β-scission:

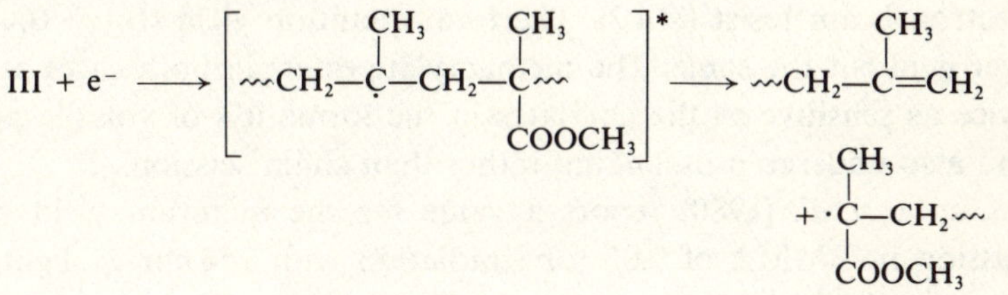

V

(14.31)

If the temperature is high enough, the tertiary radical V may split off a number of monomer units by a chain process, formally the reverse of free-radical polymerization, prior to termination by recombination, disproportionation, or hydrogen abstraction.

In the case of the phenyl (and other aromatic) esters the Fries reaction (§10.2) can also occur, via one or more of the excited states induced on the polymer chain. Thus

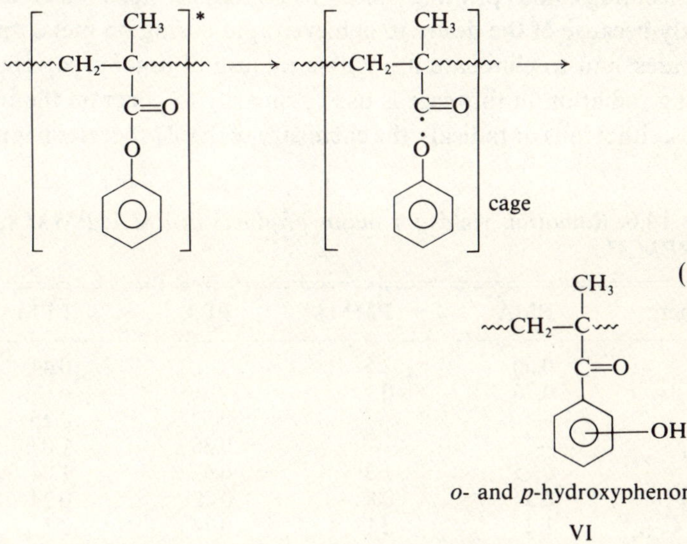

(14.32)

o- and *p*-hydroxyphenone

VI

to form *o*- and *p*-hydroxyphenone groups attached to the chain. Escape of the phenoxy radical from the cage followed by hydrogen abstraction gives rise to phenol as one of the major products.

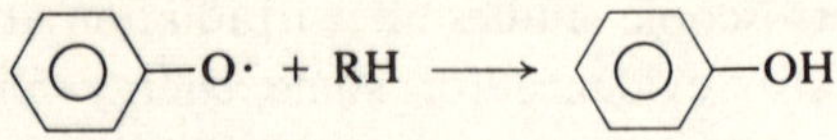

(14.33)

A summary of the *G* values for various important products for PMMA, poly(methyl acrylate) (PMA), poly(phenyl methacrylate) (PPMA) and poly(phenyl acrylate) (PPA) are given in table 14.6. It is clear that G_x for PPMA is only about one-fifth that of PMMA, making it less useful as an electron beam resist (§14.7). The total quantum yield ($G_{CO}+G_{CO_2}$) however is about the same. The methacrylate esters in both cases are about twice as sensitive as the acrylates in the formation of volatile products, and also undergo crosslinking rather than chain scission.

Gupta *et al.* (1980) report a value for the quantum yield of chain scission in PMMA of 0.05 for irradiation with 254 nm uv light. Application of (14.7) yields an estimate of G_s value for uv photons of about 1, as compared with 2.5 for γ-rays. This increase in efficiency for radical processes with γ-ray photons is similar to that observed previously by Slivinskas and Guillet (1973).

14.7 Application to photolithography

High energy radiation, including electron beams and γ-rays, has been used commercially to cause either crosslinking or polymerization of vinyl and other organic monomers and polymers in order to produce cured coatings and printing inks. These applications have increased recently because of the desire to achieve rapid curing on metal and other substrates, and to eliminate if possible the use of organic solvents. Since ionizing radiation in this case is used primarily to generate the initiating species, either ions or radicals, the chemistry of these processes has not been

Table 14.6. *Radiation yields for major products in PMA, PMMA, PPA and PPMA*[a]

Polymer	PMA	PMMA	PPA	PPMA
G_s	0.50	2.5		0.44
G_x	0.76	~0		0
G_{Fries}	—[b]	—[b]	0.52	0.25
G_{phenol}	—[b]	—[b]	0.36	1.07
G_{CO}	0.95	1.3	0.52	1.84
G_{CO_2}	0.24	0.8	0.22	0.24
$G_{CO}+G_{CO_2}$	1.2	2.1	0.74	2.1

[a] From David *et al.* (1978).
[b] Fries reaction does not occur.

dealt with in this book which, as its title suggests, is more concerned with the photochemistry of large molecules than with the processes by which such molecules may be made. Recently, however, there has been an increasing interest in the use of either photodegradable or photocross-linking polymers in the design and manufacture of microcircuits for advanced electronics. It seems likely that this new field of microlithography will generate additional fundamental information about the photochemistry and photophysics of polymer molecules and for this reason the principles will be outlined briefly here.

Photolithography dates its invention to the work of W. H. Fox-Talbot in 1852, who discovered that light exposure of films of gelatin which had been treated with bichromate induced sufficient solubility changes that unexposed regions could be selectively washed away with water. This procedure was patented for use in photoengraving and, indeed, chromated gelatin is still in use at the present time. However, for advanced applications such as in microlithography, a whole new class of photoresists has become necessary, although the principle remains the same. Resist materials are classed as either positive-working or negative-working, depending on whether the solubility of the organic polymer in the developer increases (positive) or decreases (negative) on exposure to radiation. Polymers are almost universally used for preparing photoresists because of their excellent film-forming and coating properties and the ease with which these properties can be changed by the chemist to conform with specific applications. The polymers are usually linear and have molecular weights ranging from a few to several hundred thousand, hence, when dissolved in easily evaporated solvents, films with useful thicknesses can be obtained by spin coating. The processes involved in producing a pattern are shown in figure 14.6.

14.7.1 *Negative-working resists*

A common type of negative-working photoresist used in microelectronics applications involves a synthetic polymer which is blended with a difunctional azide of the general structure:

$$N_3{-}X{-}N_3$$

VII

in which X contains aromatic groups as part of a conjugated system which provides the appropriate absorption spectrum in the near ultraviolet and visible. On excitation by light the azides lose two molecules of nitrogen to form a species known as a nitrene.

$$R-N_3 \xrightarrow{h\nu} R-\ddot{N} + N_2 \qquad (14.34)$$

These highly reactive, short-lived intermediates, in turn, react with the C–H bonds of the polymer to form the desired crosslinks:

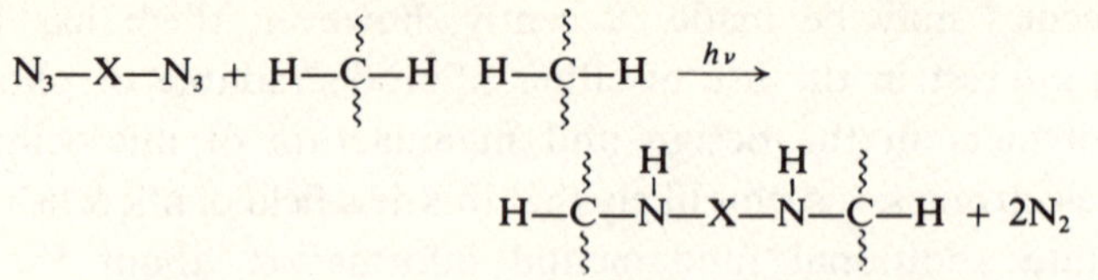

(14.35)

The polymers used for azide-based resists are usually aliphatic rubber-like materials based on polyisoprene and, in some cases, monofunctional aromatic azides have been attached to the polymer chain. The azide systems can also be sensitized by certain organic compounds which can shift the wavelength of absorption to longer wavelengths and increase the efficiency of the crosslinking process. Azide resists have certain disadvantages, particularly an inability to make really fine lines, and a sensitivity to oxygen, which reduced the efficiency of the crosslinking reaction. Typical crosslinking agents for vinyl polymers are aromatic azides (VIII) carboazides (IX) or sulfone azides (X).

Figure 14.6. Schematic diagram of the microlithographic process. (From Deckert and Ross 1980.)

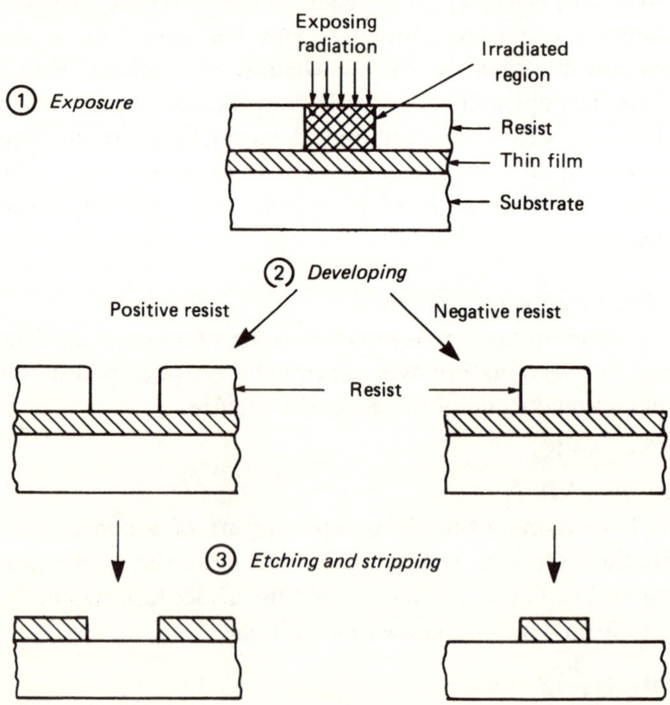

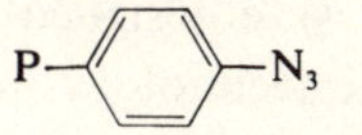

Aromatic azide VIII

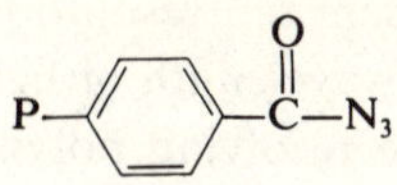

Carboazide IX

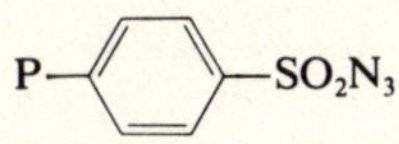

Sulfone azide X

where P represents a polymer chain.

14.7.2 *Positive-working resists*

Positive-working resists are usually based on the photochemical conversion of an *o*-diazo ketone XI to a carboxylic acid in the presence of a polymeric matrix which is usually a relatively low molecular weight hydrophobic phenolic polymer, the diazo ketone being attached directly to the polymer molecule in some cases. As a result of the association between the polymer and the diazo ketone the unexposed mixture has a low solubility in aqueous alkaline developer. However, on exposure to light, the diazo ketone liberates a molecule of nitrogen and undergoes a molecular rearrangement to the alkali-soluble acid (XII)

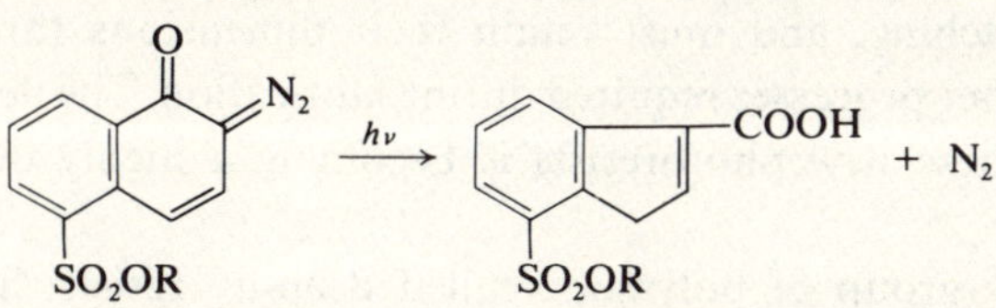

XI XII (14.36)

The formation of the polar carboxyl groups in the exposed areas makes the film much less hydrophobic and more soluble in the alkaline developer. One of the major advantages of the positive-working resists is the fact that, instead of swelling the film and leaching out the solubilized material, the developer removes the exposed areas by an etching process. As a result, no swelling occurs in the non-exposed areas, and images of extremely high resolution (less than 0.25 microns) can be formed. Because the adhesion of the substrate is not changed by the photochemistry, exposure and development conditions can be adjusted to give the desired relief pattern and geometry required. The diazo ketone resists also have the advantage that they are not affected by the presence of oxygen.

14.7.3 *High resolution photoresists*

Light sources used in graphic arts, and adapted for use in microlithography, usually have their maximum output in the region 350–

450 nm and most early photosensitive polymers were designed to have maximum sensitivity in this range. However, the technology in micro-lithography has now advanced to the stage where the resolution of the pattern on a silicon chip is now limited by the wavelength of light.

This arises from the classical formula for the resolving polymer x of any optical device:

$$x = \tfrac{1}{2}\lambda / n \sin \mathrm{u} \tag{14.37}$$

where $n \sin \mathrm{u}$ is the 'numerical aperture' of the device. $n \sin \mathrm{u}$ has a maximum value of unity, thus the theoretical limit for resolution is $\frac{1}{2}\lambda$. In order to produce electronic devices with faster response times and more components per unit area there has been an expanding effort to produce exposure devices and resists which can operate with shorter wavelength radiation. With the best optical devices commercially available at the present time, the resolution for near uv light is about 1 micron. However, for many types of advanced circuitry it is desired to reduce this to 0.3, and perhaps as low as 0.1 μm.

There are also physical requirements on photoresists, which are often coated on a chip as many as ten times, and then removed before the final device is completed. For example, they must be resistant to ion implantation and ion etching, and must retain their dimensions throughout baking and the other processes required during fabrication. The design and manufacture of these new photoresists is becoming a highly developed technology.

Recently a new group of polymers, called deep-uv resists, has been developed. These require new light sources and special quartz optics to permit exposures with wavelengths in the 200–250 nm region. While some of the conventional photoresists respond adequately to these wavelengths, many do not, because the polymers from which they are derived absorb strongly enough to block the transmission of the exciting radiation from the absorbing or photo-sensitive groups. Because of the limitations of negative-working resists in obtaining high resolution, most of the new resists are positive-working and are based on polymers that will undergo a reduction in molecular weight, and thus an increase in solubility, as a result of photo-induced main-chain scission.

14.7.4 *Electron beam and X-ray resists*

In alternative approaches to achieve higher resolution, a number of technical advances have been made in the use of electron beams and 'soft' X-rays (see table 14.1). Electrons can, depending on their energies, have extremely short wavelengths and may be focussed to very small diameters below 1000 Å. The pattern resolution obtainable with resists sensitive to electrons or X-rays no longer depends on the wavelength but on other fundamental limitations such as the scattering of electrons or, in the case of X-rays, the range of the photoelectrons produced on the absorption of radiation in the resist material, or by the substrate. The use of these new exposure devices has emphasized the differences observed earlier between uv photochemistry and radiation chemistry. For example, it seems clear that, with X-rays and electrons, energy will be absorbed more or less indiscriminately by all functional groups in the organic polymer molecules rather than being selectively absorbed by specific reactive sites. However, as discussed earlier, spatial resolution may be altered by electron or energy exchange from group to group within the polymer matrix, either by intra- or intermolecular energy transfer. It may be that the ultimate resolution of the electron beam and X-ray methods will depend more on energy transfer within the medium than on the optical and scattering effects which would normally limit the resolution.

The earliest resist used with electron beams and soft X-rays was PMMA, which on exposure to radiation, undergoes chain scission, presumably by a process analogous to reaction (14.18). This makes the exposed regions more soluble in the developer. Although PMMA has many desirable qualities as an e-beam resist, its sensitivity is low.

A negative resist which has been used commercially is a copolymer of ethylacrylate and glycidyl methacrylate developed at Bell Laboratories. The epoxy groups from the glycidyl methacrylate react under electron radiation to crosslink the polymer with high efficiency.

A new class of positive photoresists is the poly(olefin sulfone)s (Himics *et al.* 1977) which are prepared by the copolymerization of sulfur dioxide with unsaturated monomers. The copolymer of butene and sulfur dioxide, called poly(butene sulfone), is currently used in certain electronic applications. These polymers have the advantage that the initial cleavage of the backbone initiates a chain depolymerization reaction which leads to the regeneration of the original comonomers. This provides a considerable gain in the sensitivity and can produce excellent separation of molecular weights between the exposed and unexposed regions. This makes the polymer easy to develop in solvents and, in some cases, the depolymerization is sufficiently rapid that the decomposed polymer can be removed under vacuum without the need for a subsequent development step.

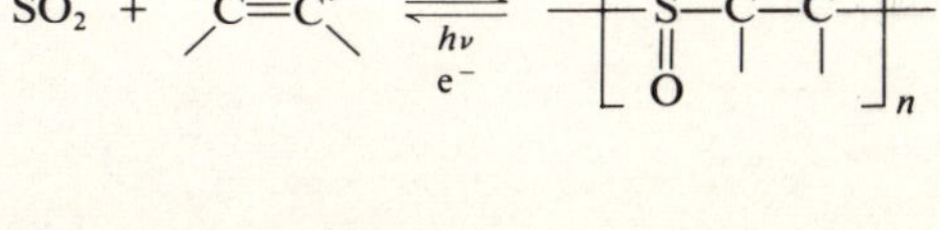

(14.38)

Much less is known or has been published about the field of photoresists for X-ray lithography. However, it is generally assumed that the sensitivity to X-rays will parallel that to electrons and γ-radiation, and in fact the usual screening process for such resists is an initial exposure to γ-radiation. The considerable commercial importance of the microelectronics industry has thus resulted in a renewed scientific interest in the field of the radiation chemistry of polymers.

References

Chapter 1

Beer, M. and Longuet-Higgins, H. C. (1955). *J. Chem. Phys.* **23**, 1390.

Berlman, I. B. (1971) *Handbook of Fluorescence Spectra of Aromatic Molecules*, 2nd edn, Academic Press, New York.

Birks, J. B. (1967). Nature, **214**, 1187.

Birks, J. B. and Kuchela, K. N. (1961). *Proc. Phys. Soc.* **77**, 1083.

Bowen, E. J. and Livingston, R. (1954). *J. Am. Chem. Soc.* **76**, 6300.

Calvert, J. G. and Pitts, J. N. (1966). *Photochemistry.* Wiley, New York.

Coulson, C. A. (1952). *Valence.* Oxford University Press.

Dexter, D. L. (1953). *J. Chem. Phys.* **21**, 836.

El-Sayed, M. A. (1963). *J. Chem. Phys.* **38**, 2834.

Förster, Th. (1948). *Ann. Phys.* **2**, 55.

Förster, Th. (1951). *Fluoresenze Organische Verbindungen.* Vandenhoech and Ruprech, Gottingen.

Gouterman, M. (1962). *J. Chem. Phys.* **36**, 2846.

Hammond, G. S. (1969). *Adv. Photochem.* **7**, 373.

Kasha, M. (1950). *Disc. Faraday Soc.* **9**, 14.

Kasha, M. (1960). *Radiation Research Suppl.* **2**, 243.

Lewis, G. N., Lipkin, D. and Magel, T. T. (1941). *J. Am. Chem. Soc.* **63**, 3005.

Lower, S. and El-Sayed, M. A. (1966). *Chem. Rev.* **66**, 199.

Parker, C. A. (1968). *Photoluminescence of Solutions.* Elsevier, Amsterdam.

Parker, C. A. and Hatchard, C. G. (1961). *Trans. Faraday Soc.* **57**, 1894.

Perrin, F. (1924). *C. R. Acad. Sci. Paris*, **178**, 1978.

Platt, J. R. (1949). *J. Chem. Phys.* **17**, 484.

Platt, J. R. (1964). *Systematics of the Electronic Spectra of Conjugated Molecules.* Wiley, New York.

Robinson, G. W. (1974). In *Excited States*, vol. 1, ed. E. Lim. Academic Press, New York.

Robinson, G. W. and Frosch, R. P. (1963). *J. Chem. Phys.* **38**, 1187.

Siebrand, W. (1967). *J. Chem. Phys.* **47**, 2411.

Stern, O. and Volmer, M. (1919). *Phys. Z.* **20**, 183.

Turro, N. J. (1978). *Modern Molecular Photochemistry.* Benjamin/Cummings, Menlo Park.

Wagner, P. J. (1972). In *Creation and Detection of the Excited State*, vol. 1, ed. A. A. Lamola. Marcel Dekker, New York.

Wagner, P. J. and Kochevar, I. (1968). *J. Am. Chem. Soc.* **90**, 2232.

Wen, W. Y. and Hsu, R. (1963). *J. Phys. Chem.* **66**, 1353.

Chapter 2

Billmeyer, F. W. Jr (1962). *Textbook of Polymer Science.* Interscience, New York.
Bryant, W. M. D. (1947). *J. Polym. Sci.* **2**, 547.
Debye, P. and Bueche, F. (1952). *J. Chem. Phys.* **20**, 1337.
Einstein, A. (1906). *Ann. Physik*, **19**, 289.
Einstein, A. (1911). *Ann. Physik*, **34**, 591.
Elias, H.-G. (1977). *Macromolecules I*, p. 80. Plenum, New York.
Flory, P. J. (1953). *Principles of Polymer Chemistry.* Cornell University Press, Ithaca, New York.
Flory, P. J. (1969). *Statistical Mechanics of Chain Molecules.* Interscience, New York.
Flory, P. J., Allcock, H. R., Eisenberg, H., Feldman, R. J. and Yoon, D. Y. (1982). In *Outlook for Science and Technology: the next five years.* W. H. Freeman, San Francisco.
Geil, P. H., Baer, E. and Wada, Y. (eds.) (1974). *The Solid State of Polymers, U.S.–Japan Joint Seminar.* Marcel Dekker, New York.
Guillet, J. E. (1977). *Pure Appl. Chem.* **49**, 249.
Haward, R. (1975). In *Molecular Behaviour and the Development of Polymer Materials*, ed. A. Ledwith and A. M. North. Chapman and Hall, London.
Hoseman, R. and Bonart, R. (1957). *Kolloid-Z.* **152**, 53.
Keller, A. (1957). *Phil. Mag.* **2**, 1171.
McCrum, N. G., Read, B. E. and Williams, G. (1967). *Anelastic and Dielectric Effects in Polymeric Solids.* Wiley, New York.
Natta, G., Corradini, P. and Bassi, I. W. (1955). *Rend. accad. nazl. Lincei VIII.* **19**, 404.
North, A. M. (1975). In *Molecular Behaviour and the Development of Polymer Materials*, ed. A. Ledwith and A. M. North. Chapman and Hall, London.
Schlesinger, C. E. and Leeper, H. M. (1953). *J. Polym. Sci.* **11**, 203.
Shen, M. C. (1972). *Chem. Space Res.* 319.
Slivinskas, J. A. and Guillet, J. E. (1973). *J. Polym. Sci. Polym. Chem. Ed.* **11**, 3043.
Wunderlich, B. (1973). *Physics of Macromolecules*, vol. 1, *Crystal Structure, Morphology, Defects.* Academic Press, New York.

Chapter 3

Brandrup, J. and Immergut, E. H. (eds.) (1975). *Polymer Handbook*, 2nd edn. Wiley, New York.
Barrer, R. M. and Skirrow, G. (1948). *J. Polym. Sci.* **3**, 549.
Braun, J. M. and Guillet, J. E. (1976). *Advances in Polymer Science*, vol. 21. Springer-Verlag, Berlin.
Brockmeier, N. F., McCoy, R. W. and Meyer, J. A. (1972*a*). *Macromolecules*, **5**, 130.
Brockmeier, N. F., McCoy, R. W. and Meyer, J. A. (1972*b*). *Macromolecules*, **5**, 464.
Brockmeier, N. F., McCoy, R. W. and Meyer, J. A. (1973). *Macromolecules*, **6**, 176.
Bueche, F. (1962). *Physical Properties of Polymers.* Interscience, New York.
Calvert, J. G. and Pitts, J. N. Jr (1966). *Photochemistry.* Wiley, New York.
Crank, J. and Park, G. S. (1968). *Diffusion in Polymers.* Academic Press, London.
Debye, P. (1942). *Trans. Electrochem. Soc.* **82**, 265.
Frisch, H. L. (1965). *J. Polym. Sci.* **C10**, 11.
Gray, D. G. and Guillet, J. E. (1973). *Macromolecules*, **6**, 223.
Guillet, J. E. (1973). In *New Developments in Gas Chromatography*, ed. H. Purnell. Wiley, New York.
Heskins, M. and Guillet, J. E. (1970). *Macromolecules*, **3**, 224.

Lipson, J. E. G. and Guillet, J. E. (1981). In *Polymer Characterization*, vol. 1, ed. J. V. Dawkins. Applied Science, Barking.

Liu, D. D. and Prausnitz, J. M. (1976). *Ind. Eng. Chem. Fundam.* **15**, 330.

MacCallum, J. R. and Rudkin, A. L. (1978). *Eur. Polym. J.* **14**, 655.

Nowakowska, M., Najbar, J. and Waligora, B. (1976). *Eur. Polym. J.* **12**, 387.

Turro, N. J. (1978). *Modern Molecular Photochemistry*. Benjamin-Cummings, Menlo Park.

van Deemter, J. J., Zuiderweg, F. J. and Klinkenberg, A. (1956). *Chem. Eng. Sci.* **5**, 271.

Chapter 4

Amerik, Y. and Guillet, J. E. (1971). *Macromolecules*, **4**, 375.

Bandrup, J. and Immergut, E. H. (eds.) (1966). *Polymer Handbook*, 1st edn. Interscience, New York.

Bandrup, J. and Immergut, E. H. (eds.) (1975). *Polymer Handbook*, 2nd edn. Wiley, New York.

Charlesby, A. (1960). *Atomic Radiation and Polymers*. Pergamon Press, New York.

Charlesby, A. and Pinner, S. H. (1959). *Proc. R. Soc. London, Ser. A.* **249**, 367.

Debye, P. (1944). *J. Appl. Phys.* **15**, 338.

Debye, P. (1947). *J. Phys. Colloid Chem.* **51**, 18.

Flory, P. J. (1953). *Principles of Polymer Chemistry*, p. 380. Cornell University Press, Ithaca, New York.

Flory, P. J. and Rehner, J. Jr (1943). *J. Chem. Phys.* **11**, 521.

Guillet, J. E. (1963). *J. Polym. Sci., Part A*, **1**, 2869.

Graessley, W. (1968). In *Characterization of Macromolecular Structure*, ed. D. McIntyre, p. 371. National Academy of Sciences Publication No. 1573, Washington, DC.

Inokuti, M. (1963). *J. Chem. Phys.* **38**, 2999.

Kells, D. I. C. and Guillet, J. E. (1969). *J. Polym. Sci., Part A-2*, **7**, 1895.

Kells, D. I. C., Koike, M. and Guillet, J. E. (1968). *J. Polym. Sci., Part A-1*, **6**, 595.

Kilp, T. and Guillet, J. E. (1977). *Macromolecules*, **10**, 90.

Kuhn, W. (1946). *J. Polym. Sci.* **1**, 380.

Lindenau, D., Beavan, S. W., Beck, G. and Schnabel, W. (1977). *Eur. Polym. J.* **13**, 819.

Nemzek, T. L. and Guillet, J. E. (1977). *Macromolecules*, **10**, 94.

Pecora, A. (1972). In *Light Scattering from Polymer Solutions*, ed. M. B. Huglin. Academic Press, London.

Reiser, A. and Edgerton, P. L. (1979). *Macromolecules*, **12**, 670.

Roberts, G. P., Budzol, M. and Dole, M. (1971). *J. Polym. Sci., Part A-2*, **9**, 1729.

Saito, O. (1958). *J. Phys. Soc. Jpn.* **13**, 198.

Saito, O. (1972). In *The Radiation Chemistry of Macromolecules*, vol. 1, ed. M. Dole, p. 223. Academic Press, New York.

Tanford, C. (1961). *Physical Chemistry of Macromolecules*, Ch. 5. Wiley, New York.

Scott, K. W. (1974). *J. Polym. Sci., Part C*, **46**, 321.

Staudinger, H. and Heuer, W. (1930). *Berichte*, **63B**, 222.

Treloar, J. R. G. (1958). *The Physics of Rubber Elasticity*, 2nd edn. Clarendon Press, Oxford.

Wall, F. T. (1943). *J. Chem. Phys.* **11**, 67.

Zimm, B. H. and Kilb, R. W. (1959). *J. Polym. Sci.* **37**, 19.

Chapter 5

Aviv, G., Sagiv, J. and Yogev, A. (1976). *Mol. Cryst. Liq. Cryst.* **36**, 349.

Bott, C. C. and Kurucsev, T. (1975). *J. Chem. Soc., Faraday Trans. 2*, **71**, 749.

Boustead, I. (1970). *Eur. Polym. J.* **6**, 731.

Buettner, A. V. (1964). *J. Phys. Chem.* **68**, 3253.
Charlesby, A. and Partridge, R. H. (1963). *Proc. R. Soc. London, Ser. A*, **283**, 312.
Farid, S., Martic, P. A., Thompson, D. R., Specht, D. P., Hartman, S. E. and Williams, J. L. R. (1979). *Pure Appl. Chem.* **51**, 241.
Gailite, V. and Freimanis, J. (1973). *Zh. Obshch. Khim.* **43**, 348.
Gardlund, Z. G. (1968). *J. Polym. Sci., Part B*, **6**, 57.
Gardlund, Z. G. and Laverty, J. J. (1969). *J. Polym. Sci., Part B*, **7**, 719.
Geacintov, N., Oster, G. and Cassen, T. (1968). *J. Opt. Soc. Am.* **58**, 1217.
Gegiou, D., Muszkat, K. A. and Fischer, E. (1968). *J. Am. Chem. Soc.* **90**, 12.
Guseva, L. N., Mikheev, Yu, A. and Toptygin, D. Ya. (1973). *Dokl. Akad. Nauk. SSSR*, **213**, 871.
Heppel, G. E. (1965). *Photochem. Photobiol.* **4**, 7.
Hirshberg, Y. and Fischer, E. (1954). *J. Chem. Soc.* 297.
Joussot-Dubien, J. and Lesclaux, R. (1973). *Pure Appl. Chem.* **34**, 265.
Kamogawa, H., Kato, M. and Sugiyama, H. (1968). *J. Polym. Sci., Part A-1*, **6**, 2967.
Kryszewski, M., Lapienis, D. and Nadolski, B. (1973). *J. Polym. Sci., Polym. Chem. Ed.* **11**, 2423.
Kryszewski, M. and Nadolski, B. (1975). *J. Polym. Sci., Polym. Chem. Ed.* **13**, 345.
Lawrie, N. G. and North, A. M. (1973). *Eur. Polym. J.* **9**, 345.
Lewis, G. N. and Lipkin, D. (1942). *J. Am. Chem. Soc.* **64**, 2801.
Li, S. K. L. and Guillet, J. E. (1977). *Macromolecules*, **10**, 840.
Li, S. K. L. and Guillet, J. E. (1980). *J. Polym. Sci., Polym. Chem. Ed.* **18**, 2221.
Martic, P. A., Daly, R. C., Williams, J. L. R. and Farid, S. Y. (1977). *J. Polym. Sci., Polym. Lett. Ed.* **15**, 295.
Martin, T. F. and Kalantar, A. H. (1968). *J. Phys. Chem.* **72**, 2265.
Priest, W. J. and Sifain, M. M. (1971). *J. Polym. Sci., Part A-1*, **9**, 3161.
Tabek, D. and Morawetz, H. (1970). *Macromolecules*, **3**, 403.
Smets, G. (1972). *Pure. Appl. Chem.* **30**, 1.
Smets, G. (1975*a*). *Polym. Prepr., Am. Chem. Soc., Div. Polym. Chem.* **16** (2), 174.
Smets, G. (1975*b*). *Pure Appl. Chem.* **42**, 509.
Smets, G. J. and Matsumoto, S. (1976). *J. Polym. Sci., Polym. Chem. Ed.* **14**, 2983.
Smets, G. J., Thoen, J. and Aerts, A. (1975). *J. Polym. Sci., Polym. Symp.* **51**, 119.
Somersall, A. C., Dan, E. and Guillet, J. E. (1974). *Macromolecules*, **7**, 233.
Vainshtok, B. A. and Rakhimbaeva, N. U. (1976). *Dozim. Ioniz. Izluch*, 111.
Vasilenko, V. V., Klinshpont, E. R. and Milinchuk, V. K. (1976). *Khim. Vys. Energ.* **10**, 428.

Chapter 6

Bentz, J. P., Beyl, J. P., Beinert, G. and Weill, G. (1975). *Eur. Polym. J.* **11**, 711.
Biddle, D. and Nordstrom, T. (1970). *Ark. Kemi*, **32**, 359.
Chapoy, L. L., Spaseska, D., Rasmussen, K. and DuPré, D. B. (1979). *Macromolecules*, **12**, 680.
Duportail, G., Froelich, D. and Weill, G. (1971). *Eur. Polym. J.* **7**, 977.
Einstein, A. (1916). *Deut. Phys. Ges. Verhandl.* **18**, 318.
Einstein, A. (1917). *Phys. Z.* **18**, 121.
Fuhrmann, J. and Hennecke, M. (1980). *Makromol. Chem.* **181**, 1685.
Heskins, M. and Guillet, J. E. (1968). *Macromolecules*, **1**, 97.
Heskins, M. and Guillet, J. E. (1970*a*). *Macromolecules*, **3**, 224.
Heskins, M. and Guillet, J. E. (1970*b*). In *Photochemistry of Macromolecules*, ed. R. F. Reinisch. Plenum Press, New York.

Hirota, K. and Nishijima, Y. (1980). *Rep. Prog. Polym. Sci. Jpn.* **23**, 555.

Holden, D. A. and Guillet, J. E. (1980). In *Developments in Polymer Photochemistry*, vol. 1, ed. N. S. Allen. Applied Science, Barking.

Holden, D. A. (1982). Private communication.

Jarry, J. P. and Monnerie, L. (1979). *Macromolecules*, **12**, 927.

Kirsh, Yu E., Pavlova, N. R. and Kabanov, V. A. (1975). *Eur. Polym. J.* **11**, 495.

Liao, T.-P. and Morawetz, H. (1980). *Macromolecules*, **13**, 1228.

MacCallum, J. R. and Rudkin, A. L. (1978). *Eur. Polym. J.* **14**, 655.

Meierstein, G., Rabani, J., Mathieson, M. S. and Meisel, D. (1978). *J. Am. Chem. Soc.* **82**, 1879.

Meisel, D. and Mathieson, M. S. (1977). *J. Am. Chem. Soc.* **99**, 6577.

Moldovan, L. and Weill, G. (1971). *Eur. Polym. J.* **7**, 1023.

Monnerie, L. (1981). *Polym. Prepr., Am. Chem. Soc., Div. Polym. Chem.* **22** (1), 96.

Morawetz, H. (1979). *Science*, **203**, 405.

Nishijima, Y. (1962). In *Luminescence of Organic and Inorganic Materials*, ed. H. Kallman and G. S. Spruch. Wiley, New York.

Nishijima, Y. (1964). *Kobunshi*, **13**, 166.

Nishijima, Y. (1970). *J. Polym. Sci., Part C*, **31**, 353.

Nishijima, Y. (1973*a*). *Progr. Polym. Sci. Jpn.* **6**, 199.

Nishijima, Y. (1973*b*). *J. Macromol Sci., Phys.* **8**, 407.

Nishijima, Y. and Mito, Y. (1967). *Rep. Progr. Polym. Phys. Jpn.* **10**, 139.

Nishijima, Y. and Saito, M. (1967). *Rep. Progr. Polym. Phys. Jpn.* **10**, 135.

Nishijima, Y., Midorikawa, T. and Taki, F. (1965). *Rep. Progr. Polym. Phys. Jpn.* **8**, 139.

Nishijima, Y., Onogi, Y. and Asai, T. (1967*a*). *Rep. Progr. Polym. Phys. Jpn.* **10**, 461.

Nishijima, Y., Teramoto, A., Yamamoto, M. and Hiratsuka, S. (1967*b*) *J. Polym. Sci., Part A-2*, **5**, 23.

North, A. M. (1975). In *Molecular Behavior and the Development of Polymer Materials*, ed. A. Ledwith and A. M. North, Ch. 11, Halsted Press, New York.

Perrin, F. (1926). *J. Phys.* **7**, 390.

Perrin, F. (1929). *Ann. Phys. (Paris)*, **12**, 169.

Perrin, F. (1934). *J. Phys. Radium*, **5**, 497.

Perrin, F. (1936). *J. Phys. Radium*, **7**, 1.

Riseman, J. and Kirkwood, J. G. (1949). *J. Chem. Phys.* **17**, 442.

Somersall, A. C. and Guillet, J. E. (1972). *Macromolecules*, **5**, 410.

Somersall, A. C. and Guillet, J. E. (1975). *J. Macromol. Sci. C*, **13**, 135.

Steiner, R. F. and Edelhoch, H. (1962). *Chem. Rev.* **62**, 457.

Taha, I. A. and Morawetz, H. (1971). *J. Polym. Sci., Part A-2*, **9**, 1669.

Valeur, B. and Monnerie, L. (1976*a*). *J. Polym. Sci., Polym. Phys. Ed.* **14**, 11.

Valeur, B. and Monnerie, L. (1976*b*). *J. Polym. Sci., Polym. Phys. Ed.* **14**, 29.

Valeur, B., Jarry, J.-P., Geny, F. and Monnerie, L. (1975*a*). *J. Polym. Sci., Polym. Phys. Ed.* **13**, 667.

Valeur, B., Monnerie, L. and Jarry, J.-P. (1975*b*). *J. Polym. Sci., Polym. Phys. Ed.* **13**, 675.

Valeur, B., Rempp, P. and Monnerie, L. (1974). *C. R. Acad. Sci., Ser. C*, **279**, 1099.

Wahl, Ph., Meyer, G. and Parrod, J. (1970). *Eur. Polym. J.* **6**, 585.

Weber, G. (1952). *Biochem. J.* **51**, 145.

Weber, G. (1953). *Adv. Protein Chem.* **8**, 415.

Weill, G. (1971). *C. R. Acad. Sci. Paris Ser. B*, **272**, 116.

Yardley, J. T. and Specht, L. T. (1976). *Chem. Phys. Lett.* **37**, 543.

Chapter 7

Aladekomo, J. B. and Birks, J. B. (1965). *Proc. R. Soc. London, Ser. A*, **284**, 551.

Alexandru, L. and Somersall, A. C. (1977). *J. Polym. Sci., Polym. Chem. Ed.* **15**, 2013.

Aspler, J. and Guillet, J. E. (1979). *Macromolecules*, **12**, 1082.

Avouris, P. and El-Bayoumi, M. A. (1973). *Chem. Phys. Lett.* **20**, 59.

Bairamov, Y. Y., Amerik, Y. B. and Krentsel, B. A. (1971). *Preprints, III IUPAC Conference on Chemical Transormation of Polymers.*

Basile, L. J. (1962). *J. Chem. Phys.* **36**, 2204.

Birks, J. B. (1970). *Photophysics of Aromatic Molecules.* Wiley–Interscience, London.

Birks, J. B. and Christophorou, L. G. (1963). *Spectrochim. Acta*, **19**, 401.

Brimage, D. R. G. and Davidson, R. S. (1972). *J. Photochem.* **1**, 79.

Bokobza, L., Jasse, B. and Monnerie, L. (1977). *Eur. Polym. J.* **13**, 921.

Braun, H. and Förster, Th. (1966). *Ber. Bunsenges. Phys. Chem.* **70**, 1091.

Castro, G. and Hochstrasser, R. M. (1966). *J. Chem. Phys.* **45**, 4252.

Chandross, E. A. and Dempster, C. J. (1970). *J. Am. Chem. Soc.* **92**, 3586.

Chandross, E. A. and Thomas, H. T. (1971). *Chem. Phys. Lett.* **9**, 393.

Cram, D. J. and Steinberg, H. (1951). *J. Am. Chem. Soc.* **73**, 5691.

Cuniberti, C. and Perico, A. (1977). *Eur. Polym. J.* **13**, 369.

David, C., Lempereur, M. and Geuskens, G. (1973). *Eur. Polym. J.* **9**, 1315.

David, C., Piens, M. and Geuskens, G. (1972*a*). *Eur. Polym. J.* **8**, 1019.

David, C., Piens, M. and Geuskens, G. (1972*b*). *Eur. Polym. J.* **8**, 1291.

David, C., Putman de Lavarielle, N. and Geuskens, G. (1974). *Eur. Polym. J.* **10**, 617.

De Schryver, F. C., Demeyer, K., van der Auweraer, M. and Quanten, E. (1981). *Ann. N.Y. Acad. Sci.* **366**, 93.

Farid, S., Martic, P. A., Daly, R. C., Thompson, D. R., Specht, D. P., Hartman, S. E. and Williams, J. L. R. (1979). *Pure Appl. Chem.* **51**, 241.

Förster, Th. and Kasper, K. (1954). *Z. Phys. Chem.* **1**, 275.

Förster, Th. and Kasper, K. (1955). *Z. Electrochem.* **59**, 976.

Fox, R. B. (1972). *Pure Appl. Chem.* **30**, 87.

Fox, R. B., Price, T. R., Cozzens, R. F. and McDonald, J. R. (1972). *J. Chem. Phys.* **57**, 534.

Frank, C. W. and Harrah, L. A. (1974). *J. Chem. Phys.* **61**, 1526.

Frank, C. W. (1975). *Macromolecules*, **8**, 305.

Frank, C. W. and Gashgari, M. A. (1979). *Macromolecules*, **12**, 163.

Frank, C. W. and Gashgari, M. A. (1981). *Ann. N.Y. Acad. Sci.* **366**, 387.

Ghiggino, K. P., Roberts, A. J. and Phillips, D. (1978*a*). *J. Photochem.* **9**, 301.

Ghiggino, K. P., Wright, R. D. and Phillips, D. (1978*b*). *Chem. Phys. Lett.* **53**, 552.

Ghiggino, K. P., Wright, R. D. and Phillips, D. (1978*c*). *J. Polym. Sci., Polym. Phys. Ed.* **16**, 1499.

Ghiggino, K. P., Wright, R. D. and Phillips, D. (1978*d*). *Eur. Polym. J.* **14**, 567.

Ghiggino, K. P., Wright, R. D. and Phillips, D. (1979). *J. Polym. Sci., Polym. Phys. Ed.* **16**, 1499.

Goldenberg, M., Emert, J. and Morawetz, H. (1978). *J. Am. Chem. Soc.* **100**, 7171.

Harrah, L. A. (1972). *J. Chem. Phys.* **56**, 385.

Hirayama, F. (1965). *J. Chem. Phys.* **42**, 3163.

Holden, D. A. and Guillet, J. E. (1980). In *Developments in Polymer Photochemistry*, vol. 1, ed. J. V. Dawkins. Applied Science, Barking.

Holden, D. A., Wang, P. Y.-K. and Guillet, J. E. (1980). *Macromolecules*, **13**, 295.

Hoyle, C. E. and Guillet, J. E. (1979). *Macromolecules*, **12**, 956.

Hoyle, C. E., Nemzek, T. L., Mar, A. and Guillet, J. E. (1978). *Macromolecules*, **11**, 429.

Ishii, T., Handa, T. and Matsunaga, S. (1978). *Macromolecules*, **11**, 40.

Ito, S., Yamamoto, M. and Nishijima, Y. (1977). *Rep. Prog. Polym. Phys. Jpn.* **20**, 401.

Iwatawa, S., Tanaka, J. and Nagakura, S. (1967). *J. Chem. Phys.* **47**, 2203.

Johnson, G. E. (1975). *J. Chem. Phys.* **62**, 4697.

Keyanpour-Rad, M., Ledwith, A., Hallam, A., North, A. M., Breton, M., Hoyle, C. E. and Guillet, J. E. (1978). *Macromolecules*, **11**, 1114.

Klöpffer, W. (1969*a*). *Chem. Phys.* **4**, 193.

Klöpffer, W. (1969*b*). *J. Chem. Phys.* **50**, 2337.

Klöpffer, W. (1973). In *Organic Molecular Photochemistry*, vol. 1, ed. J. B. Birks. Wiley, London.

Knibbe, H. and Weller, A. (1967). *Z. Phys. Chem.* (*NF*), **56**, 99.

Kühnle, W. and Zachariasse, K. (1976*a*). *J. Photochem.* **5**, 165.

Kühnle, W. and Zachariasse, K. (1976*b*). *Z. Phys. Chem.* 101, 267.

Lewis, C. and Ware, W. R. (1973). *Mol. Photochem.* **5**, 261.

Liao, T.-P., Okamoto, Y. and Morawetz, H. (1979). *Macromolecules*, **12**, 535.

Longworth, J. W. and Bovey, F. A. (1966). *Biopolymers*, **4**, 1115.

Masaki, S., Okada, T., Mataga, N., Sakata, Y. and Misumi, S. (1976). *Bull. Chem. Soc. Jpn.* **49**, 1277.

MacCallum, J. R. (1981). *Eur. Polym. J.* **17**, 797.

McConnell, H., Ham, J. S. and Platt, J. R. (1953). *J. Chem. Phys.* **21**, 66.

Merle-Aubry, L., Holden, D. A., Merle, Y. and Guillet, J. E. (1980). *Macromolecules*, **13**, 1138.

Morawetz, H. (1979). *Science*, **203**, 405.

Mulliken, R. S. (1950*a*). *J. Am. Chem. Soc.* **72**, 600.

Mulliken, R. S. (1950*b*). *J. Am. Chem. Soc.* **72**, 4493.

Mulliken, R. S. (1951). *J. Chem. Phys.* **19**, 514.

Mulliken, R. S. (1952). *J. Am. Chem. Soc.* **74**, 811.

Mulliken, R. S. (1955). *J. Chem. Phys.* **23**, 397.

Nakahira, T., Maruyama, I., Iwabuchi, S. and Kojima, K. (1979). *Makromol. Chem.* **180**, 1853.

Ng, D. and Guillet, J. E. (1981). *Macromolecules*, **14**, 405.

Okada, T., Fujita, T., Kubota, M., Masaki, S. and Mataga, N. (1972). *Chem. Phys. Lett.* **14**, 563.

Phillips, D., Roberts, A. J. and Soutar, I. (1980). *J. Polym. Sci., Polym. Phys. Ed.* **18**, 2401.

Redpath, A. E. C. and Winnik, M. A. (1982). *J. Am. Chem. Soc.* **104**, 5604.

Reid, R. F. and Soutar, I. (1978). *J. Polym. Sci., Polym. Phys. Ed.* **16**, 231.

Roberts, A. J., Cureton, C. G. and Phillips, D. (1980). *Chem. Phys. Lett.* **72**, 554.

Selinger, B. (1966). *Austral. J. Chem.* **19**. 825.

Semerak, S. N. and Frank, C. W. (1981). *Macromolecules*, **14**, 443.

Somersall, A. C. and Guillet, J. E. (1973). *Macromolecules*, **6**, 218.

Soutar, I. (1981). *Ann. N. Y. Acad. Sci.* **366**, 24.

Tazuke, S. and Matsuyama, Y. (1977). *Macromolecules*, **10**, 215.

Vala, M. T. Jr, Haebig, J. and Rice, S. A. (1965). *J. Chem. Phys.* **43**, 886.

Wang, Y.-C. and Morawetz, H. (1975). *Makromol. Chem., Suppl.* **1**, 283.

Weller, A. (1970). *Z. Phys. Chem.* **69**, 183.

Winnik, M. A., Redpath, A. E. C., Svirskaya, P. and Mar, A. (1983). *Polymer* **24**, 473.

Yaneri, S. S., Bovey, F. A. and Lumry, R. (1963). *Nature*, **200**, 242.

Yokoyama, M., Tamamura, T., Atsumi, M., Yoshimura, M., Shirota, Y. and Mikawa, H. (1975). *Macromolecules*, **8**, 101.

Chapter 8

Andrews, M. (1978). M.Sc. Thesis, University of Toronto.
Albrecht, A. (1963). *J. Chem. Phys.* **38**, 354.
Birks, J. B. (1970). *Photophysics of Aromatic Molecules.* Wiley–Interscience, London.
Boustead, I. (1970). *Eur. Polym. J.* **6**, 731.
Braun, J. M. and Guillet, J. E. (1976). *Adv. Polym. Sci.* **21**, 107.
Carslaw, H. S. and Jaeger, J. C. (1959). *Conduction of Heat in Solids.* Oxford University Press.
Dan, E., Somersall, A. C. and Guillet, J. E. (1973). *Macromolecules,* **6**, 228.
David, C., Baeyens-Volant, D., Macedo de Abreu, P. and Geuskens, G. (1977). *Eur. Polym. J.* **13**, 841.
David, C., Lempereur, M. and Geuskens, G. (1972). *Eur. Polym. J.* **8**, 417.
Guillet, J. E. and Andrews, M. Unpublished work.
Horie, K. and Mita, I. (1978). *Macromolecules,* **11**, 1175.
Jones, E. F. (1968). *J. Polym. Sci., Part D,* **6**, 487.
Kilp, T. and Guillet, J. E. (1981). *Macromolecules,* **14**, 1680.
Mar, A., Fraser, S. and Winnik, M. A. (1981). *J. Am. Chem. Soc.* **103**, 491.
Miller, L. J. and North, A. M. (1975). *J. Chem. Soc. Faraday Trans. 2,* **71**, 1233.
Oster, G., Geacintov, N. and Khan, A. V. (1962). *Nature (London),* **196**, 1089.
Perrin, F. (1929). *Ann. Phys.* **12**, 169.
Shaw, G. (1967). *Trans. Faraday Soc.* **63**, 2181.
Somersall, A. C. (1973). Ph.D. Thesis, University of Toronto.
Somersall, A. C., Dan, E. and Guillet, J. E. (1974). *Macromolecules,* **7**, 233.
Stannet, V. (1968). In *Diffusion in Polymers,* ed. J. Crank and G. S. Park. Academic Press, London.
Unterleitner, F. C. and Hormats, E. I. (1969). *J. Phys. Chem.* **69**, 2516.
Weber, G. (1966). In *Fluorescence and Phosphorescence Analysis: Principles and Applications,* ed. D. M. Hercules. Interscience, New York.
Winnik, M. A. (1981). *Chem. Rev.* **81**, 491.

Chapter 9

Adamczyk, A. and Phillips, D. (1974). *Trans. Faraday Soc.* **70**, 537.
Aspler, J. S., Hoyle, C. E. and Guillet, J. E. (1978). *Macromolecules,* **11**, 925.
Bensasson, R. V., Ronfard-Haret, J. C., Land, E. J. and Webber, S. E. (1979). *Chem. Phys. Lett.* **68**, 438.
Birks, J. B. (1964). *The Theory and Practice of Scintillation Counting.* Pergamon Press, Oxford.
Burkhart, R. D., Aviles, R. G. and Magrini, K. (1981). *Macromolecules,* **14**, 91.
Cozzens, R. F. and Fox, R. B. (1969*a*). *J. Chem. Phys.* **50**, 1532.
Cozzens, R. F. and Fox, R. B. (1969*b*). *Macromolecules,* **2**, 181.
Cundall, R. B., Lewis, C., Llewellyn, P. J. and Phillips, G. O. (1970). *J. Phys. Chem.* **74**, 4172.
Cuniberti, C. and Perico, A. (1977). *Eur. Polym. J.* **13**, 369.
Dan, E., Somersall, A. C. and Guillet, J. E. (1973). *Macromolecules,* **6**, 228.
David, C., Baeyens-Volant, D. and Geuskens, G. (1976). *Eur. Polym. J.* **12**, 71.
David, C., Demarteau, W. and Geuskens, G. (1970). *Eur. Polym. J.* **6**, 1397.
David, C., Lempereur, M. and Geuskens, G. (1972). *Eur. Polym. J.* **8**, 417.
David, C., Naegelen, V., Piret, W. and Geuskens, G. (1975). *Eur. Polym. J.* **11**, 569.
David, C., Putman de Lavareille, N. and Geuskens, G. (1977). *Eur. Polym. J.* **13**, 15.

Förster, T. (1948). *Ann. Phys.* **2**, 55.
Fox, R. B. (1972). *Pure Appl. Chem.* **30**, 87.
Fox, R. B. and Cozzens, R. F. (1969). *Macromolecules*, **2**, 181.
Fox, R. B., Price, T. R. and Cozzens, R. F. (1971). *J. Chem. Phys.* **54**, 79.
Fox, R. B., Price, T. R., Cozzens, R. F. and Echols, W. H. (1974). *Macromolecules*, **7**, 937.
Fox, R. B., Price, T. R., Cozzens, R. F. and MacDonald, J. R. (1972). *J. Chem. Phys.* **57**, 534.
Guillet, J. E. (1979). *Polym. Prepr., Am. Chem. Soc., Div. Polym. Chem.* **20**(1), 395.
Guillet, J. E. and Norrish, R. G. W. (1954). *Nature*, **173**, 625.
Guillet, J. E. and Norrish, R. G. W. (1955*a*). *La Ricerca Scientifica*, **25**, 1.
Guillet, J. E. and Norrish, R. G. W. (1955*b*). *Proc. R. Soc. London, Ser. A*, **233**, 153.
Guillet, J. E., Hoyle, C. E. and MacCallum, J. R. (1978). *Chem. Phys. Lett.* **54**, 337.
Haas, E., Wilchek, M., Katchalski-Katzir, E. and Steinberg, I. Z. (1975). *Proc. Natl. Acad. Sci. U.S.A.* **72**, 1807.
Heskins, M. and Guillet, J. E. (1968). *Macromolecules*, **1**, 97.
Heskins, M. and Guillet, J. E. (1970). *Macromolecules*, **3**, 224.
Holden, D. A. (1980). Ph.D. Thesis, University of Toronto.
Holden, D. A. and Guillet, J. E. (1980). *Macromolecules*, **13**, 289.
Holden, D. A., Rendall, W. A. and Guillet, J. E. (1981). *Ann. N.Y. Acad. Sci.* **366**, 11.
Horie, K. and Mita, I. (1978). *Macromolecules*, **11**, 1175.
Hoyle, C. E. and Guillet, J. E. (1978). *J. Polym. Sci., Polym. Lett. Ed.* **16**, 185.
Ishii, T., Handa, T. and Matsunaga, S. (1978). *Macromolecules*, **11**, 40.
Ishii, T., Ytena, Y., Handa, T., Mori, S. and Tagaki, K. (1977). *Rep. Prog. Polym. Phys. Jpn.* **20**, 415.
Itaya, A., Okamoto, K. and Kusabayashi, S. (1979). *Bull. Chem. Soc. Jpn.* **52**, 3737.
Jablonski, A. (1955). *Acta Phys. Poland*, **14**, 295.
Jablonski, A. (1958). *Acta Phys. Poland*, **17**, 481.
Johnson, G. E. (1980). *Macromolecules*, **13**, 145.
Keyanpour-Rad, M., Ledwith, A., Hallam, A., North, A. M., Breton, M., Hoyle, C. and Guillet, J. E. (1978). *Macromolecules*, **11**, 1114.
Kilp, T. (1979). Ph.D. Thesis, University of Toronto.
Kilp, T. and Guillet, J. E. (1981). *Macromolecules*, **14**, 1680.
Kilp, T., Guillet, J. E., Galin, J. C. and Roussel, R. (1982). *Macromolecules*, **15**, 980.
Kiwi, J. and Schnabel, W. (1976). *Macromolecules*, **9**, 468.
Klöpffer, W. (1969*a*). *J. Chem. Phys.* **50**, 2337.
Klöpffer, W. (1969*b*). *Ber. Bunsenges. Phys. Chem.* **73**, 864.
Lashkov, G. I. and Ermolaev, V. L. (1967). *Opt. Spektrosk.* **22**, 462.
Longworth, J. L. and Battista, M. D. C. (1970). *Photochem. Photobiol.* **11**, 207.
MacCallum, J. R. and Rudkin, L. (1977). *Nature*, **266**, 338.
MacCallum, J. R., Hoyle, C. E. and Guillet, J. E. (1980). *Macromolecules*, **13**, 1647.
Maharaj, U., Winnik, M. A., Dors, B. and Schafer, H. J. (1979). *Macromolecules*, **12**, 905.
McInally, I., Reid, R. F., Rutherford, H. and Soutar, I. (1979). *Eur. Polym. J.* **15**, 723.
Merle-Aubry, L., Holden, D. A., Merle, Y. and Guillet, J. E. (1980). *Macromolecules*, **13**, 1138.
Morawetz, H. and Amrani, F. (1978). *Macromolecules*, **11**, 281.
Ng, D. and Guillet, J. E. (1982*a*). *Macromolecules*, **15**, 724.
Ng, D. and Guillet, J. E. (1982*b*). *Macromolecules*, **15**, 728.
North, A. M. and Ross, D. A. (1976). *J. Polym. Sci., Polym. Symp.* **55**, 259.
North, A. M. and Treadaway, M. F. (1973). *Eur. Polym. J.* **9**, 609.
North, A. M., Ross, D. A. and Treadaway, M. F. (1974). *Eur. Polym. J.* **10**, 411.

North, A. M., Ross, D. A., Guillet, J. E. and Nemzek, T. L. (1977). *J. Chem. Res. S*, **49**, 49.
Nowakowska, M., Najbar, J. and Waligora, B. (1976). *Eur. Polym.J.* **12**, 387.
Ore, A. (1959). *J. Chem. Phys.* **31**, 442.
Partridge, R. H. (1970*a*). *J. Chem. Phys.* **52**, 2485.
Partridge, R. H. (1970*b*). *J. Chem. Phys.* **52**, 2491.
Partridge, R. H. (1970*c*). *J. Chem. Phys.* **52**, 2501.
Philpott, M. R. (1975). *J. Chem. Phys.* **63**, 485.
Powell, R. C. (1971). *J. Chem. Phys.* **55**, 1871.
Rauh, R. D. and Leermakers, P. A. (1968). *J. Am. Chem. Soc.* **90**, 2246.
Reid, R. F. and Soutar, I. (1977). *J. Polym. Sci., Polym. Lett. Ed.* **15**, 153.
Reid, R. F. and Soutar, I. (1978). *J. Polym. Sci., Polym. Phys. Ed.* **16**, 231.
Samedova, T. G., Karpacheva, G. P. and Davydov, B. E. (1972*a*). *Eur. Polym. J.* **8**, 599.
Samedova, T. G., Karpacheva, G. P. and Davydov, B. E. (1972*b*). *Izv. Acad. Nauk SSSR, Ser. Fiz.* **36**, 1134.
Schneider, F. (1979). *Z. Naturforsch.* **24a**, 863.
Schneider, F. and Springer, J. (1971). *Makromol. Chem.* **146**, 181.
Sisido, M., Imanishi, Y. and Higashimura, T. (1979). *Macromolecules*, **12**, 975.
Slivinskas, J. A. and Guillet, J. E. (1974). *J. Polym. Sci., Polym. Chem. Ed.* **12**, 1469.
Somersall, A. C. and Guillet, J. E. (1973). *Macromolecules*, **6**, 218.
Sugita, K., Kilp, T. and Guillet, J. E. (1976). *J. Polym. Sci., Polym. Chem. Ed.* **14**, 1901.
Taha, I. A. and Morawetz, H. (1971). *J. Polym. Sci. Part A-2*, **9**, 1669.
Webber, S. E. and Avots-Avotins, P. E. (1980). *Macromolecules*, **12**, 708.
Webber, S. E., Avots-Avotins, P. E. and Deumie, M. (1981). *Macromolecules*, **14**, 105.

Chapter 10

Alexandru, L. and Guillet, J. E. (1976). *J. Polym. Sci., Polym. Chem. Ed.* **13**, 483.
Alexandru, L. and Guillet, J. E. (1976). *J. Polym. Sci., Polym. Chem. Ed.* **14**, 2791.
Anderson, J. C. and Reese, C. B. (1960). *Proc. Chem. Soc.* 217.
Anderson, J. C. and Reese, C. B. (1963). *Proc. Chem. Soc.* 1781.
Bamford, C. H. and Norrish, R. G. W. (1935). *J. Chem. Soc.* 1504.
Bamford, C. H. and Norrish, R. G. W. (1938*a*). *J. Chem. Soc.* 1521.
Bamford, C. H. and Norrish, R. G. W. (1938*b*). *J. Chem. Soc.* 1544.
Baum, E. J., Wan, J. K. S. and Pitts, J. N. Jr (1966). *J. Am. Chem. Soc.* **88**, 2652.
Bays, J. P., Encinas, M. V. and Scaiano, J. C. (1980). *Macromolecules*, **13**, 815.
Berger, M. and Steel, C. (1975). *J. Am. Chem. Soc.* **97**, 4817.
Bonnans-Plaisance, C., Casals, P. F. and Levesque, G. (1978). *Makromol. Chim.* **179**, 2577.
Burnett, G. M. (1954). *Mechanism of Polymer Reactions.* Interscience, New York.
Dan, E. and Guillet, J. E. (1973). *Macromolecules*, **6**, 230.
Dan, E., Somersall, A. C. and Guillet, J. E. (1973). *Macromolecules*, **6**, 228.
David, C., Demarteau, W. and Geuskens, G. (1970*a*). *Eur. Polym. J.* **6**, 537.
David, C., Demarteau, W. and Geuskens, G. (1970*b*). *Eur. Polym. J.* **6**, 1405.
David, C., Lempereur, M. and Geuskens, G. (1972*b*). *Eur. Polym. J.* **8**, 417.
David, C., Putman, N., Lempereur, M. and Geuskens, G. (1972*a*). *Eur. Polym. J.* **8**, 409.
Dhanraj, J. and Guillet, J. E. (1968). *J. Polym. Sci., Part C*, **23**, 433.
El-Sayed, M. A. (1963). *J. Chem. Phys.* **38**, 2834.
Encinas, M. V., Funabashi, K. and Scaiano, J. C. (1979). *Macromolecules*, **12**, 1167.
Golemba, F. J. and Guillet, J. E. (1972). *Macromolecules*, **5**, 212.

Grassie, N. (1972). In *Polymer Science*, vol. 2, ed. A. D. Jenkins, p. 1443, North-Holland, Amsterdam.

Guillet, J. E. (1980). *Pure Appl. Chem.* **52**, 285.

Guillet, J. E. and Norrish, R. G. W. (1954). *Nature*, **173**, 625.

Guillet, J. E. and Norrish, R. G. W. (1955). *Proc. R. Soc. London, Ser. A*, **233**, 153.

Gupta, A., Rembaum, A. and Moacanin, J. (1978). *Macromolecules*, **11**, 1285.

Gupta, A., Liang, R., Moacanin, J., Goldbeck, R. and Kliger, D. (1980). *Macromolecules*, **13**, 262.

Hartley, G. H. and Guillet, J. E. (1968*a*). *Macromolecules*, **1**, 165.

Hartley, G. H. and Guillet, J. E. (1968*b*). *Macromolecules*, **1**, 413.

Heskins, M. and Guillet, J. E. (1976). In *Encyclopedia of Polymer Science and Technology*, suppl. vol. 1. Interscience, New York.

Heskins, M., McAneney, T. B. and Guillet, J. E. (1976*b*). In *Ultraviolet Light Induced Reactions in Polymers*, ed. S. S. Labana, ACS Symposium Series 25. American Chemical Society, Washington, DC.

Heskins, M., Reid, W. J., Pinchin, D. J. and Guillet, J. E. (1976*a*). In *Ultraviolet Light Induced Reactions in Polymers*, ed. S. S. Labana, ACS Symposium Series 25. American Chemical Society, Washington, DC.

Hrdlovic, P. and Lukac, I. (1982). In *Developments in Polymer Photodegradation*, ed. N. S. Allen, Ch. 4. Applied Science, Barking.

Kasha, M. (1960). *Radiation Research Suppl.* **2**, 243.

Kilp, T. and Guillet, J. E. (1977). *Macromolecules*, **10**, 90.

Kilp, T. and Guillet, J. E. (1981). *Macromolecules*, **14**, 1680.

Kiwi, J. and Schnabel, W. (1975). *Macromolecules*, **8**, 430.

Kobsa, H. (1962). *J. Org. Chem.* **27**, 2293.

Kobsa, H. (1964). *Abstracts of Papers.* A.C.S. **148**, 12E.

Li, S. K. L. and Guillet, J. E. (1977). *Macromolecules*, **10**, 840.

Li, S. K. L. and Guillet, J. E. (1980). *J. Polym. Sci., Polym. Chem. Ed.* **18**, 2221.

Li, S. K. L. and Guillet, J. E. (1984). *Polym. Photochem.* **4**, 21.

Lukac, I., Hrdlovic, P., Manasek, Z. and Bellus, D. (1971). *J. Polym. Sci., Part A-1*, **9**, 69.

McKellar, J. F. and Allen, N. S. (1979). *Photochemistry of Man-Made Polymers.* Applied Science, London.

Naito, I., Koga, K., Kinoshita, A. and Schnabel, W. (1980). *Eur. Polym. J.* **16**, 109.

Nakanishi, H., Jones, W., Thomas, J. M., Hasegawa, M. and Rees, W. L. (1980). *Proc. R. Soc. London A*, **369**, 307.

Pavlik, J. O., Plooard, P. I., Somersall, A. C. and Guillet, J. E. (1973). *Can. J. Chem.* **51**, 1435.

Plooard, P. I. and Guillet, J. E. (1972). *Macromolecules*, **5**, 405.

Salvin, R., Meybeck, J. and Faure, J. (1976). *J. Photochem.* **6**, 9.

Schnabel, W. (1979). *Pure Appl. Chem.* **51**, 2373.

Shultz, A. R. (1960). *J. Polym. Sci.* **47**, 267.

Sitek, F., Guillet, J. E. and Heskins, M. (1976). *J. Polym. Sci., Symposium*, **57**, 343.

Small, R. D. and Scaiano, J. C. (1978). *Macromolecules*, **11**, 840.

Somersall, A. C., Dan, E. and Guillet, J. E. (1974). *Macromolecules*, **7**, 233.

Sugita, K., Kilp, T. and Guillet, J. E. (1976). *J. Polym. Sci., Polym. Chem. Ed.* **14**, 1901.

Tazuke, S. and Banba, F. (1976). *Macromolecules*, **9**, 451.

Tazuke, S. and Matsuyama, Y. (1975). *Macromolecules*, **8**, 280.

Tazuke, S. and Matsuyama, Y. (1977). *Macromolecules*, **10**, 215.

Tazuke, S., Sato, K. and Banba, F. (1975). *Chem. Lett.* 1321.
Tsunooka, M., Kotera, K. and Tanaka, M. (1977). *J. Polym. Sci., Polym. Chem. Ed.* **15**, 107.
Ueno, A. and Osa, T. (1978). *J. Polym. Sci., Polym. Lett. Ed.* **16**, 539.
Ueno, A., Osa, T. and Toda, F. (1977). *Macromolecules*, **10**, 130.
Wagner, P. J. (1971). *Acc. Chem. Res.* **4**, 168.
Wagner, P. J. (1976). *Top. Curr. Chem.* **66**, 1.
Wagner, P. J., Kelso, P. A., Kemppainen, A. E., McGrath, J. M., Schott, H. N. and Zepp, R. G. (1972). *J. Am. Chem. Soc.* **94**, 7506.
Weber, G. (1954). *Trans. Faraday Soc.* **50**, 552.
Weber, G. (1966). In *Fluorescence and Phosphorescence Analysis: Principles and Applications*, ed. D. M. Hercules. Interscience, New York.
Wiles, D. M. (1973). *Polym. Eng. Sci.* **13**, 74.
Wiles, D. M. and Carlsson, D. J. (1980). *Polymer Degradation and Stabilization*, **3**, 61.
Winnik, M. A. (1977). *Acc. Chem. Res.* **10**, 173.
Wissbrun, K. F. (1959). *J. Am. Chem. Soc.* **81**, 58.
Yang, N. C. and Yang, D. H. (1958). *J. Am. Chem. Soc.* **80**, 2913.
Yokata, M. and Tanimoto, O. (1967). *J. Phys. Soc. Jpn.* **22**, 779.
Yokoyama, M., Tamamura, T., Atsumi, M., Yoshimura, M., Shirota, Y. and Mikawa, H. (1975). *Macromolecules*, **8**, 101.

Chapter 11

Bevington, J. C. (1961). *Radical Polymerization.* Academic Press, New York.
Burnett, G. M. (1954). *Mechanism of Polymer Reactions.* Interscience, New York.
De Schryver, F. C. (1973). *Pure Appl. Chem.* **34**, 213.
De Schryver, F. C., Anand, L., Smets, G. and Switten, J. (1971*b*). *J. Polym. Sci., Part B*, **9**, 777.
De Schryver, F. C., Boens, N. and Smets, G. (1974*a*). *J. Am. Chem. Soc.* **96**, 6463.
De Schryver, F. C., Boens, N. and Smets, G. (1974*b*). *Macromolecules*, **7**, 399.
De Schryver, F. C., Feast, F. C. and Smets, G. (1970). *J. Polym. Sci., Part A-1*, **8**, 1939.
De Schryver, F. C., Tran, Van T. and Smets, G. (1971*a*). *J. Polym. Sci., Part B*, **9**, 425.
De Schryver, F. C., Tran, Van T. and Smets, G. (1975*a*). *J. Polym. Sci., Polym. Chem. Ed.* **13**, 215.
De Schryver, F. C., Tran, Van T., Toppet, S. and Smets, G. (1975*b*). *J. Polym. Sci., Polym. Chem. Ed.* **13**, 227.
Enkelmann, V., Schleyer, G., Wegner, G., Eichele, H. and Schwoerer, M. (1977). *Chem. Phys. Lett.* **52**, 314.
Grätzel, M. (1981). *J. Chem. Soc. Chem. Commun.* 183.
Hasegawa, M. and Suzuki, Y. (1967). *J. Polym. Sci., Part B*, **5**, 813.
Higgins, J., Johannes, A. H., Jones, J. F., Schultz, R., McCombs, D. A. and Menon, C. S. (1970). *J. Polym. Sci., Part A-I*, **8**, 1987.
Hirshfeld, F. L. and Schmidt, G. M. J. (1964). *J. Polym. Sci., Part A*, **2**, 2181.
Hub, H. H., Hupfer, B., Koch, H. and Ringsdorf, H. (1981). *J. Macromol. Sci. Chem.* **A15**, 701.
Ledwith, A. M. (1979). *Pure Appl. Chem.* **49**, 431.
McCombs, D. A., Menon, C. S. and Higgins, J. (1971). *J. Polym. Sci., Part A-1*, **9**, 1261.
Nakanishi, F. and Hasegawa, M. (1970). *J. Polym. Sci., Part A*, **8**, 2151.
Thomas, J. M. (1974). *Phil. Trans. Roy. Soc. London*, **A277**, 251.
Wegner, G. (1969). *Z. Naturforsch.* **24b**, 824.
Wegner, G. (1977*a*). *Pure Appl. Chem.* **49**, 443.
Wegner, G. (1977*b*). *Nato Adv. Study Inst., Ser. B*, 297.

Wegner, G. (1979). *Nato Conf. Ser.* **6**, 209.
Yeè, K. C. and Chance, R. R. (1978). *J. Polym. Sci., Polym. Phys. Ed.* **16**, 431.

Chapter 12

Andrews, D. J. and Feast, W. J. (1976*a*). *J. Polym. Sci., Polym. Chem. Ed.* **14**, 331.
Andrews, D. J. and Feast, W. J. (1976*b*). *J. Polym. Sci., Polym. Chem. Ed.* **14**, 319.
Anet, R. (1962). *Can. J. Chem.* **40**, 1249.
Angus, H. F. and Bryce-Smith, D. (1959). *Proc. Chem. Soc.* 326.
Arnold, D. R. (1968). *Adv. Photochem.* **6**, 301.
Braun, A. M., Cassidy, H. G., Schulz, R. C. and Tanaka, H. (1971). *Makromol. Chem.* **146**, 195.
Delzenne, G. A. (1971). *Eur. Polym. J. Suppl.* 55.
Delzenne, G. A. (1974). *Ind. Chim. Belg.* **39**, 249.
Holden, D. A. and Guillet, J. E. (1979). *J. Polym. Sci., Polym. Lett. Ed.* **17**, 15.
Holden, D. A. and Guillet, J. E. (1980). *J. Polym. Sci., Polym. Chem. Ed.* **18**, 565.
Hrdlovic, P. and Lukac, I. (1974). *J. Polym. Sci., Polym. Symposia*, **47**, 319.
Hyde, P., Kricka, L. J. and Ledwith, A. (1973). *J. Polym. Sci., Polym. Lett. Ed.* **11**, 415.
Lyalikov, K. S., Gaeva, G. L. and Evlasheva, N. A. (1970). *Tr. Leningrad Inst. Kinoinzh.* **16**, 42.
Ng, H. and Guillet, J. E. (1977). *Macromolecules*, **11**, 221.
Schulz, R. C. (1973). *Pure Appl. Chem.* **34**, 305.
Schulz, R. C., Rohe, L. and Adler, H. (1969). *Eur. Polym. J. Suppl.* 309.
Williams, J. L. R. (1969). *Fortschr. Chem. Forsch.* **13**, 227.

Chapter 13

Agolini, F. and Gay, F. P. (1970). *Macromolecules*, **3**, 349.
Aviram, A. (1978). *Macromolecules*, **11**, 1275.
Bauser, H. and Klöpffer, W. (1970). *Chem. Phys. Lett.* **7**, 137.
Bieth, J., Vratsanos, S. M., Wasserman, N. and Erlanger, B. F. (1969). *Proc. Natl. Acad. Sci. USA*, **64**, 1103.
Braun, C. L. (1968). *Phys. Rev. Lett.* **21**, 215.
Damen, J. and Neckers, D. C. (1980). *J. Am. Chem. Soc.* **102**, 3265.
Eisenbach, C. D. (1980). *Ber. Bunsenges. Phys. Chem.* **84**, 680.
Fourny, J., Delacote, G. and Schott, M. (1968). *Phys. Rev. Lett.* **21**, 1085.
Fox, R. B. (1967). In *Progress in Polymer Science*, vol. 1, ed. A. B. Jenkins. Pergamon, London.
Gill, W. D. (1972). *J. Appl. Phys.* **43**, 5033.
Grassie, N. and Weir, N. A. (1965). *J. Appl. Polym. Sci.* **9**, 975.
Goodwin, F. and Lyons, L. E. (1967). *Organic Semiconductors.* Wiley, New York.
Guillet, J. E. (1978). *Adv. Chem. Series No. 169*, 1.
Heller, J. (1972). In *The Enzymes*, vol. 6, ed. P. D. Boyer, p. 573. Academic Press, New York.
Holtzman, P., Morris, R., Jarnagin, R. C. and Silver, M. (1967). *Phys. Rev. Lett.* **19**, 506.
Irie, M., Menju, A. and Hayashi, K. (1979). *Macromolecules*, **12**, 1176.
Kepler, R. G. (1967). *Phys. Rev. Lett.* **18**, 951.
Klöpffer, W. (1969). *J. Chem. Phys.* **50**, 2337.
Lardon, M., Lell-Daller, E. and Weigl, J. W. (1967). *Mol Crystals*, **2**, 241.
Lovrien, R. (1967). *Proc. Natl. Acad. Sci. USA*, **57**, 236.
Martin, J. T. and Norrish, R. G. W. (1953). *Proc. R. Soc. London, Ser. A*, **220**, 322.
Melz, P. J. (1972*a*). *Bull. Am. Phys. Soc.* **17**, 353.
Melz, P. J. (1972*b*). *J. Chem. Phys.* **57**, 1694.

Ng, H. C. and Guillet, J. E. (1978). *Macromolecules*, **11**, 929.
Onsager, L. (1938). *Phys. Rev.* **54**, 554.
Patsis, A. V. and Seanor, D. A. (ed.) (1976). *Photoconductivity in Polymers.* Technomic Publishing, Westport, Conn.
Rabek, J. F. and Ranby, B. (1974). *J. Polym. Sci., Part A-1*, **12**, 273.
Ranby, B. and Rabek, J. F. (1975). *Photodegradation, Photooxidation, and Photostabilization of Polymers*, Ch. 5. Wiley–Interscience, New York.
Regensburger, P. J. (1968). *Photochem. Photobiol.* **8**, 429.
Reucroft, P. J. (1976). In *Photoconductivity in Polymers, an Interdisciplinary Approach*, ed. A. V. Patsis and D. A. Seanor. Technomic Publishing Co., Westport, Connecticut.
Schaffert, R. M. (1971). *IBM J. Res. Dev.* **15**, 75.
Shattuck, M. D. and Vahtra, U. (1969). US Patent 3,484,237.
Smets, G. (1972). *Pure Appl. Chem.* **30**, 1.
Strome, F. C. Jr (1968). *Phys. Rev. Lett.* **20(3)**, 1.
Trozzolo, A. M. and Winslow, F. H. (1968). *Macromolecules*, **1**, 98.
van der Veen, G. and Prins, W. (1971). *Nature Phys. Sci.* **230**, 70.
Wald, G. (1968). *Nature*, **219**, 800.
Weir, N. A. (1973). *J. Appl. Polym. Sci.* **17**, 401.
Weir, N. A. and Milkie, T. H. (1979). *J. Polym. Sci., Polym. Chem. Ed.* **17**, 3735.
Weir, N. A., Milkie, T. H. and Nicholas, D. (1979). *J. Appl. Polym. Sci.* **23**, 609.

Chapter 14.

Ausloos, P. (1968). *Fundamental Processes in Radiation Chemistry*, vols. 1 and 2. Interscience, New York.
Bovey, F. A. (1958). *The Effects of Ionizing Radiation on Natural and Synthetic High Polymers.* Polymer Reviews, vol. 1. Interscience, New York.
Chapiro, A. (1962). *Radiation Chemistry of Polymer Systems*, vol. 15. Interscience, New York.
Charlesby, A. (1960). *Atomic Radiation and Polymers.* Pergamon Press, Oxford.
David, C. (1975). In *Degradation of Polymers*, ed. C. H. Bamford and C. F. H. Tipper. Elsevier, Amsterdam.
David, C., DeBie, F. and Geuskens, G. (1975). *Makromol. Chem.* **176**, 3493.
David, C., DeBie, F., Pastiels, P. and Geuskens, G. (1978). *Int. J. Radiat. Phys. Chem.* **11**, 275.
David, C., Fuld, D. and Geuskens, G. (1970). *Makromol. Chem.* **130**, 269.
David, C., Fuld, D. and Geuskens, G. (1972*a*). *Makromol. Chem.* **160**, 135.
David, C., Fuld, D. and Geuskens, G. (1972*b*). *Makromol. Chem.* **160**, 347.
David, C., Janssen, P. and Geuskens, G. (1972*c*). *Int. J. Radiat. Phys. Chem.* **4**, 51.
Deckert, C. A. and Ross, D. L. (1980). *J. Electrochem. Soc.* **127**, 45c.
Delzenne, G. A. (1969). *Eur. Polym. J., Suppl.* 55.
Dole, M. (ed.) (1972). *The Radiation Chemistry of Macromolecules*, vol. 1. Academic Press, New York.
Dole, M. (ed.) (1973). *The Radiation Chemistry of Macromolecules*, vol. 2. Academic Press, New York.
Gandhi, S. K. (1968). *The Theory and Practice of Microelectronics.* Wiley, New York.
Geuskens, G. and David, C. (1973). *Makromol. Chem.* **165**, 273.
Guillet, J. E. (1974). US Patent 3,812, 025.
Gupta, A., Liang, R., Tsay, F. D. and Moacanin, J. (1980). *Macromolecules*, **13**, 1696.
Hartley, G. H. and Guillet, J. E. (1968). *Macromolecules*, **1**, 413.
Himics, R. J., Kaplan, M., Desai, N. V. and Poliniak, E. S. (1977). *Polym. Eng. Sci.* **17**, 406.

Kang, H. Y., Saito, O. and Dole, M. (1967). *J. Am. Chem. Soc.* **89**, 1980.

Kells, D. I. C. and Guillet, J. E. (1968). *J. Polym. Sci., Part A-1*, **6**, 595.

Kells, D. I. C., Koike, M. and Guillet, J. E. (1968). *J. Polym. Sci., Part A-1*, **6**, 595.

Partridge, R. H. (1970*a*). *J. Chem. Phys.* **52**, 2485.

Partridge, R. H. (1970*b*). *J. Chem. Phys.* **52**, 2491.

Partridge, R. H. (1970*c*). *J. Chem. Phys.* **52**, 2501.

Pitts, J. N. and Osborne, A. D. (1961). *J. Am. Chem. Soc.* **83**, 3011.

Schultz, A. R. (1966). *Encycl. Polym. Sci. & Tech.* **4**, 398.

Slivinskas, J. A. and Guillet, J. E. (1973*b*). *J. Polym. Sci., Polym. Chem. Ed.* **11**, 3043.

Slivinskas, J. A. and Guillet, J. E. (1973*a*). *J. Polym. Sci., Polym. Chem. Ed.* **11**, 3057.

Slivinskas, J. A. and Guillet, J. E. (1974). *J. Polym. Sci., Polym. Chem. Ed.* **12**, 1469.

Index